Composite Materials

VOLUME 1

Interfaces in Metal Matrix Composites

COMPOSITE MATERIALS

Edited by

LAWRENCE J. BROUTMAN   AND   RICHARD H. KROCK

Illinois Institute of   *P. R. Mallory & Co., Inc.*
Technology   *Laboratory for Physical Science*
Chicago, Illinois   *Burlington, Massachusetts*

1. Interfaces in Metal Matrix Composites
   Edited by ARTHUR G. METCALFE

2. Mechanics of Composite Materials
   Edited by G. P. SENDECKYJ

3. Engineering Applications of Composites
   Edited by BRYAN R. NOTON

4. Metallic Matrix Composites
   Edited by KENNETH G. KREIDER

5. Fracture and Fatigue
   Edited by LAWRENCE J. BROUTMAN

6. Interfaces in Polymer Matrix Composites
   Edited by EDWIN P. PLUEDDEMANN

7. Structural Design and Analysis, Part I
   Edited by C. C. CHAMIS

8. Structural Design and Analysis, Part II
   Edited by C. C. CHAMIS

VOLUME 1

Interfaces in Metal Matrix Composites

Edited by

ARTHUR G. METCALFE

Solar Division of International Harvester Company
San Diego, California

ACADEMIC PRESS   New York and London   1974
A Subsidiary of Harcourt Brace Jovanovich, Publishers

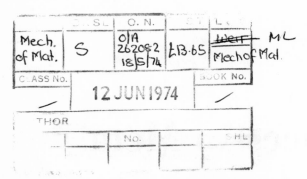
ACADEMIC PRESS, INC.
111 Fifth Avenue, New York, New York 10003

United Kingdom Edition published by
ACADEMIC PRESS, INC. (LONDON) LTD.
24/28 Oval Road, London NW1

Library of Congress Cataloging in Publication Data

Metcalfe, Arthur G.
    Interfaces in metal matrix composites.

    (Composite materials, v. 1)
    Includes bibliographies.
    1.   Metallic composites.   I.   Title.   II.   Series.
TA481.M47   620.1'6   73-5303
ISBN 0–12–136501–8

D
620.16
INT

Contents

4   Effect of the Interface on Longitudinal Tensile
    Properties

Arthur G. Metcalfe and Mark J. Klein

5   Effect of the Filament–Matrix Interface on
    Off-Axis Tensile Strength

Mark J. Klein

6   Role of the Interface on Elastic–Plastic Composite
    Behavior

A. Lawley and M. J. Koczak

7   Effect of Interface on Fracture

Elliot F. Olster and Russel C. Jones

List of Contributors

Numbers in parentheses indicate the pages on which the authors' contributions begin.

W. BONFIELD (363), Department of Materials, Queen Mary College, London, England

L. J. EBERT (31), Division of Metallurgy and Materials Science, Case Western Reserve University, Cleveland, Ohio

RUSSEL C. JONES (245), Department of Civil Engineering, Ohio State University, Columbus, Ohio

RICHARD W. HERTZBERG (329), Department of Metallurgy and Materials Science and Mechanical Behavior Laboratory, Materials Research Center, Lehigh University, Bethlehem, Pennsylvania

MARK J. KLEIN (125, 169), Solar Division of International Harvester Company, San Diego, California

M. J. KOCZAK (211), Department of Metallurgical Engineering, Drexel University, Philadelphia, Pennsylvania

A. LAWLEY (211), Department of Metallurgical Engineering, Drexel University, Philadelphia, Pennsylvania

ARTHUR G. METCALFE (1, 65, 125), Solar Division of International Harvester Company, San Diego, California

ELLIOT F. OLSTER* (245), Advanced Composite Applications, Avco Corporation, Lowell, Massachusetts

RICHARD E. TRESSLER (285), Department of Material Sciences, The Pennsylvania State University, University Park, Pennsylvania

P. KENNARD WRIGHT† (31), Division of Metallurgy and Materials Science, Case Western University, Cleveland, Ohio

* Present address: Plastics Development Center, Ford Motor Company, Detroit, Michigan.
† Present address: General Electric Company, Lamp Business Division, Lighting Research and Technical Services Operation, Cleveland, Ohio.

Foreword

The development of composite materials has been a subject of intensive interest for at least 15 years, but the concept of using two or more elemental materials combined to form the constitutent phases of a composite solid has been employed ever since materials were first used. From the earliest uses, the goals for composite development have been to achieve a combination of properties not achievable by any of the elemental materials acting alone; thus a solid could be prepared from constituents which, by themselves, could not satisfy a particular design requirement. Because physical, chemical, electrical, and magnetic properties might be involved, input from investigators of various disciplines was required. In the various volumes of this treatise references to specific materials have generally only included the man-made or synthetic composites, but certainly the broad definition of composite materials must include naturally occurring materials such as wood. Chapters dealing with analytical studies of course can apply equally to synthetic or naturally occurring composites.

While composites have been used in engineering applications for many years, the severe operating conditions at which materials have to function (in the space age) led to the science of composite materials as we know it today. The efforts of scientists and engineers working on government research and development programs created entirely new materials, fabrication techniques, and analytical design tools within a short span of time to serve a limited market, but one with constantly demanding requirements. At end of the 1960s, a sharp reduction in the level of government expenditures in these areas and redistribution or re-emphasis of much of the personnel and institutions that had been involved in the development of composite materials raised the possibility that no complete reference work would be available to record many of these important developments and techniques. It was also apparent that this great bulk of technology, if properly digested and evaluated, could be employed in industrial and consumer applications for advantages of economy, performance, and design simplicity. For these reasons, the Editors and Academic Press have prepared this treatise detailing the major aspects of the science and technology of composite materials. We believe that the wide representation of contributors and the diversity of subject matters contained in the treatise assure the complete coverage of this field.

We intend that the volumes be used for reference purposes, or for text supplement purposes, but particularly to serve as a bridge in transferring the bulk of composite materials technology to industrial and consumer applications.

The Editors are indebted for the cooperation and enthusiasm they have received from the Editor of each volume and the individual contributors who worked diligently and as a unit to complete this task. We are also grateful to the staff at Academic Press who provided constant support and advice for the project.

Finally, the Editors wish to thank the management of P. R. Mallory and Co., Inc. and the administration of Illinois Institute of Technology for providing a key element in the successful completion of this work through their support and encouragement.

LAWRENCE J. BROUTMAN
RICHARD H. KROCK

Preface to Volume 1

The main body of research on study of interfaces in metal matrix composites took place in the five years prior to the publication of this book. The volume of activity began to decrease as research focused on the development of useful materials based on the available knowledge. This book provides a summary of the position of the science of interfaces, and thereby, the necessary background for the effort in progress to apply these materials.

One indication of the newness of this field is given by consideration of the topics that are missing from this volume. Although there are chapters on the effects of interface on longitudinal and transverse strength, and on fracture, there was insufficient information for chapters on the relationship between the interface and properties such as fatigue and creep. For this reason, there could be little discussion of the question of the ideal interface. All of the authors recognize this point, and yet there seems to be an awareness that a science of interfaces is yet to fully develop. Indeed, the science may be adequate to carry the new technology through the preliminary stages of application, but, at the same time, the need for continuing development of the science is clearly recognized.

It may at first appear that we have emphasized the work of the group at Solar, Division of International Harvester Company. The studies on interfaces were supported largely by the United States Air Force but very little of it was published previously. One reason for this was that concepts such as the oxide bond in aluminum matrix composites were several years in the formulation and verification stages and have only recently been firmly established. However, other major centers of interface study have joined in this publication and provide a balanced view of this field. In nearly every case, however, the authors of these chapters critically review work in the field, including much of their own research, rather than present unpublished work.

This book would not have been possible without the continued support of many whose names do not appear within its covers. Foremost is the support and encouragement given by the author's company, and particularly that of Mr. John V. Long, Director of Research, for his continued personal encouragement and of Mrs. Patricia J. Lind for her painstaking care in preparing the manuscript.

ARTHUR G. METCALFE

1

Introduction and Review

ARTHUR G. METCALFE

Solar Division of International Harvester Company
San Diego, California

Scientific and technological contributions to our knowledge of interfaces in metal matrix composites have been made only in the last few years. Development of this knowledge has not been uniform in all areas, and the chapters of this volume are limited to those where sufficient information was available. Several important areas, including the effect of the interface on fatigue and creep properties, could not be discussed. Another result of this uneven development is that it has not been possible to conclude the book by summarizing the requirements of the ideal interface. However, the editor has felt strongly that there was need for a discussion to bring together the topics reviewed individually in the chapters. This first chapter attempts to perform this function and presents both a short introduction as well as a general review of the overall subject. The reader, new to the field, may prefer to return to the review after reading the individual chapters, although the content of the review is quite general rather than specialized.

I. Introduction

The history of metal matrix composites reveals why interest in the interface has grown rapidly in the last few years. Early work on composites was directed to demonstration of the principles governing their performance. Simple model systems sufficed for this purpose. The model systems were chosen largely for compatibility of the reinforcement and matrix, and had matrices of little reactivity, such as silver and copper with reinforcements such as aluminum oxide and tungsten. The importance of the interface was recognized in this early work, but the model systems made it relatively easy to obtain the type of interface required for the necessary load transfer from one constituent of the composite to the other. More practical systems have matrices of the common structural metals: aluminum, titanium, iron, and nickel. Such matrices are more reactive and stronger than the matrices of the model systems. The increased reactivity made control of the interface more difficult, and the higher strength required the interface to carry greater loads. Hence, control of the interface has become much more critical as interest has turned from the model systems to those with attractive engineering properties.

The problems associated with control of the interface are not unique to metal matrix composites. Finishes are applied to freshly drawn glass filaments to control the interface in glass reinforced plastics. Optimization of the finish is recognized to be a difficult compromise between many requirements such as protection of the filament from handling damage, enhancement of bonding of the glass to plastic, and preservation of the bond in service, especially in the presence of moisture. Optimization of interfaces or bonds in metal matrix composites is believed to require similar compromises. The demands on the interface are at least as severe in metal matrix composites as in glass reinforced plastics. Mention has already been made of the chemical imcompatibility of many matrix–filament combinations, and this includes both inadequate and excessive reactivity (by inadequate is meant a system where the absence of any physical chemical effects causes no bond to form). Stability of the interface is another important requirement that is made critical by the high temperature service desired for metal matrix composites. Also, metal matrix composites must be able to withstand a wider variety of loading conditions than plastic matrix composites, because metals allow many off-axis loads to be borne by the matrix with little or no reinforcement in this direction.

The basic guidelines used to select model systems were thought to apply to the selection of matrix and filament in all systems. Jech *et al.* (1960) chose the system copper–tungsten to demonstrate that the rule of mixtures was valid for both continuous and short length filaments. Copper and tungsten are essentially insoluble in each other and are nonreactive; that is, they form no compounds. Similarly, Sutton *et al.* (1960–64) provided a convincing demonstration of the validity of whisker reinforcement by choice of the model system, silver–sapphire. The degree of interaction between silver and sapphire is even less than that between copper and tungsten because molten silver will not wet sapphire. Sutton *et al.* (1960–64) evaporated nickel on to the surface of sapphire to improve the bond to the molten silver but the nickel–sapphire bond was probably mechanical and the nickel–silver interface shows no solution. It was not surprising, therefore, that Hibbard (1964), in a review presented in the introductory paper at the American Society for Metals meeting on Fiber Composite Materials, should conclude: "There should be little or no solubility or other reaction between the matrix and the fiber, which should wet each other." This condition is satisfied most readily by one special group of composites based on directionally solidified eutectics. Many of the eutectics have small changes of terminal solubility with temperature and this is the principal source of instability, although this is offset to some extent by the special crystallographic relationship between the phases. However, the condition is not satisfied by many of the artificial combinations that are of major interest. The principal advances since the 1964 meeting have been largely in control of interfaces between reinforcements and matrices that do not meet these restrictions. Silver and copper are not candidates for structural materials. Reactions between structural matrices and available reinforcements are quite complex, and can lead to a wide variety of types of interface.

A. Types of Interface

One of the first systematic examinations of types of interface was made by Petrasek and Weeton (1964), who extended the earlier work of Jech *et al.* (1960) on copper–tungsten by study of tungsten-reinforced copper alloys. Three interface types were noted with these alloy matrices, although interpretation of the results was made somewhat difficult by the

TABLE I

CLASSIFICATION OF COMPOSITE SYSTEMS

Class I	Class II	Class III
Copper–tungsten	Copper(chromium)–tungsten	Copper(titanium)–tungsten
Copper–alumina	Eutectics	Aluminum–carbon (>700°C)
Silver–alumina	Columbium–tungsten	Titanium–alumina
Aluminum–BN coated B	Nickel–carbon	Titanium–boron
Magnesium–boron	Nickel–tungsten[a]	Titanium–silicon carbide
Aluminum–boron[b]		Aluminum–silica
Aluminum–stainless steel[b]		
Aluminum–SiC[b]		

[a] Becomes reactive at lower temperatures with formation of Ni_4W.
[b] Pseudo-Class I system.

effects of the alloying elements on the tungsten wire. The types are: those where recrystallization occurred at the periphery of the wire, those where a new phase formed at the interface, and those where mutual solution occurred between the matrix and filament. The peripheral recrystallization was caused by diffusion of cobalt, aluminum, or nickel from the copper alloy into the tungsten wire. This recrystallization caused the wires to become brittle at room temperature and resulted in both loss of strength and brittleness of the composite. Compound formation occurred principally with titanium and zirconium dissolved in the copper alloy matrix and led to loss of strength and ductility of the composite. Alloying additions in the copper that were soluble in tungsten, such as chromium and niobium, resulted in the formation of solid solutions at the interface as the alloying element was removed from the copper alloy to alloy with the tungsten. The loss of ductility was small in these cases and the filaments extracted from composites remained quite ductile (the reduction in area of the filaments remained above 7 percent). In agreement with the retention of ductility, it was found that the strengths of the composites remained close to those found in the binary copper–tungsten composites. In a later review of this work, Signorelli *et al.* (1967) concluded that property deviations from those of the baseline material could usually be attributed to interfacial reactions. This viewpoint was generally held at this time. However, evidence gradually began to be accumulated to show that interfacial reactions need not interfere with the attainment of theoretical properties if the amount of reaction were controlled and if the reaction did not damage or weaken the filament.

A general scheme for the classification of interfaces has been developed and will be presented at this point before discussion is resumed on the properties attainable with each class. The scheme is based on the type of chemical reaction occurring between the filament and matrix. The term "reactive" is restricted to those systems that result in the formation of a new chemical compound or compounds. The three classes proposed are:

Class I, filament and matrix mutually nonreactive and insoluble
Class II, filament and matrix mutually nonreactive but soluble
Class III, filament and matrix react to form compound(s) at interface

Clear-cut definitions between the classes are not always possible, but the groupings provide a systematic background against which to discuss their characteristics. Table I gives examples of each type, and Fig. 1 illustrates these classes by suitable selections from the copper alloy–

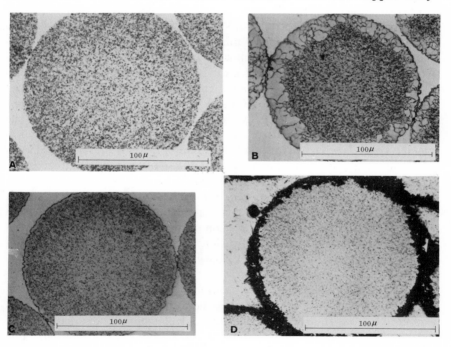

FIG. 1. Types of reaction in copper alloy–tungsten composites. (A) Cu–W, Class I system; (B) Cu(Ni)–W, Class II system; (C) Cu(Cr)–W, Class II system; (D) Cu(Ti)–W, Class III system (Petrasek and Weeton, 1964).

tungsten composite work of Petrasek and Weeton (1964). It should be noted that eutectics are included in Class II but the terminal solubility of each phase in the other may be so low that inclusion in Class I may be preferable. Similarly, the system copper (titanium)–tungsten is included in Class III because a compound forms at the interface, as shown in Fig. 1. But when the amount of titanium is small, solutions are formed with both copper and tungsten.

One of the most interesting group of system consists of the pseudo-Class I composites. The three principal members of this group are: aluminum–boron, aluminum–stainless steel, and probably aluminum–silicon carbide. Composites of this group are generally fabricated by solid state diffusion bonding. The composites appear free of interaction after fabrication by the optimum processing technique but thermodynamic data indicate that the constituents should react. Such reaction does occur when the aluminum is melted. The optimum processing was thought to result in complete absence of any interaction, in agreement with the desideratum expressed earlier for useful composites. Jones (1968) reports that a composite of 2024 aluminum with stainless steel wires should be bonded at the low temperature of 875°F for 30 min to promote bonding yet avoid reaction. Similarly, Davis (1967) reported that this type of composite suffered degradation of properties if given additional processing such as cold-rolling and annealing, because an aluminum–iron compound formed at the interface. The term, pseudo-Class I composite, describes the apparent nonreactivity when fabricated by the optimum process in contrast to the true reactivity of the constituents. Klein and Metcalfe (1971) have explained this behavior by showing that a film of oxide is preserved between the filament and matrix to account for the apparent (pseudo) nonreactivity. This oxide film has been identified in aluminum–boron composites as principally aluminum oxide. Breakdown of this film occurs readily if processing is continued beyond that regarded to be optimum, or if the aluminum is melted. Hence, fabrication of high strength aluminum–boron composites by molten metal infiltration is not possible, unless the boron is coated with suitable nonreactive coatings such as boron nitride (Camahort, 1968).

Of course, all Class III systems (and for that matter, Class II systems) may have a transitory existence as pseudo-Class I systems until the preexisting films on the surface of matrix and filament are dissipated. As a result, classification of composite systems is not always clear-cut. For example, Goddard *et al.* (1972) have identified the phase Al_4C_3 in aluminum–carbon composites formed above 700°C but the interface appears to be stable at lower temperatures. This system is included in Class III in

Table I based on the high temperature reaction. In contrast, the solution rate of films in titanium is very high at typical diffusion bonding temperatures and the transitory existence of titanium matrix composites as pseudo-Class I systems will be too short to be observed. This is believed to explain why all titanium matrix systems are of Class III; whereas solid state bonded aluminum matrix systems are included in Class I as pseudosystems. Aluminum–silica is also included as a Class III system because it has only been fabricated by infiltration with molten aluminum, and reaction is known to occur between molten aluminum and silica. Although workers at Rolls Royce, Ltd. (England) were able to reduce the amount of reaction by coating the fibers with a chemical inhibitor prior to drawing the fibers through the molten metal (Arridge et al., 1964; Cratchley, 1963), there is no evidence to suggest that reaction was completely stopped.

Three general classes of interface have been proposed for metal matrix composites but no general definition has been given for an interface. Until recently, no attempt was made to present a definition. Salkind (1968) concluded that "a precise definition of the interface is beyond our present knowledge." At that time, useful interfaces were limited to those formed between phases that wet each other; and definition for an interface of no finite width presented severe problems. It will be shown later that all three types of interface can provide useful composites when each satisfies certain conditions. This broadening of the types of useful interfaces has made it easier to define an interface. The following definition is proposed:

An interface is the region of significantly changed chemical composition that constitutes the bond between the matrix and reinforcement for transfer of loads between these members of the composite structure.

The term "significantly changed composition" would include the minor change in composition such as those discussed by Graham and Kraft (1966) in connection with the stability of eutectic composites. In this case, the solubility changes resulted from differences in curvature of the interfaces as expressed by the Thomson–Freundlich relationship. Similarly, small amounts of solution promoting recrystallization, as observed in the Cu(Ni)—W system (Petrasek, 1966; Signorelli et al., 1967), would be included in this definition. Also, segregation of elements to an interface would be included. For example, zirconium concentrates from a nickel alloy to an alumina interface to promote bonding, as shown by Sutton and Feingold (1966). Also included will be the oxide type bonds proposed for pseudo-Class I systems. These will include a sequence of phases from matrix through matrix–oxide bond, oxide films, oxide–reinforcement bond, and into the reinforcement.

Inclusion of the term "significantly changed chemical composition" in the definition of an interface is to exclude random fluctuations of composition but to include systematic changes caused by thermodynamically required effects. Solution, segregation, adsorption, and reaction are the principal systematic effects. This definition ignores the region of rapidly changing stress fields at the interface resulting from thermal contraction or applied stress, but these are transient and may be changed either by the prior history or by the ambient conditions.

B. Models of the Interface

Early models of the interface were based on the concept of no solubility or reaction at the interface. This concept leads to an interface of essentially zero thickness. Also, properties were not associated with the interface, per se. For example, the term "interfacial shear strength" was often applied to the stress in the matrix immediately adjacent to the filament. A further assumption was made that the interface was stronger than the matrix so that matrix flow limited load transfer from and to the filaments.[†] This premise forms the basis for nearly all analyses of the mechanics of composites as noted in Chapter 2.

Although quantitative analyses have been made, several workers have pointed out that the properties of the composite will be affected by the extent to which the properties of the interface differ from those of the matrix and fiber. For example, Cooper and Kelly (1968) divide the properties of the composite into those controlled principally by the tensile strength of the interface σ_i and those governed by the shear strength τ_i. Among the properties governed by the tensile strength of the interface the authors list: the transverse strength, the compressive strength, and crack arrest by tensile delamination. Properties governed principally by the shear strength were stated to include: critical or load transfer length, fracture under conditions of fiber pullout, and deformation of the matrix in fracture. This theory will be discussed further in the body of this volume.

Theories such as that due to Cooper and Kelly, and the concepts of interfacial tensile and shear strengths, were developed to fit Class I type composites. Cooper and Kelly cite cases such as copper–tungsten com-

---

† Such an assumption eliminates several of the more practical reasons to study interfaces. Analyses based on this assumption are discussed under the general heading of "Theories for Strong Interfaces" in Chapter 4, but more interest is attached to the "Theories for Weak Interfaces." The latter theories become of increasing interest as more practical systems with higher strength matrices cause the failure path to move to the interface.

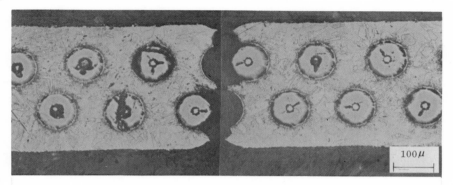

FIG. 2. Fracture of titanium–boron composite.

posites to support their analysis of the behavior of metal matrix systems. Systems of Class II and III introduce a zone of finite width with distinct properties. Ebert *et al.* (1971) have begun to analyze systems of Class II to predict the effects of interfacial diffusion on the mechanical properties of composites, using incremental methods. The initial work represents a start on analysis of nonmodel systems, but has been limited to a system with a chemical continuum (see Chapter 2). More complexity is introduced by the chemical discontinuum created by a reaction product in a Class III system, because two additional interfaces must be considered. In addition to the tensile strength of the reaction zone, two interfacial tensile strengths, and two interfacial shear strengths are required to describe the two interfaces. One of these interfaces is the matrix–interfacial phase boundary and the other is the interfacial phase–filament boundary. These strengths may be defined as σ_{iM} and σ_{iF} to define the tensile strengths of the interface in contact with the matrix and that in contract with the filament. Figure 2 reproduces an example from the system titanium-boron in which transverse failure was governed by the strength σ_{iF}.

Figure 3 defines some strengths for a simple Class III composite with a single-reaction zone product. The well-known terms for matrix and filament strengths σ_M and σ_F are used in expressions such as the rule of mixtures. The shear strength of the matrix is found in expressions for critical lengths for load transfer from matrix to filament. Some of the other terms introduced in Fig. 3 have not been used previously, and will not be used extensively in this volume because of lack of information on their values. The purpose in introducing them is to focus attention on their importance in development of a complete theory of the interface for Class III systems. One term has found quantitative use, however, and this is the strength of the interface phase σ_R. It has been pointed out by Metcalfe (1967) that

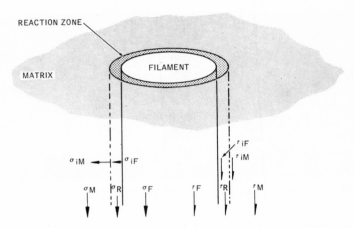

FIG. 3. Definition of some strengths for Class III composites.

the interaction compound will possess an elastic modulus different from that of the filament and matrix. In addition, the conditions of growth of this compound are such that its strength may be assumed to be that of the compound in massive form because of the presence of growth defects. For the case of longitudinal loading where the simple rule of mixtures applies, these considerations allow the strain-to-fracture of the compound to be calculated. This strain-to-fracture is less than that tolerated by the filaments in most cases, so that crack initiation starts in the interaction zone, as shown in Fig. 4 for the system titanium–boron. The theory has been developed further for three cases that depend on the thickness of the reaction zones. These cases are:

(1) *Little reaction.* Crack length equals thickness of reaction zone and is too small to initiate failure of filament. Stress concentration at filament

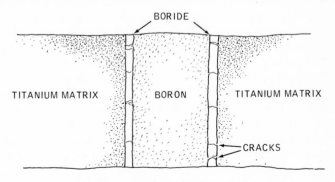

FIG. 4. Cracks in boride layer on boron.

due to crack in reaction zone is less than that due to intrinsic defects in the filament. The latter retain control of failure so that the strength of the composite is unaffected by cracks in the reaction zone. A typical thickness may extend up to 5000 Å.

(2) *Intermediate reaction.* Crack length is longer than in case (1) and creates a stress concentration greater than those due to intrinsic defects in the filament. Filaments fail prematurely at a strain that is dependent on thickness of reaction zone. Intermediate reaction may range from 5000 to 10,000 Å.

(3) *Much reaction.* Formation of crack in reaction zone leads to immediate failure of filament. Reaction zones thicker than 10,000 to 20,000 Å are typical of this grouping.

This theory led to a prediction that there is a safe limit of reaction in Class III composites below which no loss of longitudinal strength would be found. Confirmation of this prediction for titanium–boron composites provided strong support for the theory, and subsequently, other details of the theory have been partially confirmed for this and other systems. Failure strains exceeding 6000 μin./in., or a stress of nearly 360,000 psi in the boron filaments, are found with titanium–boron composites as long as the amount of reaction does not exceed a critical limit found for case (1) type behavior. This theory will be discussed in more detail in Chapter 4.

The amount of reaction permitted in Class III systems varies with many other aspects of the composite. One of the most important is the support provided at each end of the crack in the reaction zone, because this will determine the extent of crack opening and hence the stress concentration arising from the cracks. All available evidence shows that the permissible crack length increases as the elastic limit of the titanium matrix increases in the tianium–boron system. However, if the filament does not behave as a perfectly elastic member, but has some degree of plasticity, then the crack length to cause loss of strength may be much larger. This means that the thickness of reaction zone can be much larger. An example from a pseudo-Class I system is due to Jones (1968) from studies of 2024 aluminum alloy–stainless steel composites. Although the majority of specimens did not show interaction, in several instances a third phase was detected by a low angle section cut through the filaments. Figure 5 shows one such specimen in which the reaction phase is distinctly visible. The numerous cracks running through this phase are quite regularly spaced. The strain-to-fracture of this phase, given simply by σ_R/E_R, appears to be exceeded before the strains-to-fracture are attained by the filament or matrix. The

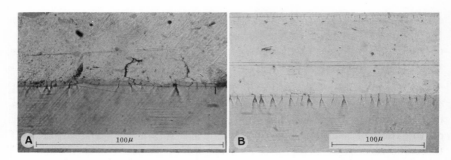

Fig. 5. Fracture in aluminum–stainless steel composite at interface compound (the 100-μ marker is on the steel filament) (Jones, 1968). Micron marker is on stainless steel filament. (A) shows cracks in aluminum matrix. (A) and (B) show cracks in reaction zone.

crack causes a stress concentration in both matrix and filament. The stress concentration in the filament is relieved by local deformation at the root of the notch that eventually extends across the full width of the steel filament. In this respect, the system differs markedly those Class III systems in which the filaments are purely elastic (e.g., Ti–B).

Demonstration that rule of mixtures type of behavior is found in composite systems of all three classes—nonreactive and insoluble, nonreactive and soluble, and reactive—has broadened the study of useful interfaces considerably in the last few years.

This general introduction and presentation of the principal theories relating to the interface should suffice as an introduction to the chapters that follow in this volume. The remainder of this chapter will review a number of special topics. Because one of the purposes of this review is to bring together all aspects of the interface problem. some knowledge of the content of the specialist chapters has been assumed. For this reason, reading of the next sections may be left until later with advantage for the reader new to this field.

II. Review

The principal topics to be reviewed in this section are the requirements of the interface, bonding in composites, and investigation of the interface in composites.

A. Requirements of the Interface

Although the broadening of the types of interface that can give theoretical mechanical properties has made the subject of interfaces more complex, the function of the interface remains unchanged. The primary function is to transfer load between the reinforcement and the matrix. This may be termed a mechanical requirement of the interface and must be satisfied for all possible types of loading, and must continue throughout the life of the composite. The latter requires a stable interface and this constitutes a physical chemical requirement of the interface.

1. Mechanical Requirements of the Interface

Nearly all mechanical studies of composites have been based either on the premise that the interface is perfect, i.e., stronger than the matrix, or that there is no bond at the interface. Such studies are important for loading conditions that do not include interface failure and are reviewed in Chapter 2. But the intermediate case is becoming of much greater importance as stronger matrices are moving failure to the interface in many composites.

Failure at the interface may be expected to occur more readily under one loading condition than under another. Longitudinal tensile loading is usually the first mechanical test condition investigated with composites, but it may not be the most demanding on the interface. Any transfer of loads between filament and matrix under longitudinal stresses can take place over extended lengths so that the shear stress imposed on the interface need not be high. On the other hand, transverse tensile loading does not have the area advantage conferred by the filament length to aid in load transfer, and may constitute a more severe loading condition on the interface. Even more severe loading conditions at the interface may arise for off-axis stresses applied to the composite, depending upon the relative strengths of the interface under different stresses and combinations of stresses such as those depicted in Fig. 3. Or, the interface may experience its most severe stress conditions under externally imposed loadings that minimize plastic deformation because the latter is able to reduce stress concentrations. Notch tensile loading and the alternating load conditions associated with fatigue may provide a more severe test of the interface than other conditions, or the conditions at the ends of broken filaments may be the most demanding. The stress distribution at the interface for some of these conditions are discussed in detail in Chapter 2.

It has been suggested that there may be cases where a weak interface is desirable. Cook and Gordon (1964) have pointed out that the stress concentration near the tip of a propagating crack includes stresses that are tending to open cracks normal to the direction of propagation in addition to the principal crack opening stress. Hence, planes of weakness that must be crossed by the principal crack may open under these subsidiary stresses. Embury *et al.* (1967) applied these concepts to the fracture of laminate composite structures. They showed that a laminate of steel sheets provided a means to arrest cracks by the delamination process. The principal effect was to cause a decrease of more than 100°C in the tempera- of the ductile–brittle transition. Almond *et al.* (1969) reviewed the subject further and presented more results for this type of crack arresting structure. There is some evidence that similar considerations may apply in the case of filamentary composites, and this is reviewed in Chapter 7 on the subject of fracture. Various mechanisms can combine to involve a considerable volume of the composite on each side of the final fracture path, and, in such cases, the fracture toughness of the composite may exceed that of the matrix metal, as shown by Adsit and Witzell (1969) in the case of aluminum–boron composites.

Two mechanical properties have been studied in sufficient detail to relate their magnitudes to the condition of the interface. These are the longitudinal and transverse tensile strengths and are the subject of Chapters 4 and 5. For the system aluminum (6061)–boron, it is shown that maxima in both tensile strengths are associated with the beginning of breakdown of the pseudo-Class I interface. Numerous but small, isolated needles of aluminum diboride grow through the original interface, but may cover less than 1% of the interfacial area. In this case, it appears that the interface needs strengthening by this additional locking of filament and matrix to realize the maximum tensile strength. The result has an interesting analogy with graphite filaments; Goan and Prosen (1969) showed that whiskers of silicon carbide grown from the surface of the graphite fiber before combination with an epoxy matrix gave much higher interlaminar shear strength.

The requirements of the interface discussed so far relate to efficient load transfer between matrix and filaments. Another important requirement is that generation of the interface must not degrade the contribution of the filaments to the strength of the composite. In general, the latter requires that the intrinsic strength of the filaments be unaffected, but this does not require that the strength of extracted filaments be unchanged. This apparent contradiction can be reconciled by considering the difference in behavior of filaments and matrix acting in concert to give true composite action, and their individual behaviors. For example, titanium and boron give true composite behavior, as discussed earlier, when the reaction is below a critical amount. However, the extracted boron filaments are apparently degraded because there is no longer support by the matrix at the ends of the cracks in the titanium diboride coating formed by reaction. Yet the intrinsic strength of the boron core is apparently unchanged. A good illustration of the latter is discussed in Chapter 4 where it is shown that fully-degraded aluminum–boron composites have each boron filament surrounded by a thick layer of aluminum diboride. The extracted filaments have the reduced strengths indicated by their behavior in the composite. However, when the aluminum diboride layer is etched away from the extracted filaments, the strengths of the latter are approximately doubled to attain essentially their original strengths.

In other cases, there is evidence that the interface reaction reduces the intrinsic strength of the reinforcement irreversibly. For example, Petrasek (1966) found that the intrinsic strengths of tungsten filaments was decreased as the amount of recrystallization increased, and the latter was markedly affected by certain alloying elements in the copper matrix. Sutton and Feingold (1966) noted the weakening effect of active alloying elements in nickel on the strength of alumina filaments after fabrication of composites by infiltration. They have based a theory of composite strength on such observations and this is discussed in Chapter 8. The weakening of the alumina is believed to be due to surface roughening, in which case removal of the reaction product will not restore the strength, although chemical or flame polishing may cause the strength to increase approaching its former value.

Where interface reactions reduce the effective strength of the reinforcement in the composite, a further consideration is the change in the strength distribution. Rosen (1965) has shown that both the average failure strength and the coefficient of variation of strength of the filaments are related to the failure stress of the composite. His analysis shows that if the average strength of the filaments is unchanged, the distribution with the larger coefficient of variation will have the higher composite strength. In other words, the coefficient of variation provides some measure of the ability of the stronger filaments to take up the load shed by the weaker filaments when they fail. Also, the larger coefficient of variation may cause the fracture energy to increase because there is a greater chance that the weak point in the filament ahead of an advancing crack may be removed from the plane of the crack. This condition will require the crack to deviate to link up with the potential failure site in the next filament, or will cause the filament to be pulled out of the matrix and increase the fracture energy by this means. Hence, the effect of reaction between the filament and matrix on the mechanical properties will depend on both the average filament strength as well as the coefficient of variation of strength after attack.

Attack on a filament that leads to weakening will reduce the strength of the composite if the strength distribution is unchanged. But uniform attack may have the effect of reducing the coefficient of variation, as is the case for boron filaments after reaction with titanium. This type of interaction may reduce the fracture energy very appreciably because the crack will not need to be deviated significantly to find a weak point on each filament.

If the reaction between the filament and matrix is severe, the bond may be weakened to the point where debonding may occur. Debonding will increase the length of filament exposed to high stresses by the advancing crack. The result may be an increase in the fracture energy, even if the coefficient of variation has been reduced by the mechanism discussed above.

The net result of interaction between matrix and filament will depend on the influence of this reaction on filament fracture, delamination, the shear strength of the interface, and many other factors. It is not surprising that complete understanding of these many factors has not been achieved for any system at the present time.

2. Physical Chemical Requirements of the Interface

In their keynote address at the American Institute of Mining and Metallurgical Engineers' symposium on Metal Matrix Composites, Burte and Lynch (1969) identified filament–matrix compatibility as the pacing area for development of the technology of these composites. Although these authors include both physiochemical as well as mechanical compatibility within this subject, the basic problem was identified as degradation resulting from chemical interactions. Three approaches were being pursued in the search for solutions to this problem:

(1) New reinforcements which are thermodynamically stable with respect to the matrix.
(2) Coatings to reduce filament–matrix interaction.
(3) Alloy additions to reduce activity of diffusing species.

Filaments of titanium diboride have been prepared and studied for compatibility with titanium by Snide (1968). Much greater compatibility was found in this system than for titanium and boron, but several factors discourage further development along these lines. Foremost amongst these factors was the low strength and high density of titanium diboride filaments. Emphasis has been directed principally to the second and third approaches listed above. Coatings represent a difficult approach, particularly for higher temperatures, because they introduce two interfaces rather than one. However, a successful coating that is compatible with the reinforcement changes the compatibility problem to one of matrix–coating rather than matrix–filament. In this regard, silicon carbide-coated boron filaments (trademark Borsic†) appear to react with titanium in the same way as silicon carbide. Hence, the physical chemical requirements of the interface are unchanged so that the discussion that follows may be restricted to the composite system type characteristic of either the matrix–coating or matrix–filament. Table I includes one example of a coated filament system (aluminum–BN coated boron) and one example of a system where the coating is believed to act in the same way as a filament (aluminum–silicon carbide-coated boron is simulated by aluminum–silicon carbide).

Stability of the interface is most readily achieved with Class I systems where the components of the composite are limited to wetting each other. But for Class II and III systems, continued solid state diffusion will cause continued growth of the interface region. Methods to reduce the rates of growth of the interface are a very important aspect of the overall development of useful composites. It has been pointed out earlier that the matrices of most practical interest tend to be more reactive than the matrices used to demonstrate the validity of the theories of composites. An additional factor is that metal matrix composites are needed most critically for use at elevated temperatures, thereby adding to the magnitude of the problem. Kinetic studies of diffusion processes and identification of diffusion mechanisms are essential for development of the science of interfaces on a sound basis, and through this science to provide answers to the problem of control of the interface. These studies of the diffusional processes and mechanisms must be performed in the range of thicknesses of interface of interest for useful composites, and often this will mean that studies should be performed within the range of thicknesses below 10,000 Å. Changes in mechanism may often occur with continued growth, particularly under the restrained conditions present in a composite. For example, Ratliff and Powell (1970) have observed a change in the mechanism of diffusion between titanium alloys and silicon carbide at a thickness of 100,000 Å and attribute the change to new reaction products. Although this thickness is above the range of interest, it does provide evidence for changes in mechanism in the later stages of growth of an interface zone. However, the

† Registered trademark of Hamilton Standard Division of United Aircraft Corporation.

change may be more subtle and result from increases in the concentration of vacancies.

Some means to control the growth of the interface have included: exhaustion of reacting species from a very low concentration matrix alloy such as Ni–0.01 at. % Ti in contact with alumina (Sutton, 1964); control of vacancies in reaction product to reduce rate of diffusion transport (Schmitz *et al.* 1970); and rejection of one solute element in the matrix ahead of the growing reaction phase (Blackburn *et al.* 1966). Coatings have been developed, in other cases, to convert a Class III system into a Class I system, such as the boron nitride coating on boron that allows infiltration by molten aluminum (Camahort, 1968).

Stability of the interface is the principal physical chemical requirement that must be satisfied if the composite is to perform satisfactorily throughout its life. Satisfaction of this requirement may depend, in turn, on the loading conditions. For example, it has been found that the thickness of titanium diboride reaction zone must be less than 5500 Å to achieve full longitudinal tensile strength in Ti(A70)–B composites, but that the transverse strength has a constant value over the reaction zone thickness range from less than 1000 to greater than 100,000 Å. However, less and less information is available as the mechanical property becomes more complex, so that almost no information is available on the subject of fatigue. Optimization of the interface for a variety of loading conditions is probably one of the most urgent problems facing the developer of composite materials.

The subject of physical chemical requirements of the interface is discussed in more detail in Chapter 3. Chapter 10 covers a special topic within this field relating to the effect of trace and impurity elements on the structure and stability of the interface.

B. Bonding in Composites

The definition given earlier for classification of composites was based primarily on phase equilibria (solubility and compound formation). It provided a useful system for classification based on the equilibrium (or apparent equilibrium) attained after fabrication of the composite. The mechanism by which bonding was achieved and the nature of the bond were ignored, but these constitute important areas in understanding interfaces.

The method of fabrication has a marked influence on the characteristics of the interface. Aluminum–boron composites made by molten metal infiltration have the characteristics of Class III composites with irregular

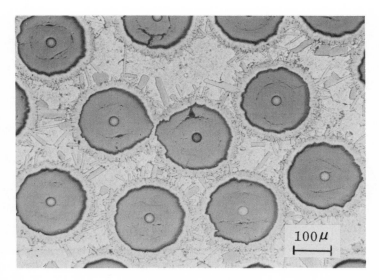

Fig. 6. Appearance of aluminum(6061)–boron composite after melting of matrix (1400°F for 1 hr).

attack on the filament and nonuniform growth of aluminum boride (Fig. 6). In contrast, composites made by the optimum solid state diffusion bonding process appear to be free of reaction, except for an occasional crystal of the boride at the interface (Fig. 7). The mechanism of solid state diffusion bonding will be discussed to explore the causes for this difference. This discussion will provide a background for review of the influence of processing on the characteristics of the interface in more general cases.

1. Fabrication Aspects

Apart from eutectic composites, fabrication generally involves the bringing together of two or more constituents. Each of these is carefully prepared to be free of contaminants, but adsorbed films will be present on any surface (unless prepared under very special conditions, such as high temperature vacuum bakeout or cathodic bombardment). Reaction with oxygen from the air is the primary source of films on metals, but reaction with water vapor may provide the films in the case of oxides and certain nonmetals. In addition to the primary source of films, other contaminants may be present in varying amounts such as oil or grease, chlorides and sulfides (from handling), dust and other foreign matter, and

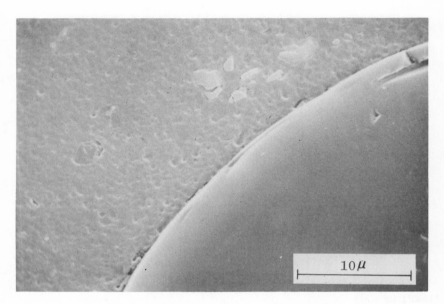

Fig. 7. Appearance of aluminum(6061)–boron composite fabricated in solid state.

reaction products of the above such as hydrated oxides. Bonding is therefore not a simple physiochemical process. In most cases, it is desired to
dissipate the films by some means so that metal-to-reinforcement bonds
can be made. However, in other cases, it may be desired to retain the films
or modified films; for example, the oxide films on aluminum and boron act
to minimize interaction in these composites.

Many composites are fabricated by solid state processes under pressure. An important variable in such processes is the amount of plastic
deformation. The American Welding Society (1971) defines "diffusion
welding" as a process involving no macroscopic deformation of the parts,
and uses the term forge welding where large amounts of deformation are
involved. However, review of diffusion bonding shows that the controlling
parameter is more descriptive, such as yield stress-controlled or creep-
controlled, with a further distinction based on the mode of application of
force by static or dynamic means. Furthermore, the most interesting area
is at the boundary between the two types of process, as defined by the
American Welding Society (1971) and, accordingly, the generally used
term diffusion bonding will be used throughout this volume to describe
this solid state method to manufacture composites.

Plastic deformation occurring under the application of the diffusion

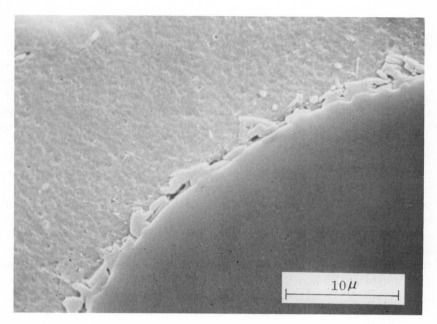

Fɪɢ. 8. Appearance of aluminum(6061)–boron composite fabricated in solid state after 50 hr at 940°F showing growth of aluminum boride through original interface film.

bonding force is the most important factor contributing to contact between constituents. Shear may take place at interfaces during this plastic deformation and lead to rupture of films. In some cases, this may be desirable to accelerate bonding, but may be undesirable if preservation of the film is required to reduce reaction as in pseudo-Class I systems. (Increase in the amount of film breakdown by shear is believed to be one reason why high‧pressures are detrimental in hot-pressing of aluminum matrix composites.) Another process by which oxide films may be disrupted is spheroidization. Thin oxide films are intrinsically unstable because they possess high surface energies. Excess temperature or prolonged times at the bonding temperature will cause loss of continuity of the oxide film by spheroidization, and permit matrix–filament reaction to occur. Figure 8 shows an example of an aluminum–boron composite in which the original oxide interface has broken down with growth of aluminum boride on each side of the spheroidized oxide.

Mechanical disruption by shear and spheroidization under thermal influence are the two principal methods of film dissipation encountered in

diffusion bonding. Other methods of fabrication introduce similar processes. For example, mechanical shear is probably replaced by erosion in
the case of composite fabrication by molten metal infiltration. A third
method of oxide film dissipation is by solution. The solubility of oxygen
in aluminum is vanishingly small, but is large enough in metals such as
nickel to aid in removal of oxides or to contribute to spheroidization through
solution and reprecipitation. The solubility in metals such as titanium and
columbium is very high so that the solution mechanism will suffice to
generate clean interfaces.

In addition to films at the interface, another factor to be considered
results from the surface roughness. Gases may be entrapped as surfaces
are pressed together and progressive deformation of asperities seals off
pockets of gas. The gas may be readily lost by diffusion through the metal
as in the case of hydrogen liberated by reaction of either hydroxyl groups
or water vapor with either matrix or filament to form an oxide. On the
other hand, it may be a noble gas such as argon and may require long
times of annealing to remove the last few atoms. Such problems become
more important in high speed processes used to make continuous tapes.

In summary, consideration of the fabrication processes shows that the
interface will be extremely complex. Oxide films, trapped gases, and high
concentrations of vacancies and dislocations are a few of the imperfections
that may occur. Imperfections are generally present in the grain boundaries of metals (e.g., a killed steel remains fine grained because aluminum
oxide particles anchor grain boundaries), but the extent is believed to be
much greater in composites.

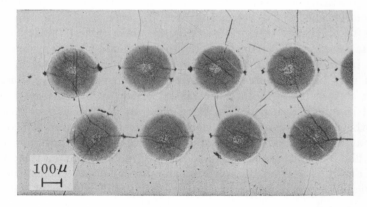

Fɪɢ. 9. Appearance of pores in columbium alloy–tungsten composite after 10 hr
at 2400°F.

The most important effect resulting from the complexity of the interface is the apparent stability of pseudo-Class I composites. This effect has already been discussed and will be examined further in other chapters of this volume. Another effect has been noted in composite systems where thermodynamic instability results in diffusion across the interface. Diffusional inbalance is often found and results in void formation by the Kirkendall mechanism. However, the high concentrations of imperfections at the interface provide nuclei for the condensation of vacancies and appear to accelerate the formation of pores. Klein *et al.* (1970) found these pores in a composite with columbium alloy matrix and tungsten filament reinforcement after heating for 10 hr at 2400°F, as shown in Fig. 9. The nucleation of pores along the site of the original interface can be seen clearly in this figure.

2. Types of Bonding

Six types of bonds are proposed in Chapter 3. These are: the mechanical bond, the dissolution and wetting bond, the oxide bond, the reaction bond, the exchange reaction bond, and mixed bonds. Figure 10 presents schematic examples of some of the principal bond types. These are discussed in more detail in Chapter 3 with fuller development of oxide bonds in Chapter 8.

Mechanical bonds require an absence of any chemical source of bonding even from van der Waals forces, and involve mechanical interlocking. However, the absence of any chemical source of bond will cause a com-

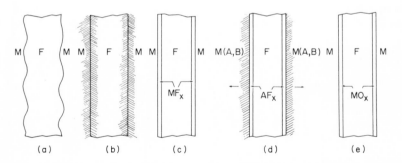

Fig. 10. Schematic of principal bond types: matrix M may contain elements A and B and filament may be single element (such as graphite) identified as F or compound (such as Al_2O_3) denoted as FO_x. (a) Mechanical bond, (b) dissolution and wetting bond, e.g., Cb–W, (c) reaction bond e.g., Ti–C, (d) exchange reaction (e.g., Ti(Al)–B), (e) pseudo-Class I oxide bond where $\Delta F_{MOx} < \Delta F_{FOy}$ (e.g., Al–B).

posite to be very weak under transverse loads and this bond is not believed to be useful in composite technology.

Composites with other than oxide reinforcements in which the matrix wets or dissolves the reinforcement, but does not form compounds with it, are included under the dissolution and wetting bond type. The oxide bond may be generated by wetting, but can also include bonds where intermediate compounds form at the interface. In general, metals with oxides that have a small free energy of formation do not form strong bonds with alumina, but traces of oxygen or active elements will enhance the bond by forming intermediate zones. All of these are included in the oxide type of bond. In addition, bonding between the oxide films on the metal matrix and on the filaments is included in the general classification of oxide bonds.

Reaction bonds, and the special case of exchange reaction bonds, are restricted to composites with other than oxide reinforcements. Titanium–boron provides a good illustration of a reaction bond. The exchange reaction bond occurs when a second element in matrix or filament begins to exchange lattice sites with the elements in the reaction product.

The final type is the mixed bond. Figure 8 showed partial breakdown of the initial bond in aluminum–boron (believed to be an oxide bond between B_2O_3 and Al_2O_3) with subsequent formation of aluminum boride AlB_2. A most interesting experiment has been reported by Klein and Metcalfe (1971) in which the 6061 aluminum alloy matrix was removed from an Al(6061)–25% B composite by solution in dilute caustic soda. The composite had been heated for up to 165 hr at 940°F to increase the amount of breakdown over that caused in fabrication. Reaction with caustic soda ceased when the interface was reached. The boron was now removed by reaction with alkaline potassium ferricyanide (Murakami's reagent). It

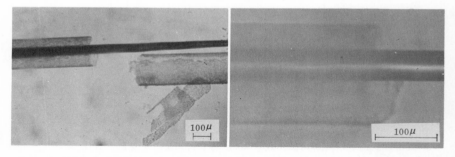

FIG. 11. Structure of extracted interface in heat-treated Al(6061)–25B composites showing the interface network around a boron filament whose diameter has been reduced by solution in Murakami's reagent.

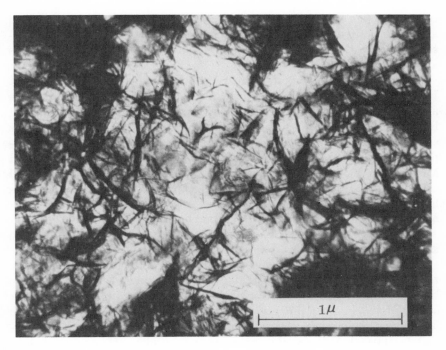

FIG. 12. Structure of extracted interface in heat-treated Al(6061)–25B composites showing the structure of this interface network as revealed by transmission electron microscopy.

appears that attack on the boron took place through holes in the interfacial zone layer. Figure 11 shows the interfacial zone layer after partial solution of the boron. This thin membrane-like layer is in the form of a tube of 0.004 in. diameter and the partially dissolved filament diameter is less than 0.002 in. When this membrane was examined in the electron microscope, the structure was found to be complex, as shown in Fig. 12. Further discussion of this interface is presented in Chapter 3.

C. *Investigation of the Interface in Composites*

In view of the complexity of some interfaces it is necessary to pay unusual attention to the methods employed to investigate interfaces. Two principal techniques have been employed. In one approach, the interface is isolated and examined, for example, in wetting studies on plates by molten metals, or by forming a diffusion couple between a pair of materials. In the other approach, typical composites have been examined after the desired interface has been generated by the fabrication procedure or postfabrication processing. Each approach has its strengths and weaknesses.

Projected lives of composites must extend to many thousands of hours. It is tempting, therefore, to reduce these times by studying interfaces after shorter exposures at higher temperatures, but this can be misleading because of changes of mechanism with temperature. This warning applies equally to the two approaches outlined above. In the case of the first approach, Brennan and Pask (1968) as well as Champion *et al.* (1969) noted a change in the wetting of alumina by aluminum that became marked above 950°C. In the case of the second approach, work in the author's laboratory has found a marked decrease in the reaction rate between the alloy Ti–13V–10Mo–5Zr–2.5Al and boron that occurs above 1400°F. This decrease results from dissolution of a molybdenum-rich boride phase above this temperature, as discussed in Chapter 3.

Another temptation in studies of the interface is to widen the interface zone by excessive reaction so that examination and measurements may be made more readily. But reference has already been made to the work of Ratliff and Powell (1970) in which it was shown that a marked change in kinetics occurred in the system titanium–silicon carbide at approximately 100,000 Å of reaction zone, whereas studies of composites show that the useful range is below 10,000 Å. However, no general guidelines can be proposed because the range of reaction providing useful properties in the composite will depend on the system, the physical size of the reinforcement, and many other factors. For example, Noone *et al.* (1969) concluded that interfacial reactions with fine alumina whiskers (a few microns in diameter) presented almost insoluble problems, but that reaction could be controlled with large diameter alumina filaments (0.010 in. diam).

In general, more difficulty has been experienced in simulation of conditions within a composite by use of isolated, single interfaces than by use of the composite. However, where true simulation can be achieved, meaningful measurements can be made more readily and accurately by this type of specimen. Reservations in connection with this type of specimen fall into two main categories:

(1) Failure or inability to control gas atmosphere.
(2) Failure to simulate geometrical conditions in a composite.

Control of the gas or vapor phase is very important to simulate the fabrication stage and the service stage in the life of a composite. Bonfield, in Chapter 10, describes the marked effect of various gas atmospheres on the wetting of silicon nitride by aluminum, so that selection of the best atmosphere for fabrication of composites can be made based on this work. On the other hand, Basche (1969) describes compatibility studies between silicon carbide coated boron filament and titanium made by heating the filaments in contact with titanium powder. Within a composite, a titanium matrix will maintain exceedingly low dissociation pressures of oxygen and nitrogen at the surfaces of the filaments. The access of reactive gas molecules to the filaments is believed to explain the low rate of reactivity reported with titanium powders.

The need to simulate geometrical conditions typical of a composite is not as generally appreciated as the other requirements in compatibility work. These requirements include:

(1) The volumetric relationship must be satisfied, particularly in a system where the reinforcement (RX) contains an interstitial element X that diffuses faster than the element R. Examples are silicon carbide SiC and alumina Al_2O_3. Rapid diffusion of the interstitial element X will eventually saturate the matrix in a composite. Simulation of the distances and volume percentages in the composite must be achieved so that the change in kinetics after saturation of the matrix occurs at a meaningful time. This point is discussed more fully in Chapter 3.

(2) The radial conditions must be simulated for reasons including: the effect of radius on solubility (Thomson–Freundlich relationship), the constraint on volume changes resulting from reaction, the influence of the fixed interface position on Kirkendall void formation, and effect of radius on concentration gradients. These effects are discussed in Chapter 3 but from the point of view of physical chemistry rather than design of experiments.

(3) The crystallographic conditions must be simulated. Champion *et al.* (1969) show a marked dependence on crystallographic orientation for the reaction between alumina and aluminum.

Although adherence to the requirements presented above will help to ensure that compatability and kinetic data are pertinent to the composite system under study, it should not be assumed that all data obtained under other conditions are invalid. Rather, caution should be exercised in accepting data obtained under conditions not satisfying these requirements. For example, both Carnahan *et al.* (1958) and Champion *et al.* (1969) observed periodic sudden contractions of molten aluminum on alumina that might be important in composite fabrication. However, the conditions of 1350°C for 1 hr are believed to be excessive for composite technology, particularly because high pumping rates allowed rapid removal of aluminum and Al_2O_3 in the vapor phase. On the other hand, the conclusion of Champion *et al.* (1969) that "there may be an optimum temperature for the incorporation of corundum whiskers into an aluminum matrix by liquid infiltration if good wetting is to be obtained without excessive attack" is of major interest to the field of composites.

Study of the mechanical properties of interfaces presents similar problems to those experienced in physical chemical examination of the interface. Isolated, single interfaces may be used, but these do not simulate the residual stress conditions in composites, and the complex stresses introduced by strain are not the same as those in a typical composite material. On the other hand, filament pullout tests also fail to simulate conditions in a composite for reasons discussed more fully in Chapter 2.

The sensitivity of the interface to fabrication conditions in many systems presents another reason to study interfaces by observation of their behavior in composites rather than as isolated, single interfaces. Results may be inferred and not be subject to quantitative measurement, but the quantitative measurements obtained on isolated interfaces do not correlate well in all cases with the conditions in a composite. It is evident that new experimental approaches will be among the most valuable tools to advance the science of interfaces.

References

Adsit, N. R., and Witzell, W. R. (1969). *SAMPE Tech. Conf. Seattle, Washington, September 9–11, 1969*, **1**.

Almond, E. A., Embury, J. D., and Wright, E. S. (1969). Fracture in laminated materials, *in* "Interfaces in Composites," STP 452, pp. 107–129. Amer. Soc. Test. Mater., Philadelphia, Pennsylvania.

American Welding Society (1971). "Welding Handbook," Sect. 3B, Chapter 52. Amer. Welding Soc., New York.

Arridge, R. G. C., Baker, A. A., and Cratchley, D. (1964). *J. Sci. Instrum.* **41** (5), 259–261.

Basche, M. (1969). Interfacial stability of silicon carbide-coated boron filament reinforced metals, *in* "Interfaces in Composites," STP 452, pp. 130–136. Amer. Soc. Test. Mater., Philadelphia, Pennsylvania.

Blackburn, L. D., Herzog, J. A., and Meyerer, W. J. *et al.* (1966). "MAMS Internal Res. on Metal Matrix Composites," MAM-TM-66-3.

Brennan, J. J., and Pask, J. A. (1968). *J. Amer. Ceram. Soc.* **58**, 569.

Burte, H. M., and Lynch, C. T. (1969). Filament–matrix compatibility, *AIME Symp. Metal Matrix Composites, Pittsburgh, Pennsylvania, May, 1969.*

Camahort, J. L. (1968). *J. Comp. Mater.* **2** (1), 104–112.

Carnahan, R. D., Johnston, T. L., and Li, C. H. (1958). *J. Amer. Ceram. Soc.*, **41**, 343.

Champion, J. A., Keene, B. J., and Sillwood, J. M. (1969). *J. Mater. Sci.* **4,** 39.

Cook, J., and Gordon, J. E. (1964). *Proc. Roy. Soc.* **282A,** 508.

Cooper, G. A., and Kelly, A. (1968). Role of the interface in the fracture of fiber–composite materials, *in* "Interfaces in Composites," STP 452, pp. 90–106. Amer Soc. Test. Mater., Philadelphia, Pennsylvania.

Cratchley, D. (1963). *Powder Met.* No. 11, 155–157.

Davis, L. W. (1967). *Met. Progr.* **91** (4), 105–114.

Ebert, L. J., Claxton, R. J., Kmieciak, H., and Wright, P. K. (1971). AFML-TR-71-133.

Embury, J. D., Petch, N. J., Wraith, A. E., and Wright, E. S. (1967). *Trans. AIME* **239,** 114–118.

Goan, J. C., and Prosen, S. P. (1969). Interfacial bonding in graphite fiber–resin composites, *in* "Interfaces in Composites," STP 452, pp. 2–36. Amer. Soc. Test. Mater., Philadelphia, Pennsylvania.

Goddard, D. M., Pepper, R. T., Upp, J. W., and Kendall, E. G. (1972). *Welding J. Res. Suppl.* **51,** 1785–1825.

Graham, L. P., and Kraft, R. W. (1966). *Trans. AIME* **236,** 94–102.

Hibbard, Jr., W. R. (1964). An introductory review, *in* "Fiber Composite Materials," pp. 1–10. Amer. Soc. Metals, ASM, Metals Park, Ohio.

Jech, R. W., McDanels, D. L., and Weeton, J. R. (1960). *Met. Progr.* **78** (6), 118–121. Also, "Stress–Strain Behavior of Tungsten-Fiber Reinforced Copper Composites" (1963). NASA TN-D-1881.

Jones, R. C. (1968). Deformation of wire-reinforced metal matrix composites, *in* "Metal Matrix Composites," STP 438, pp. 183–217. Philadelphia, Pennsylvania.

Klein, M. J., and Metcalfe, A. G. (1971), "Effect of Interfaces in Metal Matrix Composites," USAF Contract F33615-70-C-1814.

Klein, M. J., Metcalfe, A. G., and Domes, R. B. (1970). "Tungsten-Reinforced Oxidation Resistant Columbium Alloys," Naval Air Systems Command, Final Rep. on Contract N00019-69-C-0137, November.

Metcalfe, A. G. (1967). *J. Comp. Mater.* **1,** 356–365.

Noone, M. J., Feingold, E., and Sutton, W. H. (1969). The importance of coatings in the preparation of Al_2O_3 filament/metal–matrix composites, *in* "Interfaces in Composites," STP 452, pp. 59–89. Amer. Soc. Test. Mater., Philadelphia, Pennsylvania.

Petrasek, D. W. (1966). *Trans. AIME* **236,** 887–896.

Petrasek, D. W., and Weeton, J. W. (1964). *Trans. AIME,* **230,** 977–990.

Ratliff, J. C., and Powell, G. W. (1970). "Research on Diffusion in Multi-Phase Systems: Reaction Diffusion in the Ti/SiC and Ti-6Al-4V/SiC Systems," AFML-TR-70-42.

Rosen, B. W. (1965). Mechanics of composite strengthening, *in* "Fiber Composite Materials," pp. 37–75. Amer. Soc. Metals, Metals Park, Ohio.

Salkind, M. J. (1968). "Introduction to Interfaces in Composites," STP 452. Amer. Soc. Test. Mater., Philadelphia, Pennsylvania.

Schmitz, G. K., Klein, M. J., Reid, M. L., and Metcalfe, A. G. (1970). "Compatibility Studies for Viable Titanium Matrix Composites." Solar Div. of Int. Harvester Co., AFML-TR-70-237.

Signorelli, R. A., Petrasek, D. W., and Weeton, J. W. (1967). Interfacial reactions in metal–metal and ceramic–metal fiber composites, *in* "Modern Composite Materials" (Krock and Broutman, eds.) Chapter 5, pp. 146–171. Addison-Wesley, Reading, Massachusetts.

Snide, J. A. (1968). "Compatibility of Vapor Deposited B, SiC and TiB₂ Filaments with Several Titanium Matrices," Rep. AFML-TR-67-354, Air Force Mater. Lab.
Sutton, W. H. (1964). "Investigation of Bonding in Oxide-Fiber (Whisker) Reinforced Metals," General Electric Co., Final Tech. Rep. (AMRA CR 63-01/8) (DC No. AD-455, 770).
Sutton, W. H., and Feingold, E. (1966). *Mater. Sci. Res.* **3,** 577–611.
Sutton, W. H., Chorne, J., and Gatti, A. (1960–64). Development of Composite Structural Materials for High Temperature Applications, Co. Quart. Progr. Rep. 1–16, Contract Now-60-0465d, General Electric Co.

2

Mechanical Aspects of the Interface

L. J. EBERT and P. KENNARD WRIGHT†

Division of Metallurgy and Materials Science
Case Western Reserve University
Cleveland, Ohio

† Present address: General Electric Company, Lamp Business Division, Lighting
Research and Technical Services Operation, Cleveland, Ohio.

I. The Nature of the Interface

To gain an understanding of the mechanical aspects of the interface in fiber composite materials, some attention must be given to the basic nature of the interface itself. The unique nature of the interfaces in fiber composite materials, and the concomitant specific mechanical interactions produced at them, constitute one of the major factors in giving fiber composite materials their unique properties.

Fiber composite materials differ from other multiphase materials in one major way—the interfaces between the reinforcing phase and the matrix in which it is embedded are highly oriented. In addition, in certain types of fiber composites, there is an additional major difference—the existence of a chemical composition gradient across the interface. It is generally conceded that this second difference, the existence of the chemical composition gradient, is undesirable, although there may be special cases where interdiffusion of components creates favorable solid solution hardening. In rationalizing the mechanical aspects of the interfaces in fiber composite materials and their effects on the integral behavior of the composite material, the presence or absence of a continuum, both mechanically and chemically, must be taken into account.

A. The Mechanics Continuum

Any multiphase material constitutes a mechanics continuum in which the unique mechanical properties of each of the phases present contributes something to the overall properties of the material. In addition, the rheological interactions, which occur at the interfaces between the phases present because of their different elastic and plastic properties, make a significant contribution to the measured properties of the composite material when it is loaded externally. In some cases, this latter contribution is beneficial; in other cases, it is detrimental.

In composite materials in which the harder, reinforcing phase is non-fibrous, such as in the case of pearlitic steels, the rheological interactions which occur at the interfaces in the continuum are randomly directed because of the randomness of the interfaces. Thus, the effects of the reinforcement action of the harder phase are nondirectional. This is not the case for fiber composite materials, especially those in which the reinforcing fibers are highly oriented. It is this directionality of the continuum which is usually exploited in the use of fiber composite materials.

Since it is possible to apply service or use loads only to the external surfaces of the composite material, the interior members of the composite materials, both reinforcement and matrix, must receive their share of the applied load from the material which surrounds them through a shear loading action. In other words, the externally applied load is transferred from the surfaces of application to the material adjacent to it through a shearing action. This material, in turn, transmits the load further into the interior of the composite through the same kind of action.

Because of the shear-nature of the load transfer into the interior of the composite material, the externally applied load, or more precisely the externally applied strain, is not uniformly distributed over the entire cross

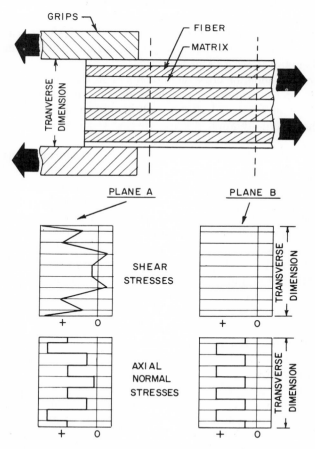

FIG. 1. The shear nature of the external loading of a fiber composite material, illustrating the internal stress distribution.

section of the composite material in the immediate vicinity of the load application. However, at some distance from this area, the externally applied strain does in fact become uniformly distributed across the cross section of the composite. At this point each component member of the composite is subjected to a stress in the direction of the externally applied load, the intensity of which stress is determined by the magnitude of the (uniform) strain, and the modulus of the component species. At this location, the value of the shear stress at the interfaces of the composite in the direction of the applied load is zero (Fig. 1).

The distance between the area of the externally applied load and the cross-sectional area of the composite upon which the strain corresponding to that load is uniform has been shown to be as little as four or five fiber diameters from theoretical considerations of Dow (1963). In actual fiber composites, the distance required for the strain distribution to become uniform is many times greater than this.

Broken fiber ends, or crossovers of fibers in the matrix, produce a discontinuum in the composite which produces a perturbation in the continuity of the stress distribution which is akin to that which exists in the area of the externally applied load. This subject is treated later in Section II,C.

Once the externally applied load, or strain, is uniformly distributed over the cross section of the composite, the differences in the stress levels of the members of the composite, together with their innately different elastic constants, produce differences in contractile tendencies among the components of the composite. Since all members of the composite are united to form a single continuum, the differences in contractile tendencies give rise to transverse stresses among the components of the composite. The sign and intensity of these stresses are a function of the species of the components, the volume fraction of the reinforcing component, the magnitude of the applied strain, and other geometric factors. The severity of the resulting complex stress state is maximum at the interfaces between the components of the composite, although the effect of this stress state complexity is reflected in the behavior of the entire integral composite material. The nature of the stress state, and its effects, are discussed more fully in following sections of this chapter.

The degree of perfection of the mechanics continuum has a major role in the performance of the composite and in maximizing the reinforcing action of the harder and stronger member of the composite. Ideally, it is hoped that the continuum is perfect, i.e., that there is a perfect bond between adjacent component members of the composite. This means that the atomic structure of the components on both sides of the interface are coherent across the interface, and that a uniform and constant interfacial strength exists all over. The continuity of all interfaces, and their constant uniform strength, are important if the interfaces are to perform their primary function of creating a continuum. Further, the functions of load distribution and resistance of the transverse tensile stresses arising because of the rheological interactions which occur at the interfaces are related to the degree of integrity of the interfaces.

In most cases, departure from the ideal continuum at the interfaces (incomplete and imperfect bonding) reduces the effectiveness of the reinforcing function of the harder and stronger of the components. Further, the degradation of the composite properties with loss of ideality of the interfacial bond occurs at a rate much greater than might be expected on the basis of a linear relation between component properties and volume fraction of the interfacial product.

B. The Chemical Discontinuum

The ideal composite material is thought to be one in which the perfect mechanics continuum (as described above) exists by virtue of perfect bonding between the components of the composite, but yet in which a perfect chemical discontinuum exists at the interface. In other words, there should be no chemical interdiffusion between the species which comprise each of the components of the composite. The reasons for the preference of the ideal chemical discontinuum are related to the nature of the problems which exist when it is not present.

Most of the preferred matrix species in metal matrix composites, those with low densities and high ductilites, have a high propensity for the formation of chemical compounds with the preferred reinforcement species (boron, silicon carbide, etc.). The resultant chemical compounds, often intermetallic in nature, are brittle and of low effective strength. Because of their characteristics, these compounds, which have a natural tendency to form at the interfaces between the components of the composites during the elevated temperature compositing process, may reduce the effectiveness of the composite interfaces in carrying out their primary functions of load distribution and resistance to fracture under the complex stress states generated at the interfaces. Metcalfe (1967) has used a model based on this effect in reacted Ti–B and Al–B composites to explain the strength degradation observed. Evidently the presence of cracks in the weak interfacial boride layer is sufficient to induce premature failure in boron filaments which by themselves are relatively insensitive to surface imperfections.

In addition to degrading the effectiveness of the interfacial functions, the chemical interactions at the interfaces have a tendency to degrade the basic properties of the reinforcement components. The interaction often creates pitted, irregular surfaces on the reinforcement components, which constitute notches. Since most of the preferred reinforcement components have a high inherent notch sensitivity, the surface irregularities greatly reduce the effective strength of the reinforcement components. Hence, the entire composite strength is thereby degraded.

Finally, the mass transport into both matrix and reinforcement, which accompanies the chemical interaction at the interface, has a tendency to reduce the desirable properties which each of the components contribute to the composite. The matrix loses some of its ductility, while the reinforcement loses some of its inherent or basic strength.

The only species of component materials which can be combined into a composite in which requirements for both a mechanical continuum and a chemical discontinuum are met (nearly) perfectly are those which are thermodynamically compatible. The outstanding example of this type of composite is the eutectic composite in which the hard component is one of the phases of the eutectic mixture. Because of the thermodynamic genesis of the hard phase, chemical interdiffusion is effectively absent, and at the same time, mechanical continuity across interfaces is virtually assured.

Oriented eutectic composites, despite their desirable features in terms of continua, have severe limitations from a practical point of view. The fiber fraction of reinforcement is set within very narrow limits by the thermodynamics of the materials systems. Further, there is little latitude on the choice of matrix and reinforcement materials, and very often the matrix component has high density. Finally, the cost of production of the oriented material is often prohibitively high, and the freedom of choice of geometric shape is severely limited.

Relatively good success has been attained in achieving the goals of chemical discontinua by using a combination of reinforcement and matrix in which the former component is virtually insoluble in the liquid phase of the latter. Copper reinforced with tungsten filaments are perhaps the first and foremost of these systems (Fig. 2). As was the case with the oriented eutectic composites, restrictions of materials combinations based on limited solubility of the hard component in the matrix component severely limit the freedom of selection of component materials for this type of composite. Hence, this class of composites find only limited usefulness.

The need for composite materials composed of thermodynamically incompatible components, the synthetic union of which would produce interfacial diffusion and its attendant damaging effects, has led to sig-

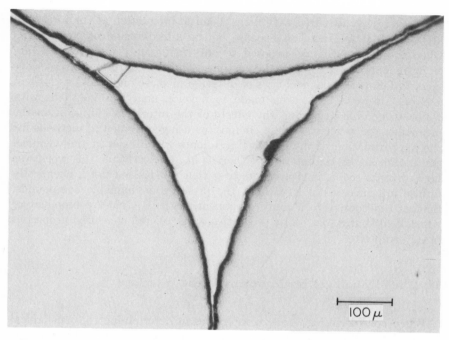

Fig. 2. Cross section of a tungsten fiber–copper matrix composite showing absence of interfacial reactions after liquid infiltration at 2200°F.

nificant activities to introduce a diffusion barrier between the components of the composite. Boron filaments coated with silicon carbide (or Borsic), and nitrided boron filaments have been used to reinforce aluminum-base alloys with a considerably reduction in the rate of reaction between the filament and matrix (see Chapter 3). This reduced rate of reaction produces composites which retain their strength much longer when exposed to elevated temperatures. It would appear that the added cost of the surface protection of these filaments has been adequately offset by the improved properties of these composites.

C. The Chemical Continuum

Despite the efforts to eliminate, or to minimize, the interfacial interaction and interdiffusion, metal matrix composite materials are still used under conditions in which a continuous chemical gradient exists across the interface between the constituent components. Even those composite materials which can be synthesized with little chemical interaction at the interfaces are subject to degradation because of the interdiffusion of components when the composites are exposed to only moderately high temperature environments for long periods of time. The interdiffusion interactions seem to follow the classical laws of diffusion in these materials.

Some attempts have been made to develop analytical methods with which to be able to predict the effects of the interfacial diffusion on the mechanical properties of these composites using incremental methods for the approximation of the chemical gradients which exist in the chemical continuum across the interface (Ebert *et al.*, 1971). While this work was limited in its scope, it clearly indicates that the presence of a chemically diffuse interface, with or without the presence of definable compounds, renders the composite a multicomponent aggregate, each component of which contributes some of its properties to the overall measured properties of the composite.

II. The Mechanical Environment of the Interface

When considering the role of interfaces in determining the mechanical properties of fiber composite materials it is necessary to keep in mind the possible dualistic approach to the phenomenon. The problem may be addressed either as finding the effect of interfacial strengths and behavior on the properties of the composite as a whole, or alternatively, as finding the effect of the mechanical behavior of the bulk of the composite on the behavior of the interface. Neither one of these approaches is singular since the two are mutually interacting; the behavior of the interface affects the behavior of the composite, which in turn affects the interfacial behavior. Since much of the material in other chapters of this volume will adopt the former approach, it will be instructive to consider the behavior of interfaces here primarily from the latter viewpoint.

Whenever possible, the approach used will be as general as possible, that is, with as few assumptions about the composite behavior as possible. While at times making the analysis more complicated, this approach will allow a more careful and accurate consideration of the mechanical behavior of the composite. Indeed, as will be shown, it will permit the establishment of the important generalization that, even for simple external composite loadings, the states of stress and strain at the interfaces between components are complex, and vary from place to place in the composite. This complexity of conditions at the interface means that the interface will respond in a manner which is different from what might be expected from simple loading conditions. Thus it is not always reliable to depend on the predictions of simplistic models of composite behavior to give the interfacial conditions.

Despite the desire to limit the number of assumptions that are made in establishing the composite response, it is necessary to make a few basic premises to simplify the mathematics of the problem. Even these may occasionally be relaxed.

(1) The components are assumed to be chemically discrete, that is, the mechanical properties of the components change discontinuously at the interface.

(2) The bonding is considered either to be perfect, of greater strength than the matrix itself, or totally debonded so that the interface can support neither tensile nor shear forces. The first case of perfect bonding has been analyzed much more extensively than that of no bonding since perfect bonding usually results in superior composite properties. The intermediate cases of limited strength bonds, or regions of good bonding mixed with regions which are unbonded, are much more difficult to analyze because of complicated boundary conditions that have to be met at the interface. As a consequence, such cases have received little attention.

(3) In the presence of very long, continuous fibers, the composite under externally applied load is assumed to be in a state of uniform strain in a plane normal to the fiber direction. In other words, in a given plane, the strains present in both fibers and the matrix in the fiber direction are identical. This follows from the second assumption of perfect bonding. The constant strain condition is relaxed when dealing with stress analyses of discontinuous fibers.

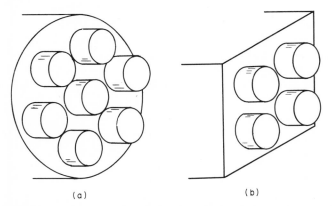

(a) (b)

Fig. 3. Idealization of random composite fiber array to a regular (a) hexagonal or (b) square array of fibers.

(4) The fibers are assumed to be distributed throughout the matrix in a regular array. This allows the isolation of a small repeating symmetry unit which is representative of the composite as a whole. As an example, square and hexagonal fiber arrays are illustrated in Fig. 3. Again, investigation of the effect of relaxing this requirement has been made for a few cases.

A. *Longitudinal Loading or Loading in a Direction Parallel to the Fibers*

This loading system is obviously of most practical interest, since by loading in the direction of the fibers, the high modulus and strength of the fiber material can be most fully exploited. Even in this simple configuration, however, the internal state of stress in the composite (including that of the interface) will deviate from uniaxial tension. This has been thoroughly demonstrated in the work of Ebert and Gadd (1965), Piehler (1967), and Bloom and Wilson (1967).

Ebert and Gadd have modeled the composite by considering a typical fiber–matrix unit to have a cylindrical configuration with a core of matrix

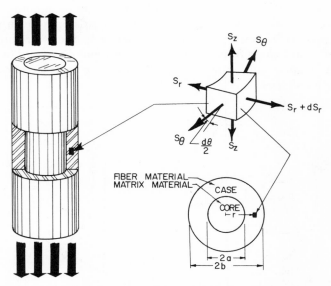

FIG. 4. Conceptual development of the concentric cylinder model for fiber composites. The high fiber fraction model is shown, with the fiber simulator as case.

and an outer sleeve of fiber simulator (Fig. 4). This arrangement was originally devised to represent a high fiber fraction composite, but was later extended (Ebert *et al.*, 1968) to low fiber fractions by simply inverting the fiber and matrix simulators.

Taking advantage of the cylindrical symmetry of the model, analytical expressions for the composite behavior could readily be developed. Because of the difference in the contractile tendencies (Poisson's ratio) of the fiber and matrix under longitudinal loading, radial and tangential stresses are set up in the fiber and matrix. These stresses arise from the existence of a sound bond between the components which requires the fiber and matrix to deform together, instead of independently. The mechanical interaction between fiber and matrix is primarily dependent upon the difference between the Poisson's ratios of the components, and to a lesser extent on the difference in Young's modulus.

The transverse stresses generated in the composite are shown schematically in Fig. 5 and reach their maximum intensity at the fiber–matrix interface. The radial interfacial stresses may be expressed in terms of the fiber fraction V_f and the component properties as follows (Ebert and Gadd, 1965)

$$\sigma_r = \frac{K_m(1 - 2\nu_m)(\nu_m - \nu_f)(V_f)(1 - V_f)\epsilon_L}{[V_f(1 - 2\nu_m)(1 - K_m/K_f)] + [1 + (1 - 2\nu_m)K_m/K_f]} \tag{1}$$

(See the List of Symbols at end of this chapter for definition of symbols.)

In order to account more exactly for the effects of neighboring fibers in

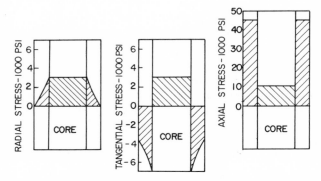

FIG. 5. Internal axial and transverse stress distributions produced by axial loading of a concentric cylinder composite into the elastic response range of both fiber and matrix. Properties are for a 25% Be–Al composite. Case: E = 45 × 10⁶, ν = 0.05; core: E = 10 × 10⁶, ν = 0.34.

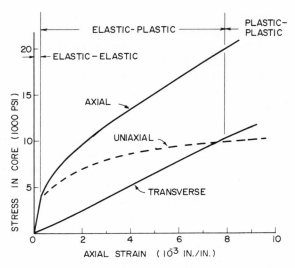

FIG. 6. Variation of interfacial transverse and axial stresses in the matrix (core of Fig. 5) of a 50% steel–copper composite with the occurrence of plastic flow. The radial stress will be the same as the tangential stress [from Hecker *et al.* (1970)].

a regular array, Piehler (1967) and Bloom and Wilson (1967) have used Fourier series and complex variable methods, respectively, to solve the problem for a hexagonal fiber array, as in Fig. 3. The distribution of stresses is very similar to that obtained by Ebert and Gadd (1965) except for a slight variation around the circumference of the fiber. Again the stress levels are most intense at the interface.

Studies of the effects of mechanical interaction into the range of plastic behavior of the matrix (and fiber) have been conducted by Ebert *et al.* (1968), Hecker *et al.* (1970), and Hamilton *et al.* (1971), using the concentric cylinder model developed by Ebert and Gadd. To account for the progression of plastic flow through the outer sleeve of matrix simulator (low fiber fraction composites), a multiring model was developed (Hecker *et al.*, 1970) which was capable of analyzing the effects of strain-hardening in both fiber and matrix components.

Aside from accurately predicting the composite stress–strain behavior when loaded in the fiber direction, the model also showed the alteration in interfacial stresses as a result of the occurrence of plastic flow. As noted above, the interfacial stresses are strongly dependent on the difference in Poisson's ratio of the components. With the onset of plastic flow in the matrix, its effective Poisson's ratio begins to increase from the elastic value toward 0.5, the ideal plastic "Poisson's ratio." This increase will increase the difference in the value of Poisson's ratio between the fiber and matrix, since fiber materials generally have lower Poisson's ratios. Hence the intensity of the transverse interfacial stresses will increase rather rapidly with the onset of plastic flow.

This change in stress distribution within the composite is shown in Fig. 6 where the internal stresses are plotted as a function of composite strain. While the composite is behaving elastically, the transverse stresses are quite small relative to the axial stresses, but with the onset of plastic flow they quickly increase to about 40% of the axial stress in the matrix.

The transverse stresses generated at the interface between the fiber and matrix can have an even more marked effect on the failure of the composite if the fibers are ductile. The effect, explained by Piehler (1965) is one of retarding the necking of the fiber at the strains at which the uncomposited fibers would ordinarily neck and fail. The transverse stresses generated at the interface act to prevent the local contraction of the fiber into a neck. This effect naturally decreases with increasing fiber content since the decreasing amount of matrix is less effective in restraining the fibers. Besides the silver–steel system used by Piehler (1965) this behavior has been noted at elevated temperatures by Mileiko (1969) in Ni–W and Kelly and Tyson (1965b) in Cu–Mo and Cu–W.

None of the above effects of transverse stresses at the fiber–matrix interface are predicted by the commonly used rule of mixtures which assumes that there are no transverse stresses generated. Indeed, this simple model implicitly assumes that there is no bonding between the fiber and matrix components. (If there were any bonding at all, the mechanical interaction described above would occur, and transverse stresses would be generated.) Hill (1964) has shown that the rule of mixtures is a lower bound of the composite properties in the fiber direction, and hence there is a tendency for this rule to underestimate composite stiffnesses and strengths. The reason that the rule of mixtures has been a useful engineering guideline to date, despite its lower bound nature, is that the transverse stresses generated in the fiber are opposite to those generated in the matrix, so that their net effect on overall composite stress–strain performance is lessened. This creates the erroneous impression that because experimental data agree with the rule of mixtures which ignores interfacial stresses, there are no significant interfacial stresses generated upon longitudinal loading.

However, compared to transverse loading (to be discussed in Section II,B) the interfacial stresses are not as severe in longitudinal loading. Hence the longitudinal strength of composite should be relatively in-

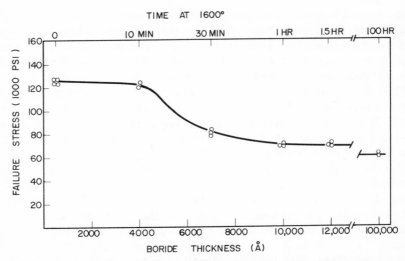

FIG. 7. Longitudinal strength of a Ti–25% B composite as affected by heat treatment at 1600°F (diboride layer growth) [from Klein *et al.* (1969)].

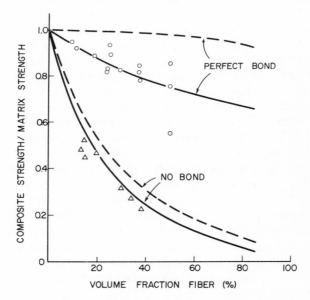

FIG. 8. Transverse strength of a fiber composite as a function of fiber fraction, fiber array, and fiber matrix bonding (–––, hexagonal array; —, square array). Included also are experimental data from 6061 Al–B (O) and Ni–Al₂O₃ (△) systems.

sensitive to interfacial bond strength (so long as the strength is enough to transfer load to the fibers). This has generally been observed when other factors are not overriding (Swenson and Hancock, 1971).

One such factor is the formation of a brittle reaction product at the interface. Here Metcalfe (1967) has shown that the strength is critically sensitive to reaction layer thickness (Fig. 7). This is postulated to be a fracture toughness effect resulting from the presence of cracks in the imperfectly formed reaction layer. At the critical thickness, these cracks (of the same order of size as the intermetallic thickness) are large enough to initiate fracture in the brittle fibers. In the system studied (Ti–B), the interfacial bond itself does not seem to have weakened; indeed, the fiber–intermetallic bond must be good to force the fracture to initiate at the crack in the intermetallic layer. The effect also does not seem to be a result of the alteration of the interfacial stress state by the introduction of a new component. Unpublished analytical work conducted in the author's laboratory relating to an investigation of the internal stress fields in composites with thin, brittle layers between the fiber and matrix shows that the interfacial stresses are but slightly affected by the addition of such a layer (prior to fracture of the layer).

B. Transverse Loading

For a loading direction perpendicular to the fiber direction, the interfacial state of stress is even more complex than that for the case of loading parallel to the fibers discussed above. Because of this complexity, the transverse loading problem has been approached chiefly with numerical techniques such as the finite difference and the finite element methods.

For extremely low fiber fractions, the stress fields are similar to those formed by an isolated rigid inclusion in an infinite plate under tension (Schuerch, 1966; Tuba, 1966). This configuration has been solved in closed form for both elastic and perfectly plastic matrix behavior, but with increasing fiber fraction, the fiber–fiber interaction makes the consideration of the entire fiber array necessary.

Solutions have been developed for two basic fiber arrays: the square and hexagonal arrays depicted in Fig. 3. The two arrays have been shown to result in slightly different composite strengths, Fig. 8, as well as somewhat different interfacial stresses as shown in Fig. 9. The effect of random fiber arrays has also been considered (Adams and Tsai, 1969; Sendeckyj and Yu, 1971) and the properties resulting from such arrays are intermediate to the two regular arrays.

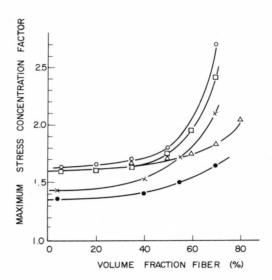

Fig. 9. Maximum interfacial normal stress concentrations produced by transverse loading (elastic behavior).

Square array: ○ $E_f/E_m = 120$ Foye (1968)
               □ $E_f/E_m = 20$ Foye (1968)
               × $E_f/E_m = 20$ Adams and Doner (1967)
               ● $E_f/E_m = 6$ Adams and Doner (1967)
Hexagonal array: △ $E_f/E_m = 20$ Foye (1968)

The interfacial stresses developed in a composite under transverse tensile loading vary significantly around the interface. An example of this variation is shown in Fig. 10 for a 50% volume fraction boron–aluminum square array composite. It is to be noted that the largest interfacial stress is the normal stress; hence failure is more likely to occur by debonding than by shear slippage. The location of this maximum normal stress is at $\theta = 0°$, where θ is measured from the leading edge of the fiber. The normal stresses will become compressive in the region of $\theta = 70°$ to $90°$, so no debonding should occur here.

The effects of plastic flow in the matrix on transverse composite behavior have been studied by Adams (1970) as well as by Wright (1971). The composite interfacial stresses are shown in Fig. 10 for an applied composite load of 2.9 times the load at which plastic flow first begins in the matrix (5400 psi for this strain hardening aluminum matrix). At this point, nearly all of the matrix has undergone some plastic flow, and the region along the interface from 0° to 80° has flowed to some extent. Never-

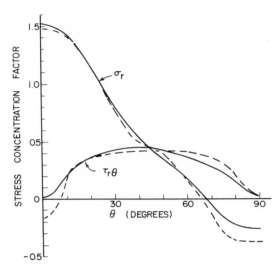

FIG. 10. Radial variation of interfacial normal and shear stresses around the circumference of the interface for transverse composite loading (Al–50% B). Both elastic (—) and plastic (– – –) behavior shown.

theless, the interfacial stress distributions are essentially the same as those prior to yielding (although the intensities are, of course, proportionately higher). Evidently the restraining effect of the stiff fiber on the material adjacent to the interface is sufficient to dominate the interfacial stress pattern for this situation.

The maximum interfacial stresses under transverse composite loading, then, are on the order of the applied composite load, as compared to the longitudinal loading case where they are somewhat less than the applied load. The exact magnitude of the interfacial stresses depends slightly on the array, the volume fraction fiber, and the component materials, as shown in Fig. 9. Below 50% volume fraction of fiber, the stress concentration factor is practically independent of the fiber volume fraction.

The case in which no bond exists between the fibers and matrix has been investigated by Chen and Lin (1969). They show that for increasing fiber content, the transverse composite strength should drop off rapidly, and that debonding of the matrix from the fiber will occur in all but a small region of the interface (Fig. 11). This type of behavior has been observed in the poorly bonded sapphire–nickel system (Mehan and Harris, 1971), as well as in stainless steel–aluminum (Lin *et al.*, 1971), and follows closely to the predicted amount of strength degradation. This degree of agreement is probably somewhat fortuitous, since Chen and Lin's analysis as-

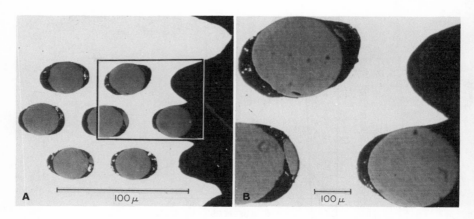

FIG. 11. Fiber–matrix debonding in a Ni–Al₂O₃ composite under transverse loading. (B) is an enlargement of the boxed area in (A). Compare debonded regions with extent of tensile radial interfacial stresses in Fig. 10 [from Mehan and Harris (1971)].

sumes a state of plane stress in the plane normal to the fibers instead of the more proper plane strain state, and neglects plastic flow effects in the ductile matrix.

Nevertheless, it is evident from studies of the transverse external loading of fiber composites that the interfacial bond strength should be a more important factor in determining the transverse strength of composites than it is for the axial loading situation. The insensitivity of some Al–B and Ti–B composites to an interfacial intermetallic layer (Klein and Metcalfe, 1971; Klein et al., 1969) may be attributed to the failure of the composites by splitting of the fibers along their length. This splitting effectively removes the interface from having a significant role when transverse loading, since the fibers can carry no transverse load whether or not the interface is capable of transmitting it.

C. Discontinuous Fibers

Considerable attention has been devoted to the study of discontinuous fiber composites, particularly for systems in which the fibers are in the form of whiskers. It has been hoped to be able to utilize the very high stiffnesses obtainable only in whisker form, as well as to improve the toughness of fiber composites. It is well understood that the interface in this situation plays a dominant role in achieving full development of the composite properties.

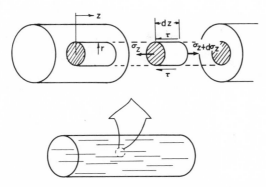

FIG. 12. Conceptual development of the shear lag model for analyzing the behavior of discontinuous fiber composites. Force balance: $\pi r^2 \, d\sigma_z = 2\pi r\tau \, dz$.

The simplest analyses of this configuration have been advanced by Kelly and Tyson (1965a) and Cox (1952). These studies all assume a simple shear lag model of stress transfer from the matrix through the fiber. In this model, the load on an isolated fiber is considered to be built up solely by the generation of a shear stresses at the fiber–matrix inter-face. Effects of neighboring fibers, of fiber end condition, of complex stress states, and of abutting fiber ends are all ignored. This simple approach, the model for which is depicted in Fig. 12, allows an elementary mechanics evaluation of some of the important characteristics of discontinuous fiber composites. The existence of a "transfer length," the length over which the discontinuous fiber is loaded to the same level as an infinitely long con-tinuous fiber, and the accompanying concept of a critical fiber length were presented by these workers. In addition, interfacial shear stress distribu-tions near the fiber ends were determined (Fig. 13).

This work, while contributing much to a conceptual understanding of discontinuous fiber composites, oversimplifies the mechanics of the situa-tion. As a result, experimental measurements of the interfacial shear stress distributions do not agree well with the predicted distributions (Fig. 13).

Unfortunately, no unifying theory of the behavior of discontinuous fiber composites has come forth. This is a result of the great complexity of the discontinuous fiber problem. Besides the difficulty imposed by the nonuniform stress states generated at the fiber ends, there are a host of relavent geometric variables which make a thorough study formidable. Some of these variables which have been investigated to date, usually in combination with one or two others, are: (1) fiber fraction, (2) modulus ratio of fiber and matrix, (3) length to diameter aspect ratio of fiber, (4)

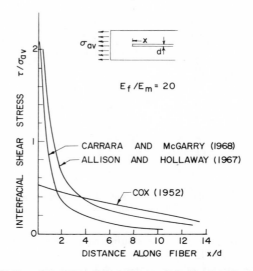

FIG. 13. Distribution of interfacial shear stress along the interface of a discontinuous fiber composite according to various investigators.

gap spacings between fiber ends, (5) fiber end shapes, (6) nearby discontinuities in other fibers, (7) debonding effects, and (8) plastic flow effects.

Most of these studies have been performed in two-dimensional (lamelar) systems because of the ease of performing analytical and experimental studies in two dimensions. A few (Haener and Ashbaugh, 1967; Carrara and McGarry, 1968; Owen et al., 1969) have considered three-dimensional systems with cylindrical symmetry, but here the geometric problems are even more formidable. The chief experimental technique has been photoelastic stress analysis; the finite element technique has been used widely, although not exclusively, for theoretical study.

In keeping with the greater generality of the finite element methods, the interfacial stress distributions along an isolated discontinuous fiber (the situation considered by Kelly and Tyson, and Cox) show good agreement with experiment. In Fig. 13 it is seen that near the fiber end, there is an additional stress concentration due to the fiber end shape which is ignored in the shear lag analysis.

Moreover, the more practical cases with adjacent unbroken fibers and abutting fiber ends can be considered by advanced analyses. These studies (Owen et al., 1969) show that decreasing the fiber spacing (increasing V_f) (MacLaughin, 1968; Barker and MacLaughlin, 1971; Iremonger and

Wood, 1969) greatly reduces interfacial shear stress concentrations, since more of the axial load can be locally transferred through the neighboring fibers. Thus, the possibility of pullout failure is decreased and failure by filament fracture (because of the stress concentrations in the adjacent fibers) is increased. Load transfer through the fiber end is generally found to be significant also when the matrix is bonded to the fiber end (Carrara and McGarry, 1968; Owen *et al.*, 1969; Iremonger and Wood, 1969). Morever, the proportion of load carried through the end is increased as the gap spacing between abutting fiber ends is decreased (Owen *et al.*, 1969; MacLaughlin, 1968; Iremonger and Wood, 1969). This again decreases the interfacial shear along the fiber.

These effects on the load transfer through the fiber ends are also reflected in the effective transfer length, since as the interfacial shear is decreased, the transfer length in general will also drecease, with more effective strengthening by shorter fibers.

Significant levels of normal interfacial stresses in the neighborhood of a fiber discontinuity have also been shown to exist (Iremonger and Wood, 1969, 1970) both at the interface of the discontinuous fiber, and at the interfaces of neighboring fibers. This normal stress arises from the local constriction of the highly stressed matrix region at the end of the discontinuous fiber. Typical levels of these stresses are shown in Figs. 14 and 15, although they are decreased somewhat by butting the fiber ends closer

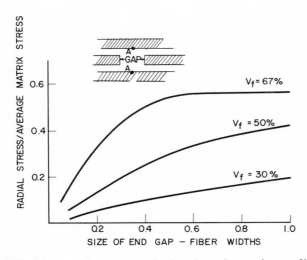

Fɪɢ. 14. Radial interfacial stresses at the interface of a continuous fiber adjacent to a fiber discontinuity.

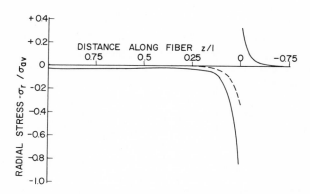

Fig. 15. Radial interfacial stresses along the interface of an isolated discontinuous fiber. — Carrara and McGarry (1968), – – – Haener and Ashbaugh (1967).

together (Iremonger and Wood, 1969), and increased appreciably by the occurrence of plastic flow. This latter effect is due to the greater contractile tendency of the matrix once plastic flow begins.

The normal interfacial stresses are tensile on the neighboring fibers and compressive on the discontinuous fiber. This additional interfacial compression at the interface of the discontinuous fiber may be beneficial in promoting shear stress transfer through a frictional mechanism, even if the bonding breaks down.

The occurrence of plastic flow has also been studied by Owen et al. (1969) and Iremonger and Wood (1970) and may be either beneficial or detrimental depending on the composite system, geometry, and interface. Plastic flow generally begins at the corners of the end of a fiber and roceeds outward and along the fiber–matrix interface. This flow, of course, lowers the stress concentrations, hence increasing the transfer length. This in conjunction with the normal compressive stress generation mentioned above, may be beneficial if the interface is strength controlled. However, if the interface is composed of a brittle compound whose fracture is strain controlled, the large strains imposed by the plastic flow are detrimental.

All of the work described above has been for two-dimensional composite models. Since fiber composites are three-dimensional structures, it may be expected that the conclusions drawn above may have some limitations when applied to three-dimensional systems. A little work has been conducted for cylindrical fibers, but it is difficult to evaluate exactly the effect of neighboring fibers in such a system. Owen et al. (1969) have made some comparative studies between planar and cylindrical models, but unfortunately, the volume fraction of fiber material was not consistent between the two. Carrara and McGarry (1968) have studied a single-filament system under elastic conditions, and have pointed out the significant role of stress transfer through the fiber ends (as much as 20% of the total fiber load) and the generation of transverse stresses near the fiber ends. These radial and tangential stresses can be much higher than those for the corresponding continuous fiber composites, and indeed, in the system examined, the radial interfacial stress reaches values of approximately 80% of the axially applied matrix stress. Apparently, this radial compressive stress is not so well-developed in planar systems.

It is well established, then, that the stresses sustained by the interface in discontinuous fiber composites are highly complex. Indeed, this may be the most severe of the interfacial loading situations, since in addition to sizable axial, radial, and tangential loads, the interface must withstand an even larger shear loading.

III. The Nature and Effects of Residual Stresses at Composite Interfaces

The role of residual stresses is often ignored in analytical and experimental considerations of interfacial effects in composite materials. This oversight is unfortunate because the resultant interpretation of properties and behaviors is usually misleading. Residual stresses are an inherent characteristic of composite materials; their absence is the exception rather than the rule.

The primary origin of residual stresses in fiber composites is twofold: thermal and mechanical. The former origin is the most prevalent, arising from the differing thermal coefficients of expansion of the component materials. Since composites are invariably used at different temperatures than those at which they are fabricated, the differing thermal expansions or contractions of the fiber and matrix set up thermally induced stresses on cool-down from the compositing temperatures. Metal matrix composites, particularly, are fabricated at temperatures which are quite high relative to ambient, and thus hold the possibility of producing very high stress levels.

The importance of residual stresses, of course, diminishes if the composite use temperature is one at which a significant amount of creep can occur. However, even under these conditions, one also must be aware that the other predictions based on the normal time-independent mechanical relations are also invalid.

The second main source of composite residual stresses is the difference in flowstress between the components. This is important when the composite is subjected to mechanical deformation at a level where one or more of the component materials begins to flow plastically. Under these conditions, the residual stresses upon loading of the composite stem from the different amounts of plastic flow which have taken place among components of the composites.

Another possible source of residual stresses in components is the occurrence of phase transformations with their accompanying volume changes. Since the transforming component is usually mechanically restrained by other components of the composite during the transformation process, the full (equilibrium) volume change cannot take place. As a result residual stresses are built up. These effects have been considered by de Silva and Chadwick (1969), and they will not be further discussed here.

A. Thermally Induced Residual Stresses

The analytical determination of residual stresses in fiber composites, along with accompanying experimental studies, has been approached in two ways: through the concentric cylinder model cited above and by studies of exact fiber arrays.

The concentric cylinder model has been the more intensely used approach (Hecker et al., 1970; de Silva and Chadwick, 1969; Karpinos and Tuchinskii, 1968) because of its simpler mathematical relations. The work by Hecker et al. (1970) has been refined to the extent that plastic flow effects in a strain hardenable matrix can be included with temperature dependent mechanical properties. For this analysis, as well as all the others, there is a certain ambiguity about the "stress-free temperature." This temperature is defined as the temperature below which creep effects can be considered to cease and significant stress levels can be generated. Hecker et al. circumvented this by comparing analytical results with experimental measurements of residual stresses in model concentric cylinder composites to establish the stress-free temperature.

Both experimental and analytical determinations of the residual stress distributions have shown that quite high levels of residual stress are generated readily. Cooling steel–copper composites from 500°F resulted in extensive plastic flow in the copper matrix (Hecker et al., 1970). Thus has also been documented in the Cu–W (Ebert et al., 1969; de Silva and Chadwick, 1969; Hoffman, 1970) and Fe–Fe$_2$B systems (de Silva and Chadwick, 1969). Use of higher strength matrices has little advantage; a drop of 1°F raises the maximum stress in a 50% Al–B composite by about 250 psi (Ebert et al., 1971) when the aluminum responds elastically. Thus

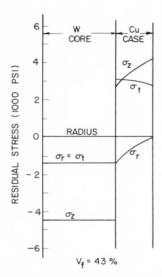

FIG. 16. Thermal residual stresses generated in a concentric copper–tungsten composite by cooling from 500°F (plastic flow effect included).

cooling from a normal fabrication temperature will create stresses well beyond the yield strength of any alloy.

A typical residual stress distribution for a tungsten fiber–copper matrix composite is shown in Fig. 16, for a low volume fraction composite. In this model the matrix is considered to surround the fiber and hence its larger thermal contraction causes it to shrink around the fiber, generating compressive radial and tensile tangential stresses. For the high fiber fraction composite the matrix is now the inside member of the model (Fig. 4), and the interfacial stresses are reversed in sign. This somewhat surprising result has also been verified by the work described below.

Thermal residual stresses have also been studied in exact array models of composites, both experimentally through the use of photoelasticity (Koufopoulos and Theocaris, 1969; Marloff and Daniel, 1969) and analytically by the finite element method (Asamoah and Wood, 1970). Koufopoulos and Theocaris (1969) have studied the plane stress case and Marloff and Daniel (1969) and Asamoah and Wood (1970) the plane strain case.

In all cases, the distributions of stresses are essentially the same, although the intensities vary. The distribution of radial stress along the interface of a hexagonal array composite is shown in Fig. 17. Of note here is that, above a certain volume fraction fiber, the radial interfacial stresses

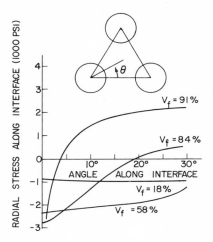

FIG. 17. Variation of radial thermal stress around the circumference of the interface in a hexagonal array composite [from Asamoah and Wood (1970)].

become tensile over a portion of the interface. This volume fraction, according to the data of Fig. 17, is 80%, but the experimental studies indicate that this changeover occurs at a lower fiber fraction, around 50%.

The high levels of residual stresses generated at the fiber–matrix interfaces may have strong effects on both interfacial and bulk composite behavior. As shown above, the residual stresses in composites may be either tensile or compressive so that the interfacial bond may appear to be either weaker (if the stress is tensile) or stronger (if compressive) than it actually is. Residual shear stresses generated at fiber discontinuities may also aggravate the bonding problem, as has been observed in glass–resin (Broutman and McGarry, 1968).

The presence of thermal residual stresses from cooling has been shown in all cases to degrade the tensile strength of fiber composites (Hecker et al., 1970), independent of their effect on the interface. This results from the overall tensile nature of the residual stress state (see Fig. 16) even though the radial stress may be compressive. When an additional axial tension is applied by axial loading, plastic flow will begin at a lower applied load (if indeed it has not already begun during cool down).

B. Deformation-Induced Residual Stresses

It is possible to alleviate the detrimental effects of thermally-induced residual stresses on the bulk mechanical behavior of composites (although not necessarily on the interfacial behavior) by using mechanical deformation (in the plastic flow range) to alter the residual stress patterns. Prestraining composites in the fiber direction before testing produces a significant improvement over the properties with thermal stresses present (Fedor and Ebert, 1973). It has been demonstrated that this effect results from the change in the relative intensities of the residual stresses in the composites and not the strain hardening effect of the prestraining. The sign of the deviatoric component of the residual stresses created by loading into the plastic range of the matrix and then unloading are opposite to those generated by cooling. Thus the net stress state remaining is less severe.

Cross rolling composites also serves to alter the composite strength beneficially (Getten and Ebert, 1969) with an increase in strength being observed that is greater than can be accounted for by work hardening of the matrix. It is also presumed that this is a residual stress effect.

IV. Interfacial Strength Tests

In order to determine the performance of the interface under the loading conditions described above, one needs to determine the strength of the interface. Ideally, such strength tests should be conducted under conditions for which the interfacial stresses are uniform over the major, if not the entire, portion of the interface. In addition, it is highly desirable that the interfacial stress state be simple, so that a simple correspondingly useful strength value is obtained. If a shear strength is desired, for example, the interface should be loaded in a state of pure shear.

Problems are encountered immediately in attempting to meet the conditions posed above. As was cited previously, whenever a system consisting of two mechanically distinct but integral phases is loaded externally, a complex state of stress arises at the interface between them. Since it is virtually impossible to duplicate this complex stress state in a simple mechanical test of the strength of the interface alone, cognizance must be taken of the stress state difference between that existing at the interface *in situ* and that used in the testing the interface outside of the actual composite.

The first criterion of constant conditions over the area of the interface is somewhat easier to achieve if interfacial areas are large enough and geometric irregularities are avoided.

Bearing in mind the above limitations on the experimental characterizations of interfacial stress states for various composite geometries and loadings, attention is directed to the various methods used to determine the

strength of interfaces. There are three general types of tests that have
been used by various investigators: (1) flat plate tests, (2) rod or fiber
pullout tests, and (3) embedded single-fiber tests.

A. Flat Plate Tests

This test consists of bonding a layer of matrix material to a flat plate of
the fiber material, as shown in Fig. 18, and then subjecting the couple to
either a lateral shear or a normal tensile force to obtain shear or tensile
strengths, respectively. One version of this test used in glass–epoxy systems is that of bonding two glass plates together with a layer of epoxy
(Broutman, 1966, 1968). Flat plate specimens have also been made by
solidifying sessile drops of the matrix on a plate of fiber material (Levitt
and Wolf, 1970).

The advantages of this type of specimen are that it is easy to fabricate
and test. Extraction of strength data would appear to be relatively simple,
also. However, thermally induced residual stresses will exist in such specimens, and their presence should be taken into account, since their intensities and signs may be different from those in the composite. Also, there
will be complex mechanically induced stresses generated at the interface
as the result of mechanical interaction effects. This effect has been well-
studied for the case of two hard plates brazed together by a thin layer
(Mar and Shepherd, 1963). When the assembly was subjected to a tensile
force normal to the braze layer, the strength of the joint was dependent on
the layer thickness, increasing greatly as the joint became thinner. This
was shown to be a direct result of the restraint of the hard plates surrounding the braze.

Nicholas *et al.* (1968) have analyzed the mechanics of the sessile drop

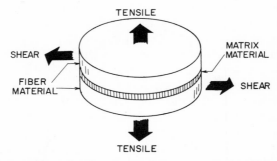

FIG. 18. Flat plate interfacial strength test specimen, used for either shear or tensile
strength measurements.

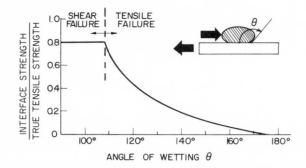

FIG. 19. Interfacial strength ("shear" loading) variation with wetting angle θ. Interfaces with $\theta > 108°$ fail by a tensile mode.

shear test, and from a strictly "strength of materials" approach (ignoring the mechanical interaction and stress concentration effects), have found that for wetting angles of greater than 108°, the specimen will not fail by shear, but by tensile forces at the interface generated by the correction of the specimen. Their analytical curve of apparent strength versus wetting angle, Fig. 19, agrees well with data obtained in the Cu–Al$_2$O$_3$ systems (see Chapter 8).

B. Rod or Fiber Pullout Tests

The interfacial shear strength of a composite system has also been measured by compositing a single fiber or a rod of fiber material in a cylinder of matrix material, and then pulling the fiber from the cylinder by exerting a tensile force along its axis (Fig. 20). This is probably the most popular strength test, especially in metal matrix composites. Proponents of the pullout test cite the great similarity of geometry and residual stresses in the specimen to these aspects in actual composites.

Kelly and Tyson (1965) have used the pullout test extensively to determine stress transfer from matrix to fiber around fiber ends. Using a simple

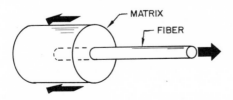

FIG. 20. Geometry of the fiber pullout test used to measure interfacial shear strength.

shear lag analysis, they determine the shear strength of the interface (for a perfectly plastic matrix) as

$$S_{\text{shear}} = \sigma_f d / (2l_c) \tag{2}$$

By embedding various lengths of the fiber into the matrix, the quantity l/d can be varied, and l_c established as the length at which the failure mode changes from pullout to fiber fracture.

There are several shortcomings of this approach. First, the stress distribution within the specimen is much more complex than that assumed by the shear lag analysis, as discussed in Section II,C. The interfacial shear stress distribution calculated by this model is in error, and hence the calculated interfacial strength will be incorrect. The gradual variation in stresses along the interface means that failure will proceed gradually and nonuniformly over a range of external loads.

Secondly, the loading situation created by the pullout specimen, despite its apparent similarity to real discontinuous fiber composites, is not the same as in such composites. This is the result of the fact that, in test, the load is applied directly to the fiber, while in composites, the load is always applied through the matrix. If the fiber in the pullout specimen has an extremely long embedded length, this difference will not be critical, since there will be enough fiber length to establish a "shared loading" between fiber and matrix. However, the high interfacial strengths of many metal matrix systems forbids this, since short lengths are needed to establish the critical fiber length l_c.

A photoelastic study by Gillan *et al.* (1969) has shown the dissimilarity of pullout specimens to an actual discontinuous fiber composite. In the former, the load in the fiber builds up linearly from the fiber end to the point of embedding, even for very long embedded lengths, while in the latter, the load builds up nonlinearly and over a confined region.

C. Embedded Single Fiber Tests

The use of this test has been confined principally to plastic matrix composites (Broutman and McGarry, 1968; Broutman, 1966, 1968). A single fiber is aligned within a block of matrix, which may be formed into several shapes, depending on the property to be measured. Normal interfacial strengths are measured by applying compression to a specimen with a curved neck (Fig. 21). A straight-sided specimen or one with a trapezoidal profile (Fig. 21) is tested under compression to determine shear strengths.

The principle of the curved-neck specimen is that of using the transverse radial tensile stresses generated at the interface under compression of the

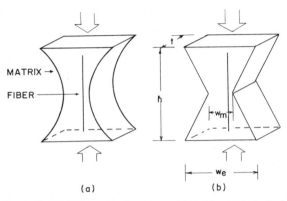

FIG. 21. Embedded single fiber specimens used to measure interfacial strengths. (a) Curve neck specimen to determine interfacial tensile strength, (b) trapezoidal specimen to determine interfacial shear strength.

specimen to cause debonding at the interface. These radial stresses are generated by the same mechanism as discussed for longitudinal loading in Section II,A. The curved neck serves to localize the failure region by creating a region of high stress in the narrow center section. The interfacial stress is then given by (Broutman and McGarry, 1968)

$$S_r = \frac{\sigma_{av}(\nu_m - \nu_f)E_f}{(1 + \nu_m)E_f + (1 - \nu_f - 2\nu_f^2)E_m} \tag{3}$$

The trapezoidal specimen measures the shear strength of the interface because the rapidly changing axial stress in the specimen causes a shear stress to be generated at the interface (Salkind, 1967)

$$S_{shear} = \frac{Pd(W_e - W_m)^3}{th^3 W_m^3} \frac{E_f}{E_m} \tag{4}$$

Straight-sided specimens have also been used to measure strengths in boron–epoxy systems (Gutfreund *et al.*, 1966). The reproducibility of these shear tests is rather poor, about +35% for the glass–resin system studied. The curved-neck specimen gives somewhat more reliable results for normal strengths.

There are two main reasons for limiting these embedded fiber tests to plastic matrix systems. First, the success of the test depends on the interface having relatively low strengths such as are found in many glass–plastic composites. The transverse stresses which are required to initiate bond failure are fairly small compared to axial stresses, and for valid data to be obtained, the interface must fail before gross composite failure. Many metal matrix systems have high bond strengths, and hence it would be impossible to determine their strength this way.

Secondly, the determination of interfacial failure is most readily done in these specimens by actually observing the fracture through the transparent matrix. Because of the stress gradients present, relatively small portions of the interface fail initially. Consequently, determination of the failure load externally would be inaccurate.

D. Evaluation of the Strength Tests

All of the strength tests discussed above suffer from the disadvantages of nonuniform or complex stress states, as well as from the problem of residual stress generation. The residual stress effects have been considered only for the embedded fiber tests, where Broutman and McGarry (1968) have studied the effect of curing temperature on the apparent interface strength. They found that there was an optimum curing temperature which gave a maximum bond strength. This behavior was explained by using a concentric cylinder composite model to calculate the residual stress states resulting from cooling from the curing temperature.

The thermally induced radial stresses at the interface were compressive and increased with increasing curing temperature. These residual compressive stresses offset the tensile radial stresses generated at the interface by the subsequent compressive loading of the specimen. Hence the interfacial strength appeared to increase for higher curing temperatures. If the temperature change with cooling became too large, however, excessive thermal shear stresses developed at the fiber ends. This caused premature failure of the specimen at the fiber end, and bond strength appeared to decrease with curing above an optimum temperature.

Although residual stresses are seen to have an important effect in the measured interfacial strengths, they have generally not been taken into account in interfacial strength measurement in metal matrix systems. This, in addition to the nonuniform distributions of interfacial stresses generated in the measurement specimens, makes any quantitative determination of interfacial strengths unreliable at the present state of the art. Even with a constant specimen geometry, residual and mechanically induced stresses will vary from system to system because of the changing mechanical and thermal properties of each system. At best, it would seem that present experimental measurement systems permit only a strictly qualitative evaluation of interfacial strengths.

List of Symbols

d	Fiber diameter	t	Specimen thickness
E	Young's modulus	w_e	Width of trapezoidal specimen at ends
h	Specimen height	w_m	Width of trapezoidal specimen at narrowest point
K	Material constant $= E/(1 + \nu)(1 - 2\nu)$	V_f	Fiber fraction by volume
l	Fiber length	ϵ_L	Longitudinally applied strain
l_c	Critical fiber length	ν	Poisson's ratio
P	Applied load	σ	Stress
S	Strength		

Subscripts

f	Fiber	r	Radial direction
m	Matrix	av	Average composite property

References

Adams, D. F. (1970). *J. Compos. Mater.* **4,** 310.

Adams, D. F., and Doner, D. R. (1967). *J. Compos. Mater.* **1,** 4.

Adams, D. F., and Tsai, S. W. (1969). *J. Compos. Mater.* **3,** 368.

Allison, I. M., and Hollaway, L. C. (1967). *Brit. J. Appl. Phys.* **18,** 979.

Asamoah, N. K., and Wood, W. G. (1970). *J. Strain Anal.* **5,** 58.

Barker, R. M., and MacLaughlin, T. F. (1971). *J. Compos. Mater.* **5,** 492.

Bloom, J. M., and Wilson, H. B., Jr. (1967). *J. Compos. Mater.* **1,** 268.

Broutman, L. J. (1966). *Polym. Eng. Sci.* **6,** 263.

Broutman, L. J. (1968). *In* "Interfaces in Composites," STP 452, pp. 27–41. Amer. Soc. Test. Mater., Philadelphia, Pennsylvania.

Broutman, L. J., and McGarry, F. J. (1968). *Mod. Plast.* **40,** No. 1, p. 161.

Carrara, A. S., and McGarry, F. J. (1968). *J. Compos. Mater.* **2,** 222.

Chen, P. E., and Lin, J. M. (1969). *Mater. Res. Std.* **9** (8), 29.

Cox, H. (1952). *Brit. J. Appl. Phys.* **3,** 72.

de Silva, A. R. T., and Chadwick, G. A. (1969). *J. Mech. Phys. Solids* **17,** 387.

Dow, N. F. (1963). General Electric Space Sci. Lab. Rep. No. R 63 SD 61.

Ebert, L. J., and Gadd, J. D. (1965). *In* "Fiber Composite Materials," p. 89. Amer. Soc. Metals, Metals Park, Ohio.

Ebert, L. J., Hecker, S. S., and Hamilton, C. H. (1968). *J. Compos. Mater.* **2,** 458.

Ebert, L. J., Fedor, R. J., Hamilton, C. H., Hecker, S. S., and Wright, P. K. (1969). Air Force Mater. Lab., AFML-TR-69-129.

Ebert, L. J., Claxton, R. J., Kmieciak, H., and Wright, P. K. (1971). Air Force Mater. Lab., AFML-TR-71-133.

Fedor, R. J., and Ebert, L. J. (1973). *J. Eng. Mater. Tech.* (*Trans. ASME Ser. H*) **95,** 69.

Foye, R. L. (1968). Air Force Mater. Lab., AFML-TR-68-91.

Getten, J. R., and Ebert, L. J. (1969). *ASM Trans. Quart.* **62,** 869.

Gillan, L. M., Hawkes, G. A., and Hill, T. G. (1969). *J. Aust. Inst. Metals* **14,** 242.

Gutfreund, K., Broutman, L. J., and Jaffee, E. (1966). *In* "Advanced Fibrous Reininforced Composites," p. E 25. 10th Nat. SAMPE Symp.

Haener, J., and Ashbaugh, N. (1967). *J. Compos. Mater.* **1,** 54.

Hamilton, C. H., Hecker, S. S., and Ebert, L. J. (1971). *J. Basic Eng. (Trans. ASME Ser D)* **93,** 661.

Hecker, S. S., Hamilton, C. H., and Ebert, L. J. (1970). *J. Mater.* **5,** 868.

Hill, R. (1964). *J. Mech. Phys. Solids* **12,** 199.

Hoffman, C. A. (1970). NASA Lewis Res. Center. NASA-TN-D-5926.

Iremonger, M. J., and Wood, W. G. (1969). *J. Strain Anal.* **4,** 121.

Iremonger, M. J., and Wood, W. G. (1970). *J. Strain Anal.* **5,** 212.

Karpinos, D. M., and Tuchinskii, L. I. (1968). *Sov. Powd. Met. Met. Cer.* No. 9, 735.

Kelly, A., and Tyson, W. R. (1965a). *In* "High Strength Materials" (V. F. Zackay, ed.), Chapter 13. Wiley, New York.

Kelly, A., and Tyson, W. R. (1965b). *J. Mech. Phys. Solids.* **13,** 329.

Klein, M. J., Reid, M. L., and Metcalfe, A. G. (1969). Air Force Mater. Lab., AFML-TR-69-242.

Klein, M. J., and Metcalfe, A. G. (1971). Air Force Mater. Lab., AFML-TR-71-189.

Koufopoulos, T., and Theocaris, P. S. (1969). *J. Compos. Mater.* **3,** 308.

Levitt, A. P., and Wolf, S. M. (1970). *In* "Whisker Technology" (A. P. Levitt, ed.), pp. 245–272. Wiley (Interscience), New York.

Lin, J. M., Chen, P. E., and DiBenedetto, A. T. (1971). *Polym. Eng. Sci.* **11,** 344.

MacLaughlin, T. F. (1968). *J. Compos. Mater.* **2,** 44.

Mar, J. W., and Sheperd, L. A. (1963). Some studies on the nature of deformation in composite materials, *AIAA Launch Space Vehicle Shell Struct. Conf., Palm Springs, California.*

Marloff, R. H., and Daniel, I. M. (1969). *Exp. Mech.* **9,** 156.

Mehan, R. L., and Harris, T. A. (1971). Air Force Mater. Lab., AFML-TR-71-150.

Metcalfe, A. G. (1967). *J. Compos. Mater.* **1,** 356.

Mileiko, S. T. (1969). *J. Mater. Sci.* **4,** 974.

Nicholas, M., Forgan, R. R. D., and Poole, D. M. (1968). *J. Mater. Sci.* **3,** 9.

Owen, D. R. J., Holbeche, J., and Zienkiewicz, O. C. (1969). *Fibre Sci. Technol.* **1,** 185.

Piehler, H. R. (1965). *Trans. Met. Soc. AIME* **233,** 12.

Piehler, H. R. (1967). *In* "Fiber-Strengthened Metallic Composites," pp. 3–26. ASTM STP 427.

Salkind, M. J. (1967). *In* "Surfaces and Interfaces" (J. J. Burke, N. L. Reed, and V. Weiss, eds.), Vol. II, Physical and Mechanical Properties, pp. 417–445. Syracuse Univ. Press, Syracuse, New York.

Schuerch, H. (1966). NASA-CR-482.

Sendeckyj, G. P., and Yu, I-W. (1971). *J. Compos. Mater.* **4,** 310.

Swanson, G. D., and Hancock, J. R. (1971). Midwest Res. Inst., TR-2. (AD-722020).

Tuba, I. S. (1966). *Appl. Sci. Res.* **16,** 241.

Wright, P. K. (1972). Ph.D. Thesis, Case Western Reserve Univ.

3

Physical Chemical Aspects of the Interface

ARTHUR G. METCALFE

Solar Division of International Harvester Company
San Diego, California

The physical chemistry of interfaces in metal matrix composites is a difficult subject and not well understood at the present time. However, an understanding of the subject is essential for an orderly development of the science of interfaces. Four topics within this field will be discussed in this chapter. These are: types of interface bonding, interfacial stability, reaction kinetics, and control of interface reactions. The text on the first topic, types of interfacial bonding, includes a review of the mechanisms of bonding and a proposed classification for the various types of bonds. Interfacial stability is an essential feature of any viable composite system. The various types of instability that may occur will be reviewed and this is followed by a discussion of reaction kinetics. The latter is of prime importance in connection with control of interfacial reactions during fabrication but has importance also with regard to stability in service. The final topic is control of interface reactions. This control may be required to moderate the bonding process and may include enhancement as well as moderation of the chemical events that occur in generating a bond. For example, control of the interface may be required to increase stability, to moderate reaction kinetics in the fabrication cycle, or to achieve optimum mechanical properties.

Many of the problems that occur in the interface in metal matrix composites have a counterpart in the interface phenomena in plastic matrix composite systems. Finishes in glass-reinforced plastics provide a transition zone from reinforcement to matrix, and it can be argued that useful metal matrix systems are those with a similar gradation of properties across the interface. Completely insoluble, nonreactive and nonwetting systems do not provide adequate bonding for useful properties. Modification of such interfaces to promote bonding leads to some measure of compositional gradients at the bondline. These considerations led to the definition of an interface proposed in Chapter 1:

An interface is the region of significantly changed chemical composition that constitutes the bond between the matrix and reinforcement for transfer of loads between these members of the composite structure.

Although this definition may bring to mind some of the similarities between plastic and metal matrix systems, there is one major difference; that of the temperature of service expected in the two types of system. The high temperatures desired for service of metal matrix composites introduce diffusional stability problems that are largely absent with plastic matrices. The questions of interfacial stability, reaction kinetics and control of interfacial reaction are all related to the high-temperature requirements of metal matrix composites. Discussion of these three topics will form the major content of this chapter.

I. Types of Interface Bonding

Composite systems were divided into three classes in the Introduction (Chapter 1). Class I included composites formed from nonreactive and insoluble constituents; Class II allowed some solubility but no reaction; and Class III included reactive constituents. These general descriptions of systems avoid consideration of the physical chemical and mechanical aspects of bonding. The latter were discussed in Chapter 2, and the former will be developed further here.

Surfaces and interfaces are not well understood for conventional materials, so that inclusion of a section on types of interface bonding may well be premature. However, its inclusion can be justified on the grounds that the status of this topic can be presented best by such an attempt. A tentative scheme for identification of bonds is based on six bond types: mechanical, dissolution and wetting, reaction, exchange reaction, oxide, and mixed.

(1) *Mechanical bond.* A purely mechanical bond requires that all chemical sources of bonding be absent. It can arise from mechanical interlocking (as in the case of boron filaments with a corncob surface) or from frictional effects arising from the contraction of the matrix on the filament.

(2) *Dissolution and wetting bond.* A contact angle of less than 90° occurs in wetting and is also characteristic of dissolution. If wetting is assumed to be accompanied by some dissolution, however small, then this bonding characteristic covers both extremes of mutual solubility. Elimination of adsorbed gases and of contaminant films must be achieved before element-to-element contact can occur and result in wetting and dissolution.

(3) *Reaction bond.* The reaction bond occurs when a new chemical compound is formed at the interface, such as the formation of titanium diboride at the interface between boron and titanium.

(4) *Exchange reaction bond.* This is a special case of the reaction bond, in which two or more reactions may occur. For example, the reaction between a titanium–aluminum solid solution (ss) and boron may be described as taking place in two steps:

$$Ti(Al)_{ss} + B = (Ti, Al)B_2$$
$$Ti + (Ti, Al)B_2 = TiB_2 + Ti(Al)_{ss}$$

(5) *Oxide bond*. The oxide bond may not involve new principles other than those enunciated earlier, but in the absence of detailed studies of bonding mechanisms, it appears desirable to introduce this bond type in a separate grouping. It may appear to be purely mechanical such as the silver–alumina whisker bonds studied by Sutton (1966). However, Moore (1969) showed that introduction of traces of oxygen converted the nickel–alumina bond to a reaction bond by formation of the $NiO \cdot Al_2O_3$ spinel. Another example may be the bond formed between the oxide-coated surfaces of aluminum and boron by solution or reaction between the two oxides. The product exists as an oxide film at the interface and constitutes the bond in this pseudostable Class I composite system.

(6) *Mixed bonds*. This may be one of the most important categories. Breakdown from one type to another will be one source of mixed bonds such as the partial transition from a pseudo-Class I system to a Class II or Class III system.

A. The Mechanical Bond

This type of bond can occur with a reinforcement that has a roughened surface, such as boron or other filaments grown by a vapor deposition process. Hill *et al.* (1969) investigated this type of bond by comparing the strength of tungsten-reinforced aluminum with various degrees of mechanical interlocking. Tungsten wires of 0.008 in. diameter were etched down to 0.0065 in. over lengths of 0.1 in., leaving the original diameter for a length of 0.025 in. Approximately 12% of filaments were incorporated into composites by vacuum infiltration of molten aluminum. Three conditions were evaluated by longitudinal tensile tests:

Condition	Theoretical Strength (%)
Smooth 0.008 in. wires	95
Smooth 0.008 in. wires, graphite coated	35
Etched wires, graphite coated	91

A reaction bond (see definitions above) forms between the uncoated tungsten and aluminum to give nearly theoretical composite strength. Graphite coating was claimed to prevent any reaction and reduced the strength to 35% of the theoretical. The etched wires were graphite coated to prevent formation of a reaction bond, but the mechanical bond was adequate to restore almost the theoretical longitudinal strength.

Hill *et al.* (1969) suggest that this type of mechanical bond might be used where the difficulties of controlling a fiber–matrix chemical interaction are considerable. Since the time of this work, major advances have been made to understand and to control reaction type bonds and the need for this type of solution is not as great. Another objection to this type of bond is that it makes no provision for load transfer in directions other than that of the reinforcement.

Some interesting observations on the mechanical type of bond have been made by Vennett *et al.* (1969) on the system brass–tungsten. Composite strengths were generally about 95% of the rule of mixtures values but the unusual effect was that the tungsten wires were necked at frequent intervals along their length, so that their total elongation was greater than for wires tested separately. Compression exerted by the matrix on the wire was ruled out as a mechanism because the necking would cause the wire to pull away from the matrix and separate in view of the weak bonds. The multiple necking was explained by local work hardening in the matrix adjacent to the neck in the tungsten. The strengthened matrix was able to take load from the tungsten until the load-bearing capability of the composite at this area exceeded the load-carrying capability elsewhere. Deformation would cease at this point and move to another position along the wire. In support of this explanation, it was found that the composite elongation decreases from 5 to 10% with less than 5% of tungsten to reach the approximate elongation of uncombined tungsten wire when 20 vol % is present. At higher volume percentages of tungsten, there will be less matrix to work harden and take up the loads.

The results described by Vennett *et al.* (1969) represent true composite behavior because properties were obtained that could not be obtained by the materials acting individually. However, the result appears to be independent of the bond at the interface and can occur with little or no bond strength, although the shear strength of the bond measured by a pullout test was reported to be appreciable, probably due to contraction of the matrix on the wire in fabrication.

Confirmation of the results has been provided by Schoene and Scala (1970) for the same system. The authors concluded that a "wetting or metallurgical" bond was not required for optimum reinforcement or multiple necking, although, in contrast to Vennett *et al.* (1969), they conclude that internal compressive forces are necessary.

It is evident from the previous discussion that a "pure" mechanical bond is unlikely to occur. It was pointed out in the presentation of classification of bond types that all chemical sources of bonding should be absent. Weak van der Waals forces will be present with all materials so that this condition can never be completely satisfied. Perhaps a better definition of the mechanical bond would be one where the mechanical interactions predominate. An interesting example of a system where the chemical forces can be varied continuously is copper–alumina. If copper oxide is heated in contact with alumina at an elevated temperature (e.g., 650°C), then a bond forms between the two oxides. Introduction of hydrogen reduces the copper oxide, first to oxygen-saturated metal, and then to metal with progressively lower dissolved oxygen. The chemical source of the bond between the reduced copper and alumina is weakened until only a mechanical bond remains with "oxygen-free" copper.

In summary, a bond formed between a matrix and reinforcement in which little or no chemical contribution can be traced is termed a mechanical bond. This type of bond may permit certain types of composite action to be demonstrated, but these appear to be limited to cases where stress is applied parallel to the interface as in a longitudinal tensile test. No data have been found to substantiate the merits of this type of bonding for any other type of loading. In consequence, it is not believed that this type of bond is an alternative for systems that are reactive.

In view of the limited application of a purely mechanical bond, the distinction between a mechanical bond and those bonds involving some chemical contributions will not be made in all cases in the subsequent pages. The term bond will mean one where chemical bonding is involved.

B. Dissolution and Wetting Bond

Bonding involves interaction of electrons on an atomic scale. Such interaction forces are quite short-range so that they do not develop until the atoms of the constituents approach within a few atom diameters of each other. The importance of the latter requirement is recognized in allied fields such as brazing where the difficulty of brazing aluminum is associated with the presence of intervening oxide films. Disruption of these films by some mechanical means (such as an ultrasonic soldering iron) leads to immediate wetting and dissolution of the substrate in the molten braze alloy. Two examples may be cited from the field of composites. Pepper *et al.* (1970) have noted that molten aluminum does not wet as-received graphite yarn until it is pretreated to desorb contaminants from the surface. Niesz (1968) made similar observations in the case of the nickel–graphite composites.

C. Reaction Bond

Although the reactions occurring in the formation of this type of bond may be very complex, the general concept of interfacial reaction forming a bond is readily understood. This type of bond occurs in Class III systems as defined in the Introduction and Review to this volume. Reaction involves transfer of atoms from one or both of the constituents to the reaction site, and these transfer processes are diffusion controlled. Hence, further discussion of this bonding process will be delayed until the section of this chapter on reaction kinetics.

Although Sutton and Feingold (1966) proposed a model for nickel–alumina in which the strength of the bond increased with amount of reaction, no other data on reaction type bonds seems to support this hypothesis. In fact, data are presented elsewhere (e.g., Chapter 4) to show that reaction often leads to weakening.

D. Exchange Reaction Bond

This is a special case of the reaction bond in which the overall reaction can be represented by two sequential reactions, although the two reactions may be indistinguishable on a more practical basis. The term exchange reaction was used widely by Rudy (1969) to describe the adjustment of equilibrium between two phases in a system containing three or more constituents. Reaction of a titanium–aluminum alloy with boron to form a boride containing both elements, followed by an exchange between the

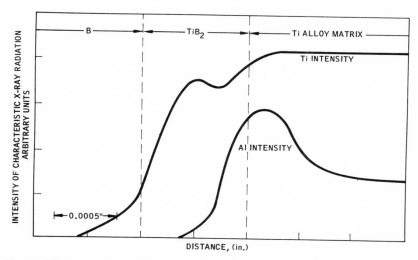

Fig. 1. Electron probe analyzer trace through reaction zone in Ti–8Al–1Mo–1V/B composite showing aluminum rejection from growing TiB$_2$ phase layer [from Blackburn *et al.* (1966)].

titanium in the matrix and the aluminum in the diboride is a good example of this special type of reaction. Figure 1 shows an electron microprobe trace across a Ti–8Al–1Mo–1V alloy interface with boron according to Blackburn *et al.* (1966). The net result of aluminum rejection ahead of the growing diboride is to concentrate the aluminum from 8 to 14%. According to Klein *et al.* (1969) this rejection by the exchange reaction leads to a reduction in the reaction rate constant at 1400°F from 5.2 to 3.4 $\times$ 10^{-7} cm sec$^{-1/2}$ for an alloy with 10% aluminum.

E. Oxide Bond

Introduction of the term oxide bond is made to group together those composites containing an oxide as reinforcement. Also included are those composites where bonding is between oxide films. It is recognized that no new principles of bonding may be involved in oxide bonds, but the classification appears desirable in view of the special problems encountered with this class of composites and the lack of a thorough study of the bonding mechanism in the majority of these systems. Indeed, oxide bonds have been studied most extensively in connection with ceramic-to-metal seals for electron tubes, enamels on metals, and applications other than composites. The most extensive work in the field of composites is that of Sutton and co-workers at the space science laboratory of General Electric Company. Sutton and Feingold (1966) noted the effect of small amounts of impurities on the bond and used high purity nickel to study bonding in nickel–alumina composites. The importance of trace impurities from any source on the bonding in composites is becoming increasingly recognized. Chapter 10 presents some results on three systems, one of which is nickel–alumina, where a detailed study of compatibility using a novel technique has shown the marked effects of impurities. In addition, Chapter 8 on interfaces in oxide reinforced metals presents a complete discussion of all aspects of the subject. The brief review to be presented here will concentrate on general aspects of the physical chemistry of the oxide bond.

From the point of view of the bond, the work summarized by Sutton and Feingold (1966) is noteworthy because they propose a model for nickel alloys and sapphire relating the effect of interaction on the strength of the bond. They propose that the bond strength increases with the amount of interaction, but the apparent bond strength may begin to decrease when excessive interaction weakens the filament. The strength aspects of this theory are discussed in more detail in Chapter 4 on the relationship of interaction to the longitudinal strength of composites. In this chapter it is pointed out that the observed bond strength changes very little as the interaction thickness increases from less than 1000 to 50,000 Å so that the data may be reinterpreted to mean that a very small amount of interaction may suffice to generate a full strength bond. The latter is in better accord with observations on the effect of reaction on bond strength in other types of system, e.g., titanium–boron.

The weakening of the sapphire by reaction with nickel alloys was proposed to occur at all amounts of reaction but only became apparent in excess of 50,000 Å of interaction when failure began to occur through the sapphire rather than at the interface. Ceramic filaments such as glass and sapphire are very susceptible to surface damage that results in weakening of the material. Removal of the damaged surface by etching or flame polishing will restore the strength. In contrast, the strength-controlling defects in boron are located largely at the filament core and the filaments do not have the same susceptibility to loss of strength by surface damage. (Surface compressive stresses in boron may also be a factor.) As a result, much greater control must be exercised to limit the amount of reaction in preparation of oxide reinforced composites than is required for composites strengthened by filaments such as boron.

Sutton and Feingold (1966) conclude that the free energy of formation of the oxide is an important criterion in determining the reactivity of an element with alumina. An element in the nickel matrix would be very reactive with alumina if it had a more negative free energy of formation for its oxide than that of aluminum oxide. Nickel alloys containing titanium and zirconium were very reactive for this reason, whereas chromium was only mildly reactive. Control of the amount of reaction could be achieved only by limiting the amount of these elements in the nickel matrix. Elements with less stable oxides than alumina were reactive only to the extent that they could obtain oxygen from other sources (e.g., the atmosphere). In a similar manner to the example cited earlier for the reaction between copper and alumina, Moore (1969) showed that the bond between pure nickel and alumina depended on the access of oxygen. The spinel $NiAl_2O_4$ formed only in the presence of oxygen.

A special case has been proposed for bonding between aluminum matrices and filaments such as boron and silicon carbide. Work in the author's laboratory has shown that many of the characteristics of the bonds in these systems can be explained if the bonding is assumed to be between the natural oxide films on the surface of aluminum and either the boron oxide films on boron, or the silicon oxide films on silicon carbide. Such an oxide film bond will explain the apparent stability of aluminum in contact with boron because these elements are highly reactive when true contact is established; for example, by infiltration with molten aluminum that causes

FIG. 2. Reaction zone (RZ) containing original interface (IF); diffraction pattern contains reflections from AlB_2, Al, and Al_2O_3.

film breakdown by erosion or some similar mechanism. The term pseudo-Class I system has already been introduced in Chapter 1 to describe composites with this type of behavior. Strong evidence in support of this model was provided by Klein and Metcalfe (1971) based on extraction of the oxide film. Examples of films were shown in Chapter 1. Further evidence for oxide bonding and the presence of an oxide film has been obtained from thin sections of an aluminum (6061)–45% boron composite prepared by Midwest Research Institute. A transmission electron micrograph and diffraction pattern is shown in Fig. 2. The diffraction pattern includes both the reaction zone and the interface region. The specimen had been annealed for 165 hr at 940°F and large amounts of aluminum diboride are identified in the reaction zone. Only aluminum oxide (Al_2O_3) was identified in the interface region. This is to be expected from considerations of the thermodynamics because any boron oxides from films on the boron filaments will be reduced by reaction with the aluminum matrix.

F. Mixed Bonds

This type of bonding will occur in pseudo-Class I systems during the stage of breakdown from a Class I to either a Class II or a Class III system.

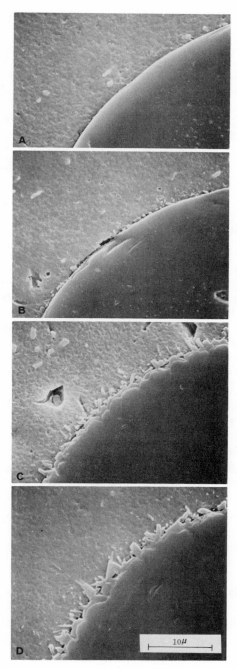

Fig. 3. Breakdown of interface in Al(6061)–25%B composite on heating at 940°F. (A) 0.5 hr, 86 ksi; (B) 5 hr, 76 ksi, (C) 12 hr, 64 ksi; (D) 165 hr, 46 ksi.

75

Figure 2 showed the oxide film at the interface of an aluminum–boron composite, in which breakdown had begun to occur with the formation of aluminum diboride. Fig. 3 supplements this view of a mixed bond by scanning electron microscopy showing the progressive breakdown with time. The film at the original site of the bond is evident in many cases. After the longer times of heating, reaction is general and the bond type may no longer be mixed, although the oxide film persists. Recently, the oxide has been shown to be alpha alumina by electron diffraction on thin sections (Gulden, 1972).

II. Interfacial Stability

One of the most demanding requirements of a metal matrix composite is that the interface be stable. The degree to which this requirement is met will determine, to a large extent, the importance of these composites as materials of the future. Metal matrix composites have their maximum potential as materials for high temperatures where plastic matrix composites are quite unstable, but economical life requirements can be met only by materials that are stable for hundreds or preferably thousands of hours.

Bayles *et al.* (1967) have defined two types of instability common to fiber composites. The first was defined as chemical instability resulting from chemical reaction between the matrix and reinforcement. The second type of instability arises in systems in which the phases are essentially chemically stable with respect to each other, and is characterized by spheroidication and/or agglomeration of the reinforcing phase. In agreement with the suggestion of Parratt (1966), based on observations of composites of nickel or cobalt alloys containing fine whiskers of silicon nitride, aluminum oxide and silicon carbide, Bayles *et al.* (1967) term this instability "physiochemical."

The field of composites has broadened considerably since this work was performed, and it is proposed that five basic types of interface instability can be identified. The first form of instability arises from the same source as that causing instability and overaging of precipitation hardened alloys. The origin of this effect is the dissolution and reprecipitation phenomenon. It is therefore identical with the physiochemical instability of Parratt (1966). The next type of instability arises from dissolution without subsequent reprecipitation. An example of such a system is columbium reinforced by tungsten filaments in the form of wires. The third type of instability arises from continued reaction at the interface in a reactive (Class

III) system. A similar type of instability arises from exchange reactions, and constitutes the fourth type of unstable system. The third and fourth types of instability are identical with the chemical instability of Bayles *et al.* (1967). The fifth type of instability is due to breakdown of a pseudo-stable interface. This type of instability is the least understood and is also the most difficult to predict. It represents a new classification compared with earlier discussions of instability.

Although these five basic types of interface instability are proposed, data are not available in all cases to distinguish between the type of reaction contributing to instability.

A. Dissolution and Reprecipitation

The driving force for this type of instability is the interfacial surface energy that is reduced as the surface area of interfaces is reduced. Spheroidization of pearlitic steels is one of the best known cases of this type of instability. Graham and Kraft (1966) have considered the factors that affect the stability of eutectic composites during high temperature exposure. They point to the special crystallographic relationship between the phases that remains unchanged throughout coarsening of the eutectic. Although dissolution of one of the phases and reprecipitation on existing grains is the mechanism of growth, the authors show that the process is diffusion-controlled rather than solution-controlled. The Thomson–Freundlich equations relating the concentration above a surface of known radius of curvature enter the analysis, but transport from point of solution to point of precipitation is shown to be the rate-controlling factor. Bayles *et al.* (1967) point out that there is little or no net change in either the composition or amount of each phase of a composite in the case of this type of instability, whereas marked changes in amount, composition and even number of phases may occur in the case of other types of instability. This subject will be discussed further in the chapter on the interfaces in eutectic composites.

An example of this type of instability from filament-reinforced composites was found by Niesz (1968) in work on nickel–graphite composites. Thermal cycling from room temperature to 1800°F led to coarsening of the graphite filaments and development of graphite bridges between other filaments. This problem may be more marked with this system because the filaments have a very irregular surface so that there are many points with acute radii of curvature. According to the Thomson–Freundlich relation, these sites will lead to higher concentrations of carbon in the matrix and more rapid transport under the higher concentration gradient.

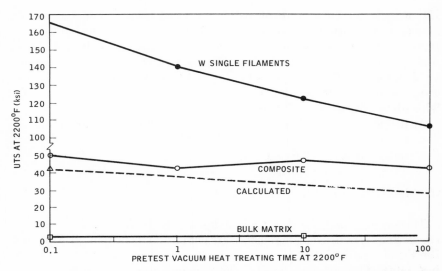

FIG. 4. Comparison of experimental and calculated composite strength for J-alloy–24%W at 2200°F as a function of pretest heat treating time at 2200°F.

B. Dissolution

The principal disadvantage of dissolution is partial loss of the reinforcement. Dissolution may occur to a large extent when molten metal infiltration is used to fabricate the composite, and is most significant when the matrix is able to dissolve large amounts of the reinforcement. Dissolution continues during high temperature exposure of the composite and may be extensive if long times of heating are necessary.

Thermal stability test results were reported on Hastelloy X reinforced by 25 to 30 vol % of 0.010 in. diam tungsten wires by Ohnysty and Stetson (1967). The composite material was made by powder metallurgical methods and interaction was small in the as-fabricated condition (Baskey, 1967). Exposure for 50 hr at 1900°F reduced the residual cross-sectional area to 81%. The same times of exposure at 2000 and 2100°F reduced the residual areas to 67 and 64%, respectively. Such a material is too unstable for use at these temperatures.

Molten metal infiltration has been used in a program at the National Gas Turbine Establishment in England to prepare nickel-base alloys with tungsten filament reinforcement. Dean (1967) reports that the problem of wire dissolution was overcome by rapidly infiltrating a bundle of tungsten wires with the molten alloy, followed by rapid solidification and cooling such that the entire process was completed in a few seconds. Shapes up to several inches in length were made by this method with alloying between wire and matrix confined to a zone of 1 μm width.

A different approach to the problem of dissolution was adopted by Brentnall *et al.* (1970) for the system columbium–tungsten. A practical limit is set to the amount of tungsten that may be added to conventional

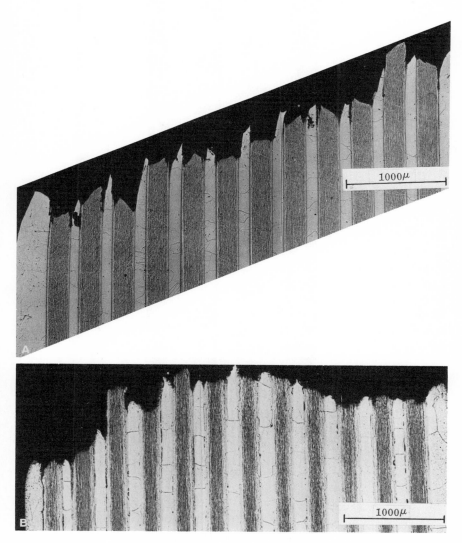

Fig. 5. Effect of 100 hr at 2200°F on the longitudinal tensile failure at 2200°F of columbium alloy–tungsten composites. (A) As-fabricated and (B) after 100 hr at 2200°F.

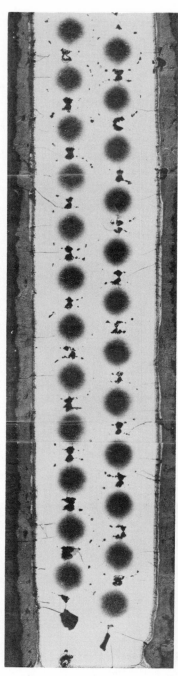

FIG. 6. Formation of voids in columbium alloy–tungsten composite after 689 hr at 2100°F under 21,000 psi.

columbium alloys by the forgeability limit encountered between 20 and 30% tungsten. A composite of tungsten wires in a columbium matrix will be unstable because of this dissolution, but the products of dissolution are the high strength alloys of columbium and tungsten that are normally regarded as being unforgeable. Generation of these alloys provides a compensating factor for the loss of the tungsten reinforcement. Figure 4 shows the effect of heat treatment at 2200°F for times up to 100 hr on the tensile strength of Cb alloy–24 vol % W3Re composites. Two weakening effects occur. The first is the loss of section of the tungsten alloy wires by dissolution; and the second is the recovery process that weakens the tungsten alloy wires. The strength of the as-received wire decreases from 170,000 psi to approximately 110,000 psi measured at a temperature of 2200°F. In spite of these weakening effects, the strength of composite remains essentially constant. It is believed that the generation of high strength columbium tungsten alloys by diffusion is responsible for this constancy of strength. Figure 5 compares the structure of failed tensile specimens of this composite in the as-fabricated condition and after 100 hr at 2200°F. The matrix is less ductile after the anneal because of the large amount of alloying with tungsten.

A problem that may arise with any instability due to a diffusion process is that inbalance of the transported species may lead to the formation of voids by the Kirkendall effect. Fig. 5 showed this porosity around each tungsten filament. A particularly serious example of this type of interface instability is shown in Fig. 6. This specimen of columbium alloy matrix composite, strengthened by 24 vol % tungsten–3 rhenium alloy wires, was tested under 21,000 psi stress for 689 hr at 2100°F without failure. Klein *et al.* (1970) suggest that the voids form because transport of the matrix elements into the filament is not balanced by the diffusion of tungsten into the matrix. However, they also find that nucleation of the voids is accelerated by the presence of residual porosity at the original interface between the filament and matrix. Reduction of this porosity delayed the formation of these voids by a factor of more than tenfold in time.

C. Reaction

Continued reaction to form a new compound at the interface is generally more degrading to the properties of the composite than simple solution. The difference arises from the mechanical properties of the layer formed at the interface. Figure 3 in Chapter 1 presented definition of some strengths for reactive composites. The strength of the reaction zone σ_R will be lower than that of the high strength reinforcement, so that, in the general case, when the compound is brittle, its strain-to-fracture will be lower than that

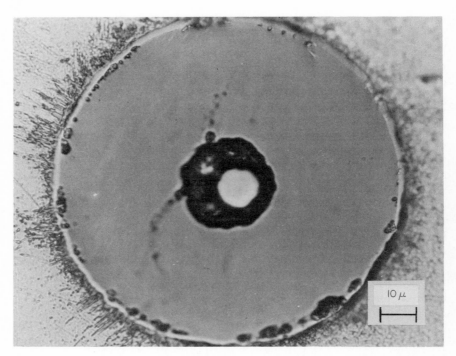

Fɪɢ. 7. Formation of voids in boron filament after reaction for 1 hr at 1700°F with unalloyed titanium.

of the filament. The cracks formed in the reaction layer at this strain will determine the subsequent mechanical behavior of the composite, as discussed in the Introduction and Review (Chapter 1) and developed further in Chapter 4 on effect of interface on longitudinal tensile properties. On the other hand, when the interface layer is a solid solution, there will be enough ductility in most cases so that the failure is initiated in the filament.

The theory of brittle interfaces advanced by Metcalfe (1967) requires that the interfacial layer remain below a critical thickness to preserve strengthening according to the rule of mixtures. These critical thicknesses are usually very small and require low rates of growth for stable systems. Because of the extreme importance of stability in such cases, many studies have been made of the growth of the reaction product(s). These results will be discussed in the next section on reaction kinetics.

Reaction between a matrix and a filament can take place at either the matrix–compound interface or at the filament–compound interface. In the first case, the atoms of the filament must diffuse through the compound, but in the second case it is the matrix atoms that diffuse through the reaction product. In some cases both processes may occur. Blackburn *et al.* (1966) and others have shown that the reaction between titanium and boron occurs by the first mechanism. The outward flow of boron from the filament leaves voids in the center of the boron filament clustered around the tungsten core, as shown in Fig. 7. Some voids may form at the filament–compound interface by this means but the origin of the majority of voids at this position is believed to have a different source. There is a volume contraction of 20% when the titanium diboride is formed from the elements and this factor is believed to contribute strongly to the voids at the inner boundary of the compound. Whatever the mechanism of formation of porosity, the instability of the interface leads to weakening of the composite. For example, reaction may change the failure mode of composites under transverse stresses from matrix to interface fracture (see Chapter 5).

An interesting example of instability resulting from reaction has been reported by workers at the materials science laboratory of the Aerospace Corporation. Aluminum–graphite composites joined by welding form aluminum carbide at the interface, in agreement with the earlier observations of Pepper *et al.* (1970). The carbide forms rapidly above 700°C, but decomposes by reaction with moisture in the air to release methane. Composites made under carefully controlled conditions do not contain the aluminum carbide.

D. Exchange Reaction

Reaction between a filament and a matrix that contains two or more elements may be considered to occur in two steps. At first, the reaction product will contain all of the matrix elements combined with the filament to form a compound. However, thermodynamic considerations will generally require that one or more of the matrix elements concentrate in the compound. Exchange of the elements between the compound and the matrix will occur in approaching equilibrium. This exchange reaction causes the composition of the matrix adjacent to the compound to become deficient in the element that concentrates in the compound, and rich in the noncompound forming element(s). Hence, reaction is slowed down to a very considerable extent, depending on the rate at which the accumulation of noncompound forming element(s) can be diffused away. The overall reaction consists of growth of a compound with rejection of one or more constituents ahead of the growth, but consideration of the two steps is more meaningful in terms of the rate-controlling reaction. Rejection of aluminum from the diboride phase and accumulation in the matrix ahead of the growing compound occurs in the system Ti–8Al–1Mo–1V/boron (Fig. 1). Interface stability will be improved by the reduction in overall reaction rate by the exchange reaction.

E. Breakdown of Pseudostable Interface

This type of instability involves a change in the interface occurring under unpredictable conditions. However, because of the statistical nature of the breakdown, and because of the very large interfacial areas, the breakdown may occur under apparently reproducible conditions.

The general mechanism of oxide bonding in aluminum–boron composites has been established in recent work although many of the details of the process remain unsettled. The oxide bond concept is consistent with erosion of the oxide films by molten aluminum but preservation of the films when composites are made by the optimum solid state diffusion process. Breakdown of this thin oxide film allows chemical reaction to begin. The mechanism of breakdown of the oxide film is believed to be complex, involving mechanical rupture and spheroidization. Mechanical rupture is the principal method of breakdown in solid state diffusion bonding, but this breakdown occurs only at local sites. Spheroidization is the long-term process for disruption of the films by thermal means and is governed by the excess surface energy of the thin oxide films.

Antony and Chang (1968) observed that the stability of aluminum–27 vol % boron composites was markedly reduced when the composite was thermally cycled. The composites were cycled between 800°F and room temperature. Both composite and filament strength decreased with increasing test duration and exhibited strength losses of about 20 and 40%, respectively, after 100 cycles. The loss of strength could be accounted for solely by the degradation noted in extracted boron filaments with no damage to matrix–filament bonding. Although the authors suggest that undefined changes in structure and/or residual stress distribution within the filaments could account for the strength loss, more recent work indicates that the source of filament degradation lies in the reaction occurring at the interface following breakdown of residual oxide films.

Stuhrke (1969) investigated the solid state compatibility of boron–aluminum composites by long term isothermal annealing at temperatures of 450, 700, and 1000°F. The decrease of strength was less than 10% after 100 hr at 700°F but increased to nearly 40% after 5000 hr at this temperature. The rate of strength decrease was much more rapid at 1000°F, and reached 50% after only 10 hr at this temperature. Figure 8 shows these re-

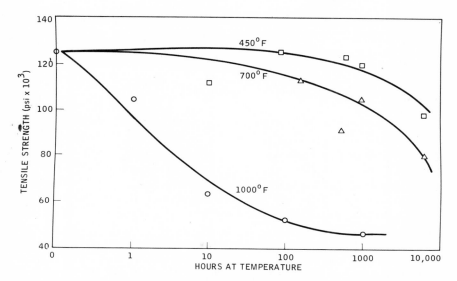

FIG. 8. Boron–aluminum composite tensile strength as a function of time at indicated temperature [□, 450°F (232°C); △, 700°F (371°C); ○, 1000°F (538°C)]; tested at room temperature [from Stuhrke, (1969)].

sults. Stuhrke advanced two explanations for the results. The first involved uncoupling of matrix and filaments, and the second included degradation due to interaction of the boron with the matrix leading to a reduction in the strain-to-fracture. Again, more recent work would support the second explanation.

Klein and Metcalfe (1972) studied the breakdown of the interface of aluminum (6061)–boron composites at temperatures above 700°F. Two volume percentages, two heat treatment conditions and four temperatures were included in the investigation with the same general results. The breakdown was followed by tensile tests, as discussed in more detail in Chapter 4. The time to 50% loss of strength was used to compare results at different temperatures by an Arrhenius type of plot, as shown in Fig. 9. Data from Fig. 8 due to Stuhrke are included, although the time for half degradation at 700°F is somewhat uncertain because the curve in Fig. 8 is incomplete. Also included is a determination by Vidoz *et al.* (1968) on the time for degradation of boron filaments in molten aluminum at 1290°F. The filaments used were low in strength (240 ksi) but confirm the rapid rate of strength loss in solid matrices. Further discussion of the relationship between interface and strength is presented in Chapters 4 and 5.

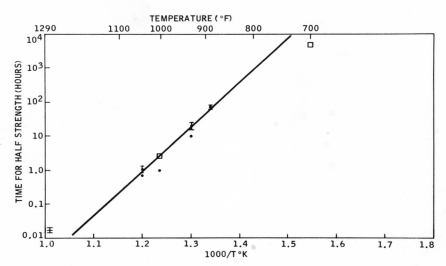

Fig. 9. Time for half strength loss of aluminum–boron composites versus reciprocal temperature. Data from: □, Stuhrke (1969); •‡, Klein and Metcalfe (1972); and ‡, Vidoz, *et al.* (1968).

The examples of breakdown given to this point are drawn from studies of the system aluminum–boron. Much effort has been expended to understand this system, but other systems with an aluminum matrix exhibit breakdown of the interface under certain conditions. Jones (1968) and Pattnaik and Lawley (1971) have studied the system aluminum–stainless steel, and observe behavior consistent with breakdown of the interface. Pinnel and Lawley (1970) describe the preparation of such composites from 0.006 in. diam stainless steel wire (Fe–15Cr–4Ni–3Mo) with a strength of 450,000 psi, and high purity aluminum. These composites were examined by Pattnaik and Lawley in the as-pressed condition and after one day at 550 and 625°C. In the as-pressed condition, these composites are described as devoid of significant chemical interaction at the matrix–reinforcement interface, but reaction was noted after the heat treatments. The tensile test specimens show that the wires separate from the matrix because necking of the wires results in significant contraction, as shown in Fig. 10. The exposed surfaces of the wires were noted to retain the original wire drawing marks although there were some isolated areas of interface reaction. In this regard, the incidence of reaction was strikingly similar to that for the as-bonded specimen of aluminum–boron shown in Fig. 3. A corncob structure formed by reaction at 550°C (1022°F) is shown in Fig. 10. The reaction is

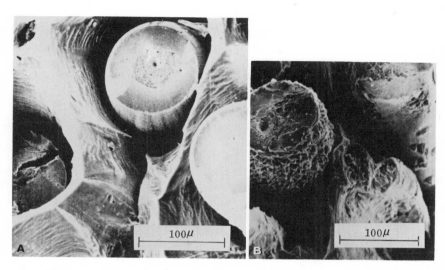

Fig. 10. Tensile fracture surfaces of aluminum–19.6% stainless steel composite: (A) As-fabricated and (B) after 24 hr at 550°C (1022°F) [from Pattnaik and Lawley, (1971)].

not uniform and again resembles that for the aluminum–boron specimens after reaction. The authors note that the reactions at the interface do not occur uniformly in a radial direction and this has been cited earlier as a characteristic of the breakdown of pseudo-Class I systems. This type of structure was more marked after a day at 625°C (1157°F) and again shows the same pattern of behavior associated with extensive breakdown and reaction in the aluminum–boron composites.

III. Reaction Kinetics

Studies of reaction kinetics are made for several purposes. Selection of matrix–filament combinations, selection of processing conditions to avoid excessive reaction, determination of service life expectancy, and development of means to control the kinetics are a few of the applications for such data. However, the point was introduced in Chapter 1 that certain limitations must be placed on the conditions under which the data are generated if the results are to be applied to these problems. The principal requirements noted in the Introduction were that the geometrical conditions should closely approach those in composites, the reaction thicknesses studied should be those of interest in application of the composite, and the temperatures should be similar to those characteristic of composite fabrication and use.

A range up to 20,000 Å of interaction product in titanium–boron and titanium–silicon carbide composites includes all thicknesses of interest in useful materials. Studies of reaction kinetics should be restricted to this range, but methods for the study of the growth of such thin layers are not well developed so that accuracy will suffer if this limitation is imposed. On the other hand, Ratliff and Powell (1970) studied reaction between titanium and silicon carbide, and found a marked change of rate at 100,000 Å. No observations were made for thicknesses below 44,000 Å. The authors show that one change of mechanism could be attributed to saturation of the titanium side of the couple by carbon from the silicon carbide. It seems clear that the very small amount of titanium present in a composite, compared with the thickness of titanium used by these authors (0.150 in.), would saturate rapidly so that changes in reaction kinetics due to this effect would occur at an earlier stage. Hence, some compromise must be sought between the reduced accuracy caused by limitation of the thicknesses investigated and the greater applicability of data generated for the appropriate thickness range.

The geometrical conditions must be those in a composite if the data are to be of immediate use. There are several reasons for this that can be illustrated by reference to the system titanium–boron. The compound titanium diboride is the principal product of this reaction. Figure 7 shows that it forms in a ring around the boron filament. The porosity in the center of the filament shows that it grows by the outward flux of boron through the diboride to react with titanium at the outer edge of the ring. Diffusion couples with flat interfaces will allow discharge of vacancies by escape to the free surfaces and by movement of the boundaries. No loss of vacancies can occur with cylindrical geometry of the filament because the inner diameter of the titanium diboride ring will be that of the original diameter of the filament and will be invariant. Also, the formation of titanium diboride from the elements is accompanied by a contraction in volume of 20%. Adjustment to take care of this volume change is difficult with the geometry of a filament in a composite, but would be accommodated readily in a typical diffusion couple between flat specimens. Additional vacancies and matrix stresses are created in the case of a composite.

The third limitation on data generation relates to the choice of temperature. There is a temptation to use higher temperatures to accelerate data collection. As in all diffusion work, this can lead to errors when the results are extrapolated to lower temperatures. Causes of such errors that have been indicated are: change of stoichiometry of reaction product with temperature (Klein *et al.*, 1969) ; solution of one reaction product in another (Schmitz and Metcalfe, 1971); and disappearance of one of two rate-controlling processes by solution (Schmitz *et al.*, 1970).

It will be noted that all of the examples cited in this section refer to the interaction of various filaments with titanium and its alloys. This is because most of the significant work at the present time has been performed for titanium matrix systems. Less complete studies on other systems have been performed and referenced in the previous section on interface stability. The systems Ti–B, Ti–SiC, Ti–Al₂O₃, and Ni–Al₂O₃ will be discussed in this section.

The method selected to determine kinetics data is important in view of the problems discussed, and the techniques will be examined before the results are presented.

A. Experimental Methods

Apart from the work of Ratliff and Powell (1970), all other studies have been made with the appropriate matrix and filament. Ratliff and Powell (1970) used small cylinders but were interested in the determination of mechanisms of reaction diffusion rather than reaction kinetics. The cylinders were diffusion bonded together under the differential expansion pressure exerted by a molybdenum container. The zero time for the start of diffusion was taken to be the time at which the temperature stabilized at the desired value. Snide (1968) in his study of boron, silicon carbide, and titanium diboride filaments in several titanium matrices, prepared his specimens by electrical resistance heating for a time of 5 to 10 sec at a temperature estimated to be 900°C. The specimens consisted of a few filaments sandwiched between sheets of the desired titanium alloy. Schmitz and Metcalfe (1968), Klein *et al.* (1969), and Schmitz *et al.* (1970) used a variation of this method for titanium alloys in which the current and pressure for bonding were applied by refractory metal anvils. Ashdown (1968) studied the reaction of silicon carbide filaments with commercial purity titanium and prepared his specimens in a manner very similar to that used by Snide. In addition, Schmitz and Metcalfe (1968) and Klein *et al.* (1969) used continuously fabricated titanium–boron and titanium–silicon carbide coated boron tapes, made by the method described by the first authors, to study the kinetics of reaction between unalloyed titanium and the filament. These tapes were made by a high speed diffusion bonding process so that the amount of reaction in the starting material was very low.

Measurement of the amount of reaction after annealing presents one of

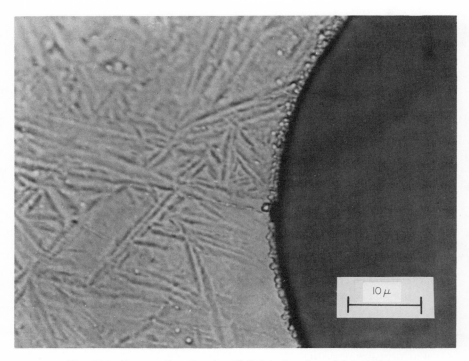

FIG. 11A. Cross section showing Ti–B interface in as-fabricated tape.

the most difficult problems in view of the desire to restrict the total amount of reaction to approximately 20,000 Å. Optical metallography was used in most of the studies. Sepcimen preparation is critical in such work because the prepared surface must be flat to avoid a step between the hard material of the filament and the much softer matrix. Each laboratory has developed its own specialized techniques but key points in this work appear to be to avoid excess pressure in polishing and to provide adequate edge support by the mounting media and by edge support additions. Taper sections were introduced by Schmitz and Metcalfe (1968), and used in later studies. In order to compensate for the lack of flatness of these taper sections, a method was developed to fit a conic section to the obliquely cut filament in order to determine the local magnification in the direction of the taper. This method was most useful for thicknesses below 3000 Å but became less reliable as the thickness became large because of errors caused by lack of flatness of the section. Electron microscopy by means of replicas is not entirely satisfactory because of the out-of-flatness problem. Scanning electron microscopy appears to be more promising. Figure 11 compares cross sections of a titan-

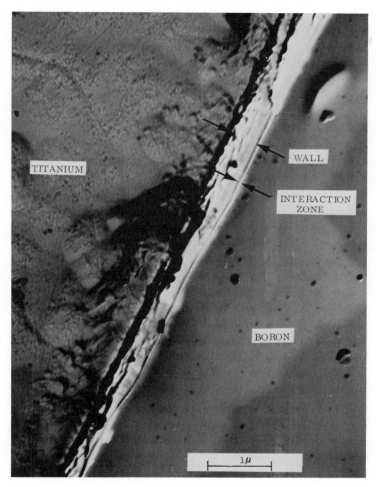

FIG. 11B.

ium–boron tape made by the high speed diffusion bonding process: (A) is the result of conventional metallographic examination at 1400× and (B) is the result derived from electron microscopy using the replica technique. No reaction zone can be seen in the optical microscope pictures but dimpling of the titanium adjacent to the filament is visible. The authors suggest that this zone is caused by solution of oxygen-rich films at the surface during the bonding, but the short time of heating (approximately 1 sec) was inadequate to take the oxygen into general solution. Interpretation of Fig. 11B is based on formation of a step in polishing between the hard

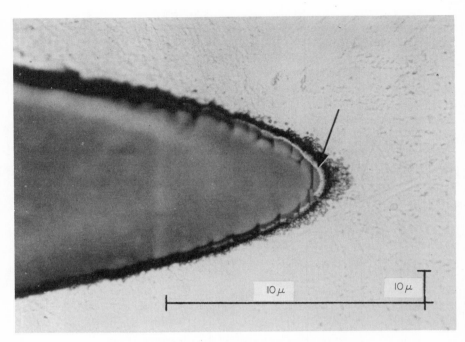

FIG. 12. Taper section of titanium–boron interface.

filament and soft matrix. This step in the pliable replica is pulled out in the view at 25,000× and is marked as the wall. The interaction zone of the hard titanium diboride is at the top of this step and is estimated to be 500 Å thick. Another example of titanium–boron is shown in Fig. 12, where the taper section has been analyzed to measure 1000 to 1500 Å of reaction.

Analysis of diffusion data by most of the workers in this field has assumed that diffusion through the reaction zone was the rate-determining process, when the reaction kinetics are described by

$$x = k\sqrt{t} \tag{1}$$

where x is the width of the reaction zone in centimeters, t is the time in seconds, and k is the reaction rate constant with units of cm sec$^{-1/2}$. This relation has been found to describe adequately most of the growth data determined between filaments and matrices of titanium and its alloys. The derivation of this relationship assumes that growth is controlled by the diffusion of one of the reactants through an interaction zone of constant interfacial area and constant boundary conditions. One of the problems encountered in use of this equation for small amounts of growth is that the amount of reaction at the start is appreciable. Klein *et al.* (1969) used the concept of equivalent time to solve this problem. By this was meant the time necessary at the annealing temperature to grow the amount of reaction formed during the preparation of the specimen. If this time was t_e and the initial thickness of reaction formed in specimen preparation was x_i, then

$$(x + x_i)^2 = K(t + t_e) \qquad (2)$$

The value of the equivalent time was determined from a plot of total thickness squared versus annealing time. The intercept on the time axis gave the equivalent time. This was then added to the annealing time to give the corrected time and plotted versus the total thickness according to Eq. (2). The equivalent time could also be calculated from the known diffusion bonding parameters, although there is uncertainty in the bonding temperature when the electrical resistance method of heating is used to bond in a period of a few seconds. However, such calculations provided a check of the equivalent time determined by the intercept method. Good agreement was obtained in most cases.

The temperature dependence of the reaction rate constant k is expressed by an Arrhenius type relationship, i.e., $A \exp(-Q/RT)$, where A and Q are constants). Since the diffusion constant is proportional to x^2/t rather than $x/t^{1/2}$, as is k, the activation energy for diffusion would be twice the value of Q. The constant Q is often called the "apparent" activation energy. This difference may lead to confusion in comparison of data and is discussed at length by Ratliff and Powell (1970).

B. Reaction Kinetics of Boron with Titanium

The principal studies of the reaction between boron and titanium and its alloys have been: Blackburn *et al.* (1966), Schmitz and Metcalfe (1968), Snide (1968), Klein *et al.* (1969), and Schmitz *et al.* (1970). Blackburn *et al.* (1966) investigated the reaction of boron with the alloy Ti–8Al–1Mo–1V and found that the compound TiB_2 was almost the only product with this alloy. The thickness of this compound is a measure of the amount of reaction, and has been used in all studies of the titanium–boron reaction. The investigation covered the range 750 to 1000°C for times up to 200 hr. The thicknesses of diboride formed varied between approximately 10,000 and 90,000 Å. Table I presents their results with results by other workers for comparison. Results for unalloyed titanium show that the rates are reduced for the alloys.

Schmitz and Metcalfe (1968) investigated the reaction with Ti–6Al–4V and Ti–5Al–2.5Sn. The work was designed to investigate the possibility

TABLE I

Constants for Titanium Alloy–Boron Reactions[a]

Alloy	750°C 1382°F	760°C 1400°F	800°C 1472°F	850°C 1562°F	871°C 1600°F	900°C 1652°F	982°C 1800°F	1000°C 1832°F	1038°C 1900°F	Reference
Ti-8Al-1Mo-1V	2.05	—	5.4	—	—	17	—	28	—	Blackburn et al. (1966)
Ti-6Al-4V	—	3.4	—	—	—	—	—	—	—	Klein et al. (1969)
	—	—	—	—	—	—	25	—	35	Schmitz and Metcalfe (1968)
	—	2.6	—	18	—	—	—	—	—	Snide (1968)
Ti-13V-11Cr-3Al	—	—	—	—	—	—	—	—	—	Klein et al. (1969)
	—	—	—	—	—	—	10	—	18	Schmitz and Metcalfe (1968)
Ti-8V-8Mo-2Fe-3Al	—	1.4	—	—	—	—	—	—	—	Klein et al. (1969)
Ti-11Mo-5Zr-5Sn	—	1.6	—	—	8.0	—	—	—	—	Schmitz et al. (1970)
Unalloyed Titanium	—	2.0	—	23	—	—	32	—	—	Klein et al. (1969)
(CP)	—	—	—	23	—	—	—	—	—	Snide (1968)
	—	5.0	—	—	18.6	—	48	—	78	Klein et al. (1969)

[a] Units are 10^{-7} cm sec$^{-1/2}$.

that an incubation period might exist before reaction commenced. This incubation period would correspond to the transition from a Class I system to a reactive, Class III system. Solution of the oxide films present on the metal and filament surfaces would provide this delay time. Specimens were bonded for very short periods and growths up to 12,000 Å were investigated for times of 960 sec at 1800°F and 2160 sec (36 min) at 1700°F. Plots of reaction thickness against the square root of the time failed to show a significant intercept on the time axis that could be related to a delay in the start of reaction. Because of the short times studied in this work, the accuracy of results was not high, but the results for Ti–6Al–4V included in Table I are in good agreement with other data.

Snide (1968) determined the parabolic rate constants for iodide grade titanium, commercial purity titanium, and the alloy Ti–6Al–4V at a single temperature of 850°C (1526°F). His results are included in Table I. Reaction with the iodide grade of titanium gave a higher value for the constant; 29×10^{-7} as compared to 23×10^{-7} cm sec$^{-1/2}$ for the commercial purity titanium included in Table I. Snide used the single time of 100 hr at a temperature such that after the reaction the specimens contained a large amount of internal porosity. He concluded that the boron–titanium interaction takes place by a predominantly one-way diffusion of boron outward, which was evidenced by the lack of recession of the initial boron–metal interface and the void formation in the filaments. No explanation

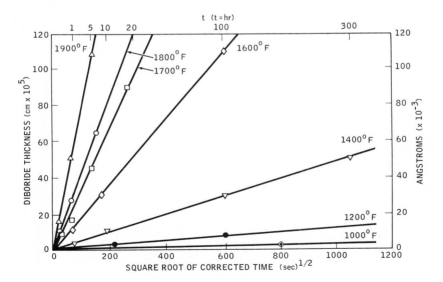

Fig. 13. Growth of diboride in Ti–B composites.

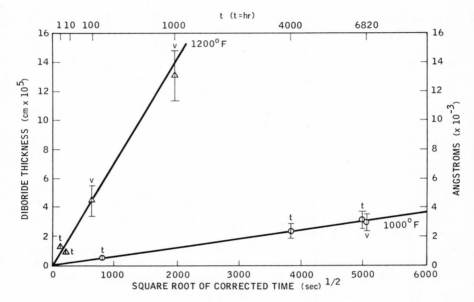

FIG. 14. Growth of diboride in Ti–B composites at 1000 and 1200°F, where t is the taper section and v the vertical section.

was advanced to explain the faster rate of reaction with the higher purity titanium.

1. *Effect of Temperature on Reaction of Boron with Titanium*

The most extensive study of the reaction of boron with commercial purity titanium was made by Klein *et al.* (1969). Plots made of total diboride thickness against the square root of the corrected time, as discussed earlier, are shown in Figs. 13 and 14. The parabolic growth law seems to hold for this system over a wide range of thicknesses. The rate constant k was determined from the slopes of these lines.

Figure 15 shows the temperature dependence of the rate constant, k, according to an Arrhenius-type plot of log k versus reciprocal of the absolute temperature. The linear plot between 1400 and the 1900°F shows that the Arrhenius law is obeyed, i.e., $k = A \exp(-Q/RT)$, where A and Q are constants. The value of the apparent activation energy, Q, is 27,000 cal/mole for the linear portion of the curve. (The true activation energy is 54,000 cal/mole.) The greater value at lower temperatures is believed to indicate some departure from the constant boundary conditions assumed

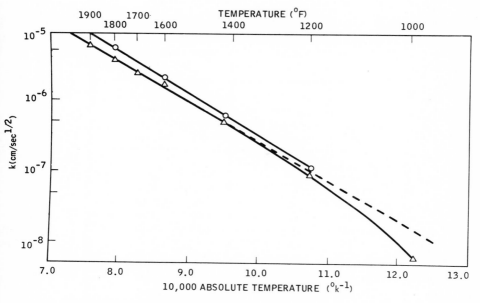

FIG. 15. Variation of rate constant k with temperature, where k = (zone width in centimeters) (time in seconds)$^{-1/2}$. Here ○ is Ti–Borsic and △, Ti–boron.

in the derivation of the parabolic law. This effect is discussed in the section on mechanism of diffusional growth of diboride.

The values for the reaction rate constant for selected temperatures were included in Table I.

2. Effect of Composition on Reaction Rate Constant

Klein *et al.* (1969) studied the effect of alloying elements in titanium on the reaction rate to provide the necessary background data for development of alloys compatible with boron. The single temperature of 1400°F was selected for measurement of rate constants for the reaction between boron and binary alloys of titanium. It was felt that this temperature was sufficiently low so that changes in mechanism at expected service temperatures would be unlikely to occur, yet sufficiently high so that measurable growths could be attained with reasonable periods of heating. These times of heating were chosen to be 10, 50, and 100 hr.

The thickness of the diboride phase formed in 100 hr at 1400°F varied from nearly 30,000 Å for unalloyed titanium to 3000 Å for the alloy Ti–30V, but the plots of thickness versus the square root of the time showed no

TABLE II

REACTION RATE CONSTANTS FOR METALS AND
BINARY TITANIUM ALLOYS WITH BORON AT
1400°F

Metal	Rate constant (10^{-7} cm sec$^{-1/2}$)	Specific rate constant for alloy element	
		(wt %)	(at. %)
Titanium	5.2	—	—
Ti–0.5Si	5.2	0	0
Ti–20Sn	5.3	0	0
Ti–2Ge	5.1	−0.05	−0.06
Ti–10Cu	4.7	−0.05	−0.06
Ti–10Al	3.8	−0.14	−0.08
Ti–30Mo	1.8	−0.10	−0.19
Ti–17V	2.0	−0.19	−0.20
Ti–22V	1.3	−0.18	—
Ti–30V	0.6	−0.15	—
Ti–70V	0.9	—	—
Vanadium	2.5	—	—
Columbium	20.0	—	—
Zirconium	4.2	—	—
Hafnium	2.7	—	—

departure from linearity. A repeat determination was made on the unalloyed titanium–boron tape material to provide a cross-reference with the earlier results. The previous study had given a value of 4.8×10^{-7} cm sec$^{-1/2}$ which compared favorably with the value determined in this study of 5.2×10^{-7} cm sec$^{-1/2}$. In view of the difficulty of controlling temperatures to within a few degrees for periods of 100 hr, the unalloyed titanium–boron specimen provided an internal standard against which the effects of alloying elements could be assessed. A specific rate constant, S_E, was introduced to evaluate the contribution of each element, defined as

$$S_E = \frac{\text{rate constant for alloy} - \text{rate constant for titanium}}{\text{alloy content (\%)}}$$

The alloy content may be expressed in weight or atomic percent. The former is more useful in alloy development but the latter provides insight into the mechanism of control of the kinetics. Table II includes both presentations. The alloying elements may be divided into three types. The first type shows no effect on the rate constant, that is, the specific rate

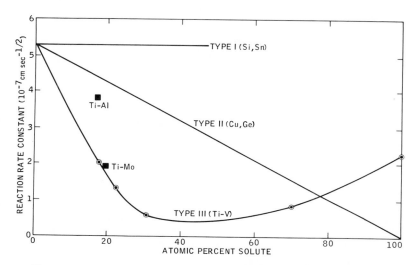

Fig. 16. Effect of alloying on rate constant for titanium–boron reaction.

constant is zero. Silicon and tin fall into this category and appear to cause no change in the behavior of titanium when present in solid solution. Copper and germanium are representative of the second type which acts as a diluent, that is to say, that the effective concentration of the titanium is reduced proportionally to the amount of element in solid solution. A perfect diluent would cause the rate constant to decrease linearly from 5.2×10^{-7} cm $\sec^{-1/2}$ to zero as the titanium content of the alloy is decreased to zero, or the specific rate constant would be -0.052×10^{-7} cm $\sec^{-1/2}$ per atomic percent. Aluminum, molybdenum, and vanadium cause a faster rate of decrease than expected for the diluent types and constitute the third type of alloying element behavior. Examination of the data in Table II shows that vanadium is more effective in dilute solutions than in concentrated solutions. Figure 16 shows the effect of different types of alloying element on the rate constant at 1400°F. The plot of the results for titanium–vanadium alloys shows the effect of concentration. Values for the specific rate constant per weight percent for certain alloying elements in dilute solution were estimated from these results: vanadium, -0.32×10^{-7} cm $\sec^{-1/2}$; aluminum, -0.14×10^{-7} cm $\sec^{-1/2}$; and molybdenum, -0.17×10^{-7} cm $\sec^{-1/2}$.

These values (termed "dilute") were used to calculate the rate constants for commercial alloys to investigate the additivity of these effects. Table III presents results for Ti–6Al–4V and Ti–8Al–1Mo–1V calculated from these values for dilute solutions. The agreement suggests additivity is a

TABLE III

| Alloy | Rate constant $(10^{-7}$ cm sec$^{-1/2})$ | | Specific rate constants used |
	Obs.	Calc.	
Ti–6Al–4V	2.6	3.08	Dilute.
Ti–8Al–1Mo–1V	3.4	3.59	Dilute.
Ti–8V–8Mo–2Fe–3Al	1.6	1.42	Dilute, vanadium from Fig. 16. S_{Fe} assumed = −0.05.
Ti–13V–11Cr–3Al	1.4	1.73	Dilute for aluminum. Vanadium from Fig. 16. S_{Cr} assumed = −0.05.
Ti–11Mo–5Sn–5Zr	2.0	3.08	Dilute. S_{Zr} assumed = −0.05.
	2.0	1.98	Dilute. S_{Zr} assumed = −0.27.

reasonable assumption. Rate constants for more concentrated alloys were calculated using the vanadium values from Fig. 16 and assuming first that any unknown element acted as a diluent so that the specific rate constant was equal to -0.05×10^{-7} cm sec$^{-1/2}$. It appears this value for iron in the alloy Ti–8V–8Mo–2Fe–3Al is valid, but that the value assumed for chromium in the alloy Ti–13V–11Cr–2.5Al may be slightly low, that is, chromium may be more effective than a diluent. On the other hand, the disagreement for the alloy Ti–11Mo–5Sn–5Zr suggests that zirconium is very effective in reducing the rate constant. To obtain agreement between calculated and observed rate constants requires that the specific rate constant for zirconium be -0.27×10^{-7} cm sec$^{-1/2}$. Subsequent work confirmed this value, and use of this element in development of compatible alloys will be discussed later. Table III includes a recalculation of the rate constant for the Ti–11Mo–5Sn–5Zr alloy, assuming the revised specific rate constant for zirconium.

3. Mechanism of Diffusional Growth of Diboride

The mechanism of growth of diboride has already been discussed at the beginning of the section on reaction kinetics. The outward diffusion of boron was first pointed out by Blackburn *et al.* (1966) based on the appearance of porosity in the center of the filament. Klein *et al.* (1969) measured the rate of growth of this porosity and found that it was approximately proportional to the square root of reaction time, indicating that it is a diffusion-controlled process.

The needle-like growth outside the diboride has been identified as titanium monoboride by Blackburn *et al.* (1966) and confirmed by later workers. The amount of titanium monoboride appears to increase with the temperature of annealing and with the time at temperature.

One effect of alloying elements has already been mentioned. This was the rejection of aluminum ahead of the growing titanium diboride phase in a matrix of Ti–8Al–1Mo–1V, and was shown in Fig. 1. Adequate thermodynamic data for a complete analysis are not available, but from general trends, it seems probable that the diborides of zirconium and hafnium are the only ones to be slightly more stable than the phase TiB_2. The diborides of the group five elements are probably less stable, while the group six elements have diborides of yet lower stability. For example, the enthalpy

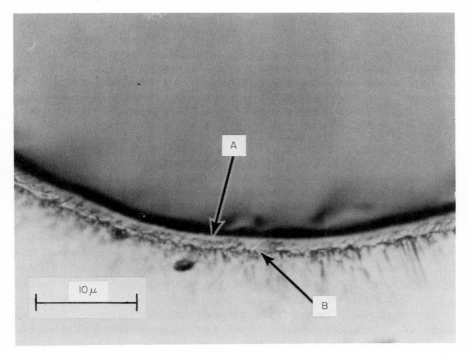

Fig. 17. Cross section of boron filament in Ti–30Mo matrix heat-treated at 1400°F for 100 hr. A is inner zone of diboride; B is outer zone believed to be molybdenum-rich boride.

of formation of the group four diborides are between -70 and -80 kcal/ mole but change to between -20 and -30 kcal/mole for the diborides of the group six metals, chromium and molybdenum. Aluminum diboride also is believed to be of much lower stability than the titanium diboride. On the basis of the considerations discussed by Rudy (1969), it can be concluded that elements that form diborides of low stability will be rejected from the diboride phase. Aluminum and molybdenum will be examples of this type of element. Figure 17 shows the appearance of a diffusion couple of Ti–30Mo/B after 100 hr at 1400°F. Klein *et al.* (1969) interpret this to indicate the rejection of molybdenum from the diboride has led to the appearance of a second zone (B) outside of the diboride (A). However, the combined thicknesses of zones A and B were used in calculation of reaction constants.

Elements that form diborides of high stability will partition between matrix and diboride, and may concentrate in the diboride in the case of the very stable diboride formers, zirconium, and hafnium. Alloying the diboride phase will change the range of composition across the reaction zone. Rudy and St. Windisch (1966) have shown that the maximum range of the titanium diboride phase is from $TiB_{1.95}$ to $TiB_{1.98}$ at the temperature of the first melt reactions, and that this range decreases with decreasing temperature. Hence, at all temperatures, the concentration of vacancies in the boron lattice sites of the titanium diboride will be much greater than the vacancies generated by thermal means alone. Diffusion of boron in titanium diboride will occur by means of these vacant lattice sites. A simple analysis has been made to calculate the mean diffusion coefficient for the composition limits TiB_{2-l} and TiB_{2-n} where $l < n$, giving

$$x^2 = BD_0 \exp(-Q_j/RT)[(n^2 - l^2)/4]t$$

where x is the reaction zone thickness, B is a constant, D_0 is the diffusion constant, Q_j is the activation energy for an atom jump, R is the gas constant, T is the absolute temperature, and t is the time.

It is assumed in the derivation of this equation that n and l are both small and are close in value, so that the concentration gradient of vacancies can be assumed to be linear. Another assumption made is that the thermal vacancies are always much smaller than the vacancies due to nonstoichiometry so that the thermal vacancies can be ignored.

The equation fits the observed parabolic relationship between thickness and time. In addition, the equation offers an explanation for the departure from linearity in the Arrhenius plot presented in Fig. 15. The reasonable assumption must be made that the range of composition of the TiB_2 phase decreases as the temperature is reduced so that the values of n and l move closer together. For example, a change in range of composition from between $TiB_{1.95}$ and $TiB_{1.98}$ to between $TiB_{1.965}$ and $TiB_{1.98}$ would explain the observed departure from linearity at 1000°F.

The change in the nonstoichiometry of the diboride caused by temperature has been used in the preceding discussion to explain the reduction in the reaction rate at 1000 and 1200°F. Alloying may be expected to have a similar effect. Klein *et al.* (1969) reviewed the data on range of composition of the diborides developed by Rudy (1969). These show that titanium, molybdenum, and hafnium diborides are hypostoichiometric at the temperature of the melt reaction, but that vanadium, niobium, and tantalum diborides extend on both sides of the stoichiometric composition. The bracketed composition for zirconium diboride was not adequate to establish its degree of stoichiometry. All of these diborides are isomorphous, so that alloying a hypostoichiometric diboride (such as titanium diboride) with one whose composition extends to the hyperstoichiometric composition will reduce the vacancies in the lattice to very low values as the composition passes through the stoichiometric. This is believed to be the explanation for the minimum in the reaction rate constant plot at 30% vanadium, as shown in Fig. 16. Residual vacancies at the stoichiometric composition will be due to thermal effects so that the equation derived earlier will no longer be valid. One additional assumption made in this analysis is that the variation of composition of the diboride occurs solely by reduction in the number of vacancies on boron lattice sites as the stoichiometric composition is approached.

C. Reaction Kinetics of Silicon Carbide with Titanium

Silicon carbide filaments, silicon carbide-coated boron filaments, and silicon carbide cylinders or pellets were used in the diffusion work included in this section. The principal contributors to the study of the system Ti–SiC have been Blackburn *et al.* (1966), Snide (1968), Schmitz and Metcalfe (1968), Ashdown (1968), Ratliff and Powell (1970), and Klein *et al.* (1969). Blackburn *et al.* (1966) investigated the reaction of silicon carbide filaments with Ti–6Al–4V alloy. Table IV includes their data. They observed that the reaction zone seemed to exhibit two phases, and also noted that the reaction consumed both filament and matrix with no void formation within the filament to indicate preferential movement outwards such as occurs in the titanium–boron system. Another important observation was that the rate of reaction of Ti–6Al–4V with SiC was less than that between Ti–8Al–1Mo–1V with boron.

TABLE IV

CONSTANTS FOR TITANIUM ALLOY–SILICON CARBIDE REACTION[a]

Alloy	650°C 1200°F	677°C 1230°F	760°C 1400°F	800°C 1472°F	850°C 1562°F	871°C 1600°F	900°C 1652°F	927°C 1700°F	982°C 1800°F	1000°C 1832°F	1038°C 1900°F	Reference
Ti-6Al-4V	—	—	—	—	—	—	7.6	—	18.7[b]	—	—	Blackburn et al. (1966)
	—	—	—	3.8	9.5	—	8.3	—	13.5[b]	—	—	Snide (1968)
	—	—	—	—	—	—	—	10	15.0	—	20	Schmitz and Metcalfe (1968)
	—	—	—	—	—	—	—	—	—	22[c]	—	Ratliff and Powell (1970)
Ti-13V–11Cr–3Al	—	—	—	—	—	—	—	—	9.0	—	8	Schmitz and Metcalfe (1968)
Titanium	1.1	—	6.0	—	—	21	—	—	70.0[d]	—	—	Klein et al. (1969)
	1.3	2.5	6.0	—	—	16	—	—	32.0[d]	—	—	Ashdown (1968)

[a] Units are 10^{-7} cm $\sec^{-1/2}$.

[b] 991°C.

[c] Stage I kinetics (see discussion).

[d] See Fig. 21.

Schmitz and Metcalfe (1968) examined the reaction of some titanium alloys with silicon carbide filaments using the same type of conditions outlined earlier for their study of titanium alloys with boron. Their results are included in Table IV. Snide (1968) investigated the reaction of silicon carbide filament w. the alloy Ti–6Al–4V at temperatures of 800, 850, 900, and 991°C for a reaction period of 100 hr, but did not use the value at 850°C for an Arrhenius plot. The results for the other three temperatures gave good agreement with the reciprocal temperature relationship, for which he calculated an activation energy of 31.3 kcal/mole. The reaction layer was observed to be multilayered in agreement with the phase diagram of Rudy (1969) which indicates that the phases TiC, Ti_5Si_3, and $TiSi_2$ should form between the diffusion couple of titanium and SiC. Microprobe work indicated a very small amount of titanium in the filament after diffusion. Snide notes the SiC titanium reaction proceeds by the simultaneous recession of the SiC–filament and growth of the reaction zone into the titanium, and attributes the absence of porosity in the filament to this mechanism of growth.

1. Effect of Temperature on Reaction of Silicon Carbide with Titanium

Ashdown (1968) studied the reaction of silicon carbide filaments with the commercial purity titanium. Each specimen contained two filaments and was diffusion bonded for 20 to 40 sec. Diffusion anneals were performed at temperatures of 650, 700, 750, 800, 850, 950, and 1050°C. The reaction zone after selected times of diffusion were between 10,000 and 100,000 Å in thickness, and gave good agreement with the parabolic growth law when plotted against the square root of the time. The thickness of the reaction zone included the two phases observed earlier by Snide (1968) and others.

Ashdown (1968) plotted his results on an Arrhenius type of presentation. He found some scatter of data but fitted a straight line to the results by means of a least squares analysis. The data for 650°C were not included because "the reaction zone was small and difficult to resolve in the photographs." However, Ashdown's data show that two determinations in good agreement were made at this temperature. The amounts of reaction were 12,000 and 15,000 Å, respectively, after the longest time of heating (140 hr). When the data for 650°C were omitted, the activation energy was found to be 36,500 cal/mole.

Klein *et al.* (1969) investigated the reaction of silicon carbide-coated boron with titanium using specimens in the form of continuously made tapes with 30 filaments and approximately 25 vol % of filaments. The experimental methods were similar to those described for titanium–boron.

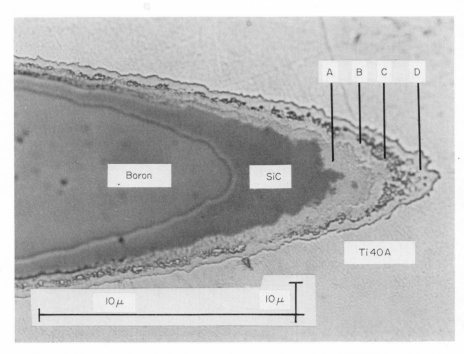

FIG. 18. Taper section of titanium-Borsic composite heated at 1600°F for 10 hr. See text for discussion of reaction zones, A, B, C, and D.

The initial thickness of the silicon carbide coating had been approximately 38,000 Å so that reaction zones of more than about 40,000 Å began to include reaction with the underlying boron. The rate of reaction begins to decrease at this time below that expected from shorter times based on parabolic growth, and voids begin to appear in the filament. The reaction zone formed between the silicon carbide coating and titanium appeared to be more complex than in the earlier studies of Blackburn *et al.* (1966), Snide (1969), and Ashdown (1968). Figure 18 shows a taper section of one of the specimens used by Klein *et al.* (1969) after 10 hr at 1600°F. Four types of reaction product are visible although it is likely that B and D are the same with a dispersion of C in this phase. Based on electron microprobe analysis, Ashdown (1968) believed that the dispersed phase was titanium carbide.

Klein *et al.* (1969) investigated the reaction at temperatures of 1200, 1400, 1600, and 1800°F for times at 1200°F that extended up to 1000 hr. Fig. 15 shows these results in comparison with the results of the same investigators for titanium–boron. The activation energy for this reaction was calculated to be 59,800 cal/mole compared with 54,000 cal/mole for the reaction of titanium with boron.

2. Mechanism of Reaction of Silicon Carbide with Titanium

Ratliff and Powell (1970) studied the reaction of silicon carbide pellets or discs with both unalloyed titanium and the alloy Ti–6Al–4V. The temperatures investigated were between 1000°C (1832°F) and 1200°C (2192°F) so that the results may not be applicable to the lower temperatures of interest in composite fabrication and service. However, interesting effects were observed that are believed to be important to interpretation of other results on this system.

The authors consider the types of binary reaction diffusion problems solved in prior studies. One of two simplifying conditions have been assumed in these analyses. These conditions were as follows:

(1) Terminal phases are semiinfinite in extent.

(2) Terminal phases are saturated with respect to adjacent intermediate phase.

Neither of these conditions is satisfied in a composite, yet assumption of constant boundary conditions is necessary for derivation of the parabolic rate law. Other complications may occur because the reaction product is not a single compound. Two cases were analyzed to develop mathematical relationships. In one of these it was assumed that the metal was initially saturated with the nonmetal, and in the other it was assumed that the metal was not saturated with the nonmetal. The first case yielded the expected parabolic growth relationship, but in the second, the relationship was complex. Two diffusion coefficients were necessary, and were for the nonmetal in both the intermediate compound as well as in the metal matrix.

The results of the experimental study of diffusion reaction between unalloyed titanium and silicon carbide containing 8% excess silicon are shown in Fig. 19, according to Ratliff and Powell (1970). Transition from Stage I to Stage II reaction kinetics occurs over a discrete period of time at 1100 and 1200°C but the transition at 1000°C is clearly a continuous process. Prior carbon saturation of the titanium with carbon causes much faster growth of the reaction zone because the matrix no longer provides a sink for the carbon. The diffusion path was investigated and found to vary with both temperature and time. At the lower temperatures of 1000 and 1100°C, a sequence of three phases marks the diffusion path at the shorter times in Stage I (Ti_3Si; Ti_5Si_3; and TiC_{1-x}) but this increased to four phases at the longer times in Stage II by the addition of a ternary

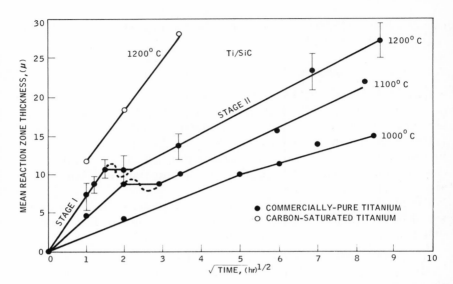

F$_{IG}$. 19. Reaction zone growth kinetics for the Ti–SiC diffusion couples [from Ratliff and Powell, (1970)]. Here ● is commercially pure titanium and ○, carbon-saturated titanium.

phase. The reader is referred to the original reports for fuller details because the results apply to conditions far removed from those of interest to the field of composites. The value of these results lies primarily in the demonstration of the effects of experimental conditions on the data generated. Both Ashdown (1968) and Snide (1968) used a few filaments in a large quantity of titanium so that saturation of the matrix with carbon was unlikely to be achieved. For example, Ashdown (1968) used specimens estimated to contain 0.03 vol % of filaments. Klein *et al.* (1968) used composites with approximately 25 vol % of filaments and noted a general hardness increase in the matrix after small amounts of reaction that was attributed to carbon solution.

Although such differences as those described in the previous paragraph make intercorrelation of results uncertain, some interesting comparisons can be made between the results for reaction of unalloyed titanium with silicon carbide. The results from Ashdown (1968), Ratliff and Powell (1970), and Klein *et al.* (1969) are assembled in Fig. 20 using averages where multiple determinations were made at any temperature. Reexamination of Ashdown's results shows that the reaction constant at 650°C (1200°F) was 1.3×10^{-7} cm sec$^{-1/2}$ for the mean of the two determinations. This compares very favorably with the result of Klein *et al.* (1969) at the

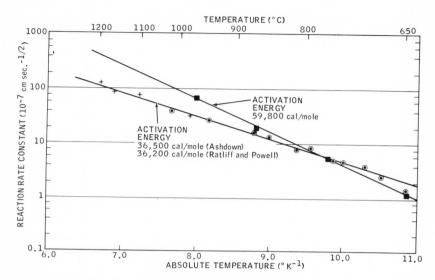

FIG. 20. Arrhenius plot for the Ti–SiC reaction. Data from: +, Ratliff and Powell (1970), stage 1; ⊙, Ashdown (1968); and ■, Klein *et al.* (1969).

same temperature where a value of 1.1×10^{-7} cm sec$^{-1/2}$ was obtained. Inclusion of this data point in the Arrhenius plot (Fig. 20) shows fair agreement with the results of Klein *et al.* (1969) at the lower temperatures and excellent agreement with the results of Ratliff and Powell (1970) at the higher temperatures. This agreement at high temperatures is believed to result from the similarity of conditions, because both Ashdown (1968) and Ratliff and Powell (1970) used a large amount of titanium relative to the silicon carbide, so that saturation of the matrix with carbon was unlikely. Once saturation of the matrix is attained, the reaction constant will increase. This occurs because the carbon from the reacted silicon carbide can no longer dissolve but will form part of the reaction product. Although Ratliff and Powell report a rate constant for carbon-saturated titanium of 110×10^{-7} cm sec$^{-1/2}$ at 1200°C, derived from the slope of the line reproduced in Fig. 19, an even faster rate must occur at shorter times between the origin and the first data point. Such a rate will be in better agreement with the results of Klein *et al.* (1969) for composites where the distances are small so that carbon saturation will be very rapid.

Table IV summarizes results for the reaction of titanium alloys with silicon carbide. Ratliff and Powell (1970) determined the reaction constants and diffusion paths for Ti–6Al–4V with silicon carbide (containing 8% silicon). The reaction followed the same stages observed in the earlier work with unalloyed titanium, with transitions following the same pattern. The activation energy was found to be 63,200 cal/mole for Stage I and 69,700 cal/mole for Stage II. Snide (1968) found 31,300 cal/mole for the same system but used a few filaments in a large volume of titanium instead of cylinders of silicon carbide. It is worth noting that Klein *et al.* (1969) found an activation energy of 59,800 cal/mole for unalloyed titanium, but investigated typical composite specimens. Much additional work will be required to understand the full significance of these results.

D. Reaction Kinetics of Alumina with Titanium

Tressler and Moore (1970) have studied this system for both unalloyed titanium and for the alloy Ti–6Al–4V over the temperature range 1200 to 1600°F. The reaction layer was identified as containing the low oxygen phase Ti_3Al, so that the study of the kinetics included both growth of this reaction layer as well as the solution of oxygen in the matrix. Details of this reaction are discussed in Chapter 8.

The growth of the gross reaction zone with time was found to follow a parabolic law indicating control by a diffusion process. The reaction constant k calculated from those plots followed the Arrhenius relationship. Reaction was more rapid with the alloy than with the unalloyed titanium, and probably reflects the presence of aluminum in the matrix in the former case so that the Ti_3Al phase formed with less transport of aluminum from the filament. The activation energies were quite similar for the two cases and were 50.3 kcal/mole for the alloy and 51.5 kcal/mole for the unalloyed titanium. Tressler and Moore (1970) point out that these values are reasonable from the standpoint of a process controlled by the diffusion of aluminum through the intermetallic phase which forms at the interaction compound–matrix interface.

The relative rate constants for matrix–filament reactions were compared by Tressler and Moore (1970). A comparison is presented below with revisions in the data to reflect the refinements since the time of this work. In addition, a temperature of 1200°F (650°C) has been chosen instead of the 1000°F (538°C) selected by Tressler and Moore (1970):

Ti/Al$_2$O$_3$	1.5×10^{-7} cm sec$^{-1/2}$	Tressler and Moore (1970)
Ti–6Al–4V/Al$_2$O$_3$	2.5×10^{-7} cm sec$^{-1/2}$	Tressler and Moore (1970)
Ti/boron	0.7×10^{-7} cm sec$^{-1/2}$	Klein *et al.* (1969)
Ti/Borsic®	1.1×10^{-7} cm sec$^{-1/2}$	Klein *et al.* (1969)
Ti/SiC	1.3×10^{-7} cm sec$^{-1/2}$	Ashdown (1968)

The relative rates of reaction seem to be such that boron has the lowest rate followed by silicon carbide and alumina. Alloying may increase or decrease the reaction rate. The elemental filament, boron, forms a reaction product that appears to be directly related to the amount of boron consumed, but this is not true for the compound filaments (or coated filaments). The small interstitial elements in the compounds, Al_2O_3 and SiC, go into solution in the matrix to cause hardening and embrittlement, yet the rate of formation of reaction product is faster than with boron. Tressler and Moore (1970) point out that titanium–alumina composites are more tolerant to interaction than composites involving reaction between titanium–boron or titanium–silicon carbide. These points will be explored further in Chapter 4 on tensile strength and Chapter 8 on oxide composites.

E. Reaction Kinetics of Alumina with Nickel

Mehan and Harris (1971) have studied the reaction between nickel and alumina by heating specimens in air. This procedure allows oxygen access to the system so that the nickel is saturated with oxygen at all times. The work of Ratliff and Powell (1970), discussed in connection with the mechanism of reaction between silicon carbide and titanium (Section C), showed the importance of matrix saturation in order to satisfy the conditions for parabolic rates of growth. Metallographic taper sections were used to measure the thickness of the reaction layer. Moore (1969) had previously identified the reaction product as the spinel $NiAl_2O_4$. and discussed the conditions necessary for formation of this compound. These conditions included the access of sufficient oxygen to continue the reaction. Nickel saturated with oxygen at 1000°C has a dissociation pressure of 4×10^{-11} atm, whereas the dissociation pressure of the spinel is 7×10^{-13} atm at the same temperature. Hence, the formation of the spinel can occur readily even if the nickel is not fully saturated with oxygen. Mehan and Harris (1971) found that the spinel growth followed the parabolic law for temperatures of 1100, 1150, and 1200°C for times up to 250 hr. The rate constants were between 4 and 17×10^{-7} cm sec$^{-1/2}$. These values are in the same range as those given by Tressler and Moore (1970) for the reaction between titanium and alumina for temperatures of 1200 to 1500°F (650 to 815°C).

The temperature dependence of reaction found by Mehan and Harris (1971) corresponded to a reaction with an activation energy for diffusional growth equal to 122 kcal/mole.

Further discussion of this topic will be presented in Chapter 8 on oxide composites.

IV. Control of Interface Reaction

Control of the amount, type, and rate of reaction occurring at the interface may be required for one or more reasons. Excess reaction is usually undesirable, as noted by Sutton and Feingold (1966) for interfacially active metals in oxide bonds and by Metcalfe (1967) for titanium matrix composites with boron and silicon carbide filaments. But, equally, inadequate reaction may be unacceptable if this leads to poor bond strength. This section will present a reveiw of the approaches that have been used to control the interface in composites.

Four main lines of effort can be identified. These are enhancement of bonding, suppression of undesirable species, reduction of differences in chemical potential, and reduction of diffusion rates controlling growth. Each of these areas will be discussed in this section.

A. Enhancement of Bonding

This need is found most acutely with alumina-reinforced composites. The oxide is not readily wetted by many metals unless these metals are very reactive, e.g., zirconium, but filament damage can result. Also, the bonding must be strong to allow load transfer between filaments, particularly if the filaments are short whiskers. Consequently, there must be a balance between reaction that increases the bond strength and attack on the filaments that causes the strength of the filaments or whiskers to be diminished. Sutton (1966) discusses this balance for the model system silver–alumina in which the alumina is not wetted by the silver and provides very little reinforcement. The lack of wetting of the whiskers by molten silver was overcome by first vapor-depositing a thin coating of metal such as nickel on the alumina whisker prior to infiltration. Noone *et al.* (1969) discuss this problem in a later paper and extend the results to development of coatings for alumina to be used with a nickel-base alloy matrix. Several coatings for the alumina were developed but either the stability of the coating or the degradation of the filament caused by the coating were not entirely acceptable. Another approach to control of the degree of reaction at the interface was proposed by Sutton and Feingold (1966). Nickel alloys containing 1% of various active metals attacked sapphire excessively, but by reducing the alloy addition to much lower values, some degree of control over reaction could be obtained. The strengths of such bonds were raised to 10,000 psi, but these values were still approximately half the strength obtained with pure nickel.

Further discussion of this subject will be delayed until Chapter 8 which deals with the subject of interfaces in oxide-reinforced metals.

B. *Suppression of Undesirable Species*

Two methods to suppress undesirable compounds that form at the interface have been examined. The first is the use of coatings on the filament, and the second is the use of alloys that reduce the tendency for the compound to form.

Coatings on the filament have been studied most extensively for boron where reaction with common matrices such as aluminum, titanium, and the ferrous metals (iron, cobalt and nickel) led to a search for suitable barriers to prevent the formation of compounds at the interface. Aluminum–boron composites form a special case. It appears that the already existing oxide films on aluminum and boron can act to delay reaction if consolidation is done in the solid state. But the oxides were useless in the presence of molten aluminum and may be inadequate if the solid state diffusion bonding parameters are not optimum. Kreider and Leverant (1966) studied aluminum–boron composites by plasma arc spraying aluminum onto aligned boron filaments. They observed a 50% decrease in composite strength after 1000 hr at 400°C (752°F) and after correspondingly shorter times at higher temperatures. This is a more serious rate of decrease than that found by Stuhrke (1969) and may be due to the rapid deterioration that takes place during the short exposure of the boron filaments to the molten aluminum spray. For example, Stuhrke (1969) found that the strength decrease was from 124,000 to 106,000 psi after 1000 hr at 700°F for aluminum–boron composites with 35 vol % of filaments. Silicon carbide coatings on the boron [trade name Borsic (see footnote Chapter 1, Section II.,A,2)] provide some protection so that plasma arc spraying with molten aluminum does not degrade the filament. Typical coatings are 0.4 mil in thickness. Basche (1969) shows that this thickness of silicon carbide coating on the boron prevented loss of strength of the filament when heated in air or in the presence of aluminum or titanium powders. Although exposure of filaments to loosely packed powders is an unreliable method of investigation because oxide films can preserve a Class I condition for an extended period in a pseudo-Class I system, subsequent investigations under more rigorously controlled conditions have confirmed the general result that the silicon carbide coating reduces reaction with an aluminum matrix.

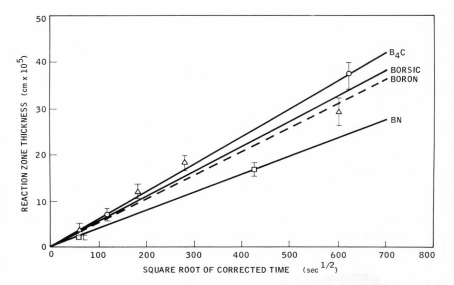

Fɪɢ. 21. Rate of reaction of coated filaments in Ti–40A matrix at 1400°F. Here ○ is B₄C coated; □, BN coated; and △, Borsic.

Boron nitride-coated boron appears to be very stable in the presence of molten aluminum. Camahort (1968) shows that this coating will allow boron to be exposed to molten aluminum for periods of several minutes at temperatures as high as 800°C without attack so long as the coating remains intact. Molten metal infiltration methods to fabricate aluminum–boron composites have been developed based on this finding. Forest and Christian (1970) investigated the shear and transverse strength of diffusion-bonded composites of nitrided boron filaments with 6061 aluminum alloy matrix. They found that the shear strength was lower than that of composites with boron and Borsic. The transverse strength was considerably lower. These low strengths may result from a poor bond between boron nitride and aluminum, but no data are provided on the fracture path. The silicon carbide-coated boron was affected to a much smaller extent. The authors concluded that the standard 6061 uncoated boron composite material, solution heat-treated at 980°F (526.5°C) followed by aging, gave the most desirable combination of mechanical properties.

Klein *et al.* (1969) investigated the usefulness of several coatings on boron for reduction of the reaction rate between titanium and boron at 1400°F. Figure 21 shows their results for boron nitride, silicon carbide, and boron carbide coatings on boron, in comparison with the reaction rate of uncoated boron with titanium. The boron carbide-coated filaments were as

reactive as uncoated boron. Figure 22 shows the reaction zone after 100 hr at
1400°F in which the residual boron carbide layer can be seen with an interaction layer, believed to be TiB₂, and needles of a phase tentatively identified as the monoboride as the principal reaction products. Reaction with
the silicon carbide-coated boron follows the path discussed previously and
presented in Fig. 18. The lowest growth rate was associated with the boron
nitride-coated filament. Figure 23 shows the filament after reaction for 50 hr
at 1400°F using polarized light to make the thin boron nitride coating
more readily visible. The reaction zone was typical of that between titanium and boron, with titanium diboride as the principal product, although the acicular phase was reported to be Ti₂B₅ for other specimens,
according to Moore (1968). The boron nitride coating appears to be intact
with a measured thickness of 2500 Å, that is close to the original thickness
of 2000 to 4000 Å indicated by the manufacturer. This result implies that
the nitrogen is not taking part in the reaction but is rejected to the filament
so that the thickness of the boron nitride layer remains unchanged. The

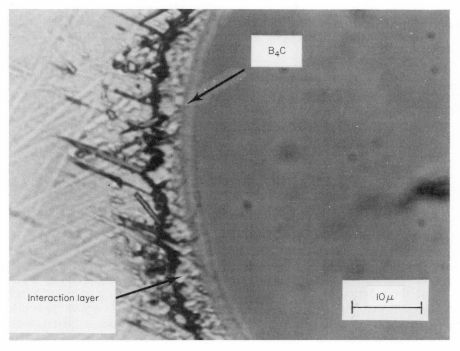

FIG. 22. Reaction between B₄C-coated boron filament and Ti–40A matrix after 100
hr at 1400°F.

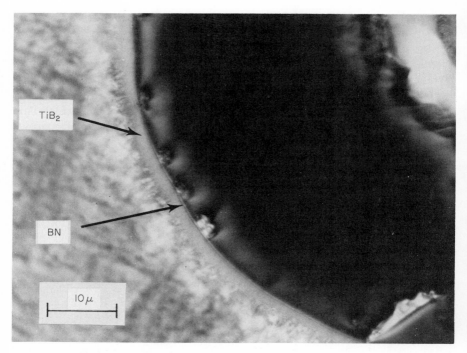

Fig. 23. Boron nitride-coated filament in Ti–40A matrix after 50 hr at 1400°F (polarized light).

reduction in the concentration of boron in the boron nitride layer compared with uncoated boron is adequate to explain the observed reduction in the reaction rate constant at 1400°F from 5.2 to 4.0 $\times$ 10^{-7} cm sec$^{-1/2}$.

A coating of silicon carbide on the boron does not reduce the rate constant at 1400°F for reaction with titanium according to Fig. 21. It is not valid to regard the silicon carbide coating as a sacrificial layer that can delay reaction between the titanium and boron until the silicon carbide layer has been consumed. This viewpoint ignores the harmful effect of the reaction products formed in the sacrificial reaction. The analysis of Metcalfe (1967) and subsequent experimental results of Klein *et al.* (1969) show that the reaction product causes the onset of degradation to occur after the same amount of reaction, although the fully-degraded condition is not so serious for the silicon carbide reaction as it is for the reaction with boron.

It can be concluded that coatings have been successful in reducing the reaction between aluminum and boron but have been unable to moderate the reaction between titanium and boron.

Coatings are not the only method to suppress reaction. Alloying of the reinforcement will also change the form of the reaction products. For example, Harden and Wright (1969) found that laminate composites of aluminum and boron reacted when bonded for various time periods at 600°C and 4000 psi. The strength and elastic moduli began to decrease after 3 to 5 hr respectively, and both were reduced considerably when extensive reaction had taken place after 8 hr. In contrast, no reaction product could be detected when B_4C–Al laminates were bonded under these conditions.

C. Control of Chemical Potential across Interface

The chemical potential gradient across an interface for any species will depend on the phase relationships, and also on the bonding type because the latter will control the activity of the reacting species. However, significant information is available for only the dissolution bond. Two types of dissolution bond have been considered. The first is the system of complete solubility, or the miscible system. Tungsten-reinforced columbium matrix composites are an excellent example of this type of system and were investigated by Brentnall *et al.* (1970) and by Klein *et al.* (1970). It is not possible to reduce the chemical potential gradient to zero in this type of system, but it can be reduced significantly. Diffusion with a matrix free of tungsten occurs initially under the highest possible gradient of chemical potential and results in Kirkendall void formation for this system, as discussed in Section B. and illustrated in Fig. 5. Increase of the tungsten content of the matrix to 10% increased the time to develop voids at 2400°F by a significant amount.

The second type of dissolution-bonded system has limited solubility. Reference to the nickel–tungsten binary system (Hansen and Anderko, 1958) shows that a nickel alloy with 38% tungsten is in equilibrium with tungsten containing a small amount of nickel (less than 0.3%) in solution. This equilibrium implies equal chemical potentials. Petrasek *et al.* (1968) used this principle in development of a compatible nickel-base alloy for nickel alloy–tungsten composites. The starting point for this development was a Ni–20Cr–25W alloy. This was modified by additions of titanium and aluminum. A second series of alloys was made with reduced content of tungsten but with other refractory metals: columbium, molybdenum, and tantalum. The alloys were more compatible with tungsten wires than commercial superalloys but the alloys with tungsten showed a marked increase in compatibility. A further significant increase in compatibility resulted from the addition of titanium and aluminum, but the mechanisms in these cases must differ from the chemical potential control. The authors concluded that 0.015 in. diam wires should be used to withstand the interdiffusion without excessive loss of section. The depth of penetration was 0.001 in. after an exposure time of 60 hr at 2000°F, and corresponded to a diffusion coefficient between 2 and 5×10^{-11} cm sec^{-1}. Because of this interdiffusion of matrix and filament with loss of filament section, the Larson–Miller stress rupture plot was markedly steeper for the composites than for the wires.

D. Reduction of Diffusion Rates

The development of special matrices for metal matrix composites has been undertaken in very few cases, but may be the best approach to control diffusion at the interface. Compatibility of boron or silicon carbide filaments (including silicon carbide-coated filaments) with titanium was reviewed by Klein et al. (1969). They concluded that coatings offered little chance to control reaction and also offered little chance for a reliable system unless thick coatings were used. On the other hand, a compatible matrix would provide a highly reliable system. The results presented in Table I show that considerable variation in the rate of reaction could be realized by variation in the matrix composition. This conclusion was supported by the results for the effects of individual alloying elements and the analysis presented in the section on the mechanism of diffusional growth of diboride (Section B.). Two separate mechanisms for control of growth were identified: one was the rejection of alloying elements ahead of the growing diboride into the titanium alloy matrix, and the other was the addition of elements to the matrix that alloy with the diboride and change its stoichiometry. The elements aluminum and molybdenum were identified as the most important representative of the first class, while vanadium and zirconium were the most effective elements of the second type. In addition, it was shown that the effects of certain combinations of alloying elements were additive. This background formed a promising starting point for the development of a compatible matrix alloy for boron.

Other considerations enter into the development of a compatible matrix alloy. The alloy must be metallurgically stable; it must be rolled readily into foils of the thicknesses required for composite manufacture (for composites to be manufactured by solid state diffusion bonding); it must be low in density; it must possess good strength at the use temperature of the composite; and it must possess the fabricability required of a commercial product. Klein et al. (1969) noted that the high alloy contents required to reduce the reaction rate with boron to low values would cause the titanium alloy to have a beta structure. This structure is favorable to ease of rolling to foil. The maximum addition of aluminum is limited by the formation of alpha or the Ti_3Al phase in the beta alloy. Based on the ternary phase diagram for the ternary system Ti–V–Al, according to Farrar and Margolin (1961) an addition of 2.5% was selected as the probable limit of solubility. In addition to aluminum, molybdenum is also rejected by the diboride. This element was investigated in more detail. It has been noted by Klein *et al.* (1969) that high molybdenum content leads to a duplex structure in the boride phase, as shown in Fig. 17. A series of beta alloys was examined to determine the effect of different amounts of molybdenum. Table V shows these results for the reaction rate constant at 1400°F. The reaction rate constant k was analyzed to determine the contribution attributable to molybdenum. The latter was expressed as a specific rate constant for this element in terms of weight percentage. The alloy Ti–17V–5Mo showed no duplex structure and had a correspondingly low value for the specific rate constant attributable to molybdenum. Once a duplex structure begins to form at $\geq 8\%$ molybdenum, the effectiveness of the addition increases markedly and approaches the extrapolated value given earlier for dilute alloys of -0.17×10^{-7} cm sec$^{-1/2}$ percent^{-1} (see Section B).

A reason for this critical percentage of molybdenum is indicated by the phase relationships in the binary phase diagrams. The binary system Ti–B has two intermediate phases. These are the diboride TiB_2 melting at 3225°C according to Rudy (1969), and stable down to room temperature; and the monoboride TiB formed by a peritectic reaction at 2190°C and also stable to room temperature. The binary system Mo–B is more complex with up to six compounds. The diboride MoB_2 forms by a peritectic reaction at 2375°C but decomposes by a eutectoid reaction to a mixture of MoB and Mo_2B_5 at 1520°C. The diborides of molybdenum and titanium are isomorphous with a hexagonal lattice [crystal structure is that of AlB_2 (C32) type]. Complete solid solution would be expected because the lattice constants for the two diborides are similar. The lattice constants (Å) are:

Diboride	a	c
TiB_2	3.03	3.23
MoB_2	3.04	3.07

However, complete solid solubility can occur only at high temperatures because the molybdenum diboride is not stable below 1520°C. Hence, the amount of molybdenum that can be dissolved in the titanium diboride lattice will be expected to decrease with temperature. Apparently the amount that can be dissolved is very small at 1400°F (760°C) because

TABLE V

Effect of Molybdenum on Rate Constant for Reaction of
Titanium Alloys with Boron at 1400°F

Alloy	Rate constant, k (10^{-7} cm sec$^{-1/2}$)	Specific rate constant for Mo	Structure of reaction zone
Ti–17V–5Mo	1.8	−0.04	Single
Ti–8V–2Fe–3Al–8Mo	1.6	−0.15	Duplex
Ti–13V–5Zr–2.5Al–10Mo	0.2	−0.155	Duplex
Ti–30Mo	1.9	−0.11	Duplex

the diboride becomes duplex when the compound grows in contact with a titanium alloy containing as little as 8% molybdenum. This percentage is dependent to some extent on the other alloying elements in the titanium because the diboride formed with the alloy Ti–11Mo–5Zr–5Sn was single-phased at 1400°F.

The effectiveness of molybdenum in slowing the rate of reaction of titanium alloys with boron derives from the rejection of molybdenum ahead of the growing diboride. The discussion in the previous paragraph suggests that the effectiveness should decrease with increasing temperature because the solubility of MoB_2 in TiB_2 increases markedly with temperature until the two borides should be miscible above 1520°C. Preliminary results have indicated that the duplex zone found for the Ti–13V–10Mo–5Zr–2.5Al alloy at 1400°F becomes single-phased at 1600°F, and the reaction rate increases unexpectedly rapidly.

The role of elements that enter into the diboride phase has been discussed in Section B. It was pointed out that the effect of composition on the rate constant for titanium–vanadium alloys presented in Fig. 16 could be explained by the movement of the composition with alloying to include the stoichiometric composition. For example, the hypostoichiometric diboride of titanium was estimated to become stoichiometric at approximately 20 at. % of vanadium, in approximate agreement with the minimum in Fig. 16. These considerations led Klein *et al.* (1969) and Schmitz *et al.* (1970) to develop alloys combining the two types of growth-controlling mechanisms. One such alloy was included in Table V and had a rate constant of 0.2 × 10^{-7} cm sec$^{-1/2}$, or 4% that for the reaction between unalloyed titanium and boron. This change in the rate constant means that the time to generate a certain amount of reaction product is increased by the square of this ratio, or the time increases to 625 times that required for a given amount of reaction with titanium.

References

Antony, K. C., and Chang, W. H. (1968). *Trans. A.S.M.* **61**, 550–558.

Ashdown, F. A. (1968). "Compatibility Study of SiC Filaments in Commercial Purity Titanium," GSF/MC/68-1 Thesis presented to the Air Force Inst. of Technol., Dayton, Ohio, June.

Basche, M. (1969). Interfacial stability of silicon carbide coated boron filament reinforced metals, *ASTM Symp. Interfaces Composites*, San Francisco, June 1968. Published as STP 452, "Interfaces in Composites." Amer. Soc. Test. Mater., Philadelphia, Pennsylvania.

Baskey, R. H. (1967). "Fiber-Reinforced Metallic Matrix Composites," Rep. AFML-TR-67-196, September.

Bayles, B. J., Ford, J. A., and Salkind, M. J. (1967). *Trans. AIME* **239**, 844–849.

Blackburn, L. D., Herzog, J. A., Meyerer, W. J., Snide, J. A., Stuhrke, W. F., and Brisbane, A. W. (1966). "MAMS Internal Research on Metal Matrix Composites," MAM-TM-66-3.

Brentnall, W. D., Klein, M. J., and Metcalfe, A. G. (1970). "Tungsten Reinforced Oxidation-Resistant Columbium Alloys," First Annual Report on Contract N00019-69-C-0137, January 1970.

Camahort, J. L. (1968). *J. Comp. Mater.* **2** (1), 104–112.

Dean, A. V. (1967). *J. Inst. Metals* **95**, 79–86.

Farrar, P. A., and Margolin, H. (1961). *Trans. AIME* **221**, 1214–1221.

Forest, J. D., and Christian, J. L. (1970). *Metals Eng. Quart.* **10** (1), 1–6.

Graham, L. P., and Kraft, R. W. (1966). *Trans. AIME* **236**, 94–102.

Gulden, M. E. (1972). Unpublished work at Solar Res. Lab., San Diego, California, on Contract F33615-70-C-1814.

Hansen, M., and Anderko, K. (1958). "Constitution of Binary Alloys." McGraw-Hill, New York.

Harden, M. J., and Wright, M. A. (1969). Development of film-reinforced laminate aluminum composite. *AIME Symp. Metal Matrix Composites, Pittsburgh, Pennsylvania, May 1969.*

Hill, R. G., Nelson, R. P., and Hellerich, C. L. (1969). "Composites Work at Battelle-Northwest." Battelle Memorial Inst., Pacific Northwest Lab., 16th Refractory Composites Working Group Meeting, Seattle, Washington, October 1969.

Jones, R. C. (1968). Deformation of wire reinforced metal matrix composites *in* "Metal Matrix Composites." STP 438, pp. 183–217. Amer. Soc. Test. Mater., Philadelphia, Pennsylvania.

Klein, M. J., and Metcalfe, A. G. (1971). "Effect of Interfaces in Metal Matrix Composites on Mechanical Properties." AFML-TR-71-189, October.

Klein, M. J., and Metcalfe, A. G. (1972). "Effect of Interfaces in Metal Matrix Composites on Mechanical Properties." AFML-TR-72-226, November.

Klein, M. J., Reid, M. L., and Metcalfe, A. G. (1969). "Compatibility Studies for Viable Titanium Matrix Composites." AFML-TR-69-242, October.

Klein, M. J., Metcalfe, A. G., and Domes, R. B. (1970). "Tungsten Reinforced Oxidation Resistant Columbium Alloys." Final Rep. on Naval Air Syst. Command Contract N00019-69-C-0137, November.

Kreider, K., and Leverant, G. R. (1966). "Boron Fiber Metal Matrix Composites by Plasma Spraying." United Aircraft Res. Lab., Contract AF33(615)-3209, Rep. AFML-TR-66-219, July.

Mehan, R. L., and Harris, T. A. (1971). "Stability of Oxides in Metal or Metal Alloy Matrices." General Electric Company, Philadelphia, Pennsylvania. AFML-TR-71-150, August.

Metcalfe, A. G. (1967). *J. Comp. Mater.* **1,** 356–365.

Moore, T. L. (1968). Unpublished information, cited in "Effectiveness of Boron Nitride Coating in Boron Composites," by A. E. Vidoz, J. L. Camahort, W. C. Coons, and A. R. Hansen, Lockheed Missiles and Space Company Rep. 6-78-68-26, July.

Moore, T. L. (1969). DMIC Memorandum 243. Battelle Memorial Institute, Columbus, Ohio.

Niesz, D. E. (1968). "Development of Carbon-Filament Reinforced Metals." Final Rep., Battelle Memorial Inst., Columbus, Ohio, Contract N00019-67-C-0342, June.

Noone, M. J., Feingold, E., and Sutton, W. H. (1969). The importance of coatings in the preparation of Al_2O_3 filament/metal matrix composites, *ASTM Symp. Interfaces Composites, San Francisco, June 1968.* Published as STP 452, "Interfaces in Composites." Amer. Soc. Test. Mater., Philadelphia, Pennsylvania.

Ohnysty, B., and Stetson, A. R. (1967). "Evaluation of Composite Materials for Gas Turbine Engines." Solar Div. Int. Harvester Co., San Diego, California, Rep. AFML-TR-66-156, December.

Parratt, N. (1966). *Chem. Eng. Progr.* **62** (3), 61–67.

Pattnaik, A., and Lawley, A. (1971). "Mechanical Behavior of Aluminum–Stainless Steel Composites Subjected to Elevated Temperature Exposure." Tech. Rep. No. 6 on Contract N00014-67-A-0406-0001, March.

Pepper, R. T., Rossi, R. C., Upp, J. W., and Riley, W. C. (1970). The room temperature mechanical behavior of aluminum–graphite composites, *A.S.M. Mater. Eng. Congr., Cleveland, Ohio, October 20, 1970.*

Petrasek, D. W., Signorelli, R. A., and Weeton, J. R. (1968). "Refractory Metal-Fiber-Nickel-Base-Alloy Composites for Use at High Temperatures." NASA TN-D-4787.

Pinnel, M. R., and Lawley, A. (1970). *Metall. Trans.* **1**(5), 1337–1348.

Ratliff, J. C., and Powell, G. W. (1970). "Research on Diffusion in Multiphase Systems, Reaction Diffusion in Ti/SiC and Ti-6Al-4V/SiC Systems." AFML-TR-70-42, March.

Rudy, E. (1969). "A Compendium of Phase Diagrams." AFML-TR-65-2, Part V.

Rudy, E., and St. Windisch (1966). "Ternary Phase Equilibria in Transition Metal Carbides." AFML-TR-65-2, Part I, Vol. VII, January.

Schmitz, G. K., and Metcalfe, A. G. (1968). "Development of Continuous Filament Reinforced Metal Tape." AFML-TR-68-41, February.

Schmitz, G. K., and Metcalfe, A. G. (1971). "Evaluation of Compatible Titanium Alloys in Boron Filament Composites." Unpublished work on Contract F33615-71-C-1150 at Solar Div. of Int. Harvester Co.

Schmitz, G. K., Klein, M. J., Reid, M. L., and Metcalfe, A. G. (1970). "Compatibility Studies for Viable Titanium Matrix Composites." AFML-TR-70-237, September.

Schoene, C., and Scala, E. (1970). *Metall. Trans.* **1,** 3466–3469.

Snide, J. A. (1968). "Compatibility of Vapor Deposited B, SiC, and TiB$_2$ Filaments with Several Titanium Matrices." AFML-TR-67-354, February.

Stuhrke, W. F. (1969). Solid state compatibility of boron–aluminum composite material, *AIME Symp. Metal Matrix Composites, Pittsburgh, Pennsylvania, May 1969.*

Sutton, W. H. (1966). "Role of the Interface in Metal–Ceramic (Whisker) Composites." Soc. of Plast. Ind., Chicago, Illinois, February.

Sutton, W. H., and Feingold, E. (1966). *Mater. Sci. Res.* **3,** 577–611.

Tressler, R. E., and Moore, T. L. (1970). *Metals Eng. Quart.* **11** (1), 16–22.

Vennett, R. M., Wolf, S. M., and Levitt, A. P. (1969). *Metall. Trans.* **1,** 1569–1575.

Vidoz, A. E., and Camahort, J. L., Coons, W. C., and Hansen, A. R. (1968). Effectiveness of boron nitride coating in boron composites. *14th Refractory Composites Working Group Meeting, Dayton, Ohio, May, 1968.*

4

Effect of the Interface on Longitudinal Tensile Properties

ARTHUR G. METCALFE and MARK J. KLEIN

Solar Division of International Harvester Company
San Diego, California

The longitudinal tensile properties of unidirectional composites have been more amenable to theoretical analysis than any other area in composite technology. The simple rule of mixtures law and the shear lag analysis for load transfer under tensile loading are two examples of theoretical analysis where theory was ahead or level with experimental verification. However, such analyses were applied primarily to ideal, nonreactive systems. These systems have been designated Class I according to the scheme proposed in the Introduction to this volume. Demonstration that reactive systems (Class III) could have theoretical properties was closely followed by a theoretical prediction of the effect of interaction on longitudinal tensile strength and strain-to-fracture. As is shown further on in this chapter, this prediction was subsequently verified. Because of this unusual development where theory led, or at least kept pace with the experimental work, it appears appropriate to follow the same order of presentation in this chapter. The theoretical work is presented first followed by the experimental work on various systems.

A requirement stipulated in much of the early theoretical work is that the interface be strong enough to transfer load from the tensile grips into the specimen and then to distribute the load equally amongst the filaments. Also, it was required that the interface be strong enough to permit load redistribution to the other filaments from the site of a filament break. Such theories applied if the interface had in excess of the minimum strength required to perform these functions. It is proposed to call these analyses "theories for strong interfaces." In general, these theories were developed for nonreactive systems with filaments insoluble in the matrix, i.e., Class I systems. The validity of these theories has been demonstrated with such systems. However, as interest grew in less ideal systems where reaction was a characteristic of the interface, and as interest was transferred from the weak matrices of model systems to the strong matrices characteristic of useful composites, it became evident that the interface was not always strong enough to satisfy the theories for strong interfaces. Analyses have been developed to cover the cases where failure initiates at the interface and these are called "theories for weak interfaces." In some cases, the theories cover the entire range from a "strong" to a "weak" interface, but these will be included under the general heading of weak interface theories.

This presentation in this chapter will follow the general lines indicated above. Theories for strong interfaces will only be briefly discussed because longitudinal tensile properties are not dependent upon the interface for this type of composite. Theories for weak interfaces will be discussed in detail and will include both the theories relating to Class III (reactive) interfaces, such as those due to Sutton and Feingold (1966) and to Metcalfe (1967), as well as the more recently developed theory relating to pseudo-Class I systems. (The latter is defined in Chapter 1 as a system that behaves as a nonreactive Class I system until breakdown of the interface occurs, revealing the true classification of the system.) The third section will cover the effect of reaction on the strength of Class III systems. Three systems with titanium matrices will be discussed and will include composites with boron, silicon carbide or silicon carbide-coated boron, and alumina filaments. The next section will cover the effect of reaction on the strength of pseudo-Class I systems and will include aluminum–boron and aluminum–stainless steel. The chapter will conclude with a review of the present status of the relationships between interface, reaction, strength, and strain-to-fracture.

Emphasis in this chapter tends to be on reaction and degradation of the interface. There are several reasons for this emphasis. The first is that the matrices of major technical interest are aluminum, titanium, and the ferrous group of metals, and these metals tend to be reactive with many of the filaments of interest. The second reason is that interest in the interface tends to focus on those conditions where the interface begins to be unable to perform in the desired manner, rather than on those conditions when the interface has more than enough strength to perform its desired function. Understanding of the causes and results of degradation of the interface will make possible the necessary controls to ensure theoretical performance is maintained.

The discussion in this chapter will be limited to unidirectional reinforced composites. The reason for this limitation is twofold. First, understanding is far from complete even for the simple, unidirectional composites and little would be gained by presentation of data on more complex materials; and second, the nonunidirectional composite involves off-axis loadings that will be reviewed in the next chapter.

I. Theories for Strong Interfaces

The simple rule-of-mixtures relates the strength of a composite to the relative volumes of each member and the stress borne by each. In the elastic range, this stress is proportional to the elastic modulus of each component, assuming the strains are equal in the reinforcement and matrix. (Residual stresses and Poisson stresses are assumed to be negligible.) At failure, the stress in the reinforcement equals the "strength" of this component. The problem is to define this "strength." In the absence of a matrix, this strength is the bundle strength of the filaments, and remains at this value when a matrix is added with zero interfacial shear strength. The effect of increasing the interfacial strength depends upon the characteristics of the reinforcement. Composites with continuous reinforcement in which the dispersion of strength is zero, that is, the average strength and bundle strength are equal, are unaffected by the interfacial strength. As the dispersion becomes greater, more and more filaments will fail at weak points located out of the plane of fracture. Under these conditions load transfer must occur between breaks in the reinforcement by a mechanism involving load transfer through the interface into the matrix. When the interface strength becomes greater than the matrix strength, matrix shear occurs rather than interface failure, and further increase of interface strength has no effect on the failure mode. This regime of interface-independent failure is described by theories for strong interfaces. Since the longitudinal tensile properties are independent of the interface for this type of composite, theories predicting their behavior fall outside the scope of this chapter. Reviews of these theories are available elsewhere, including Chapter 2 of this volume dealing with the mechanical aspects of the interface.

II. Theories for Weak Interfaces

In contrast to composites with strong interfaces that fail in either the matrix or the filaments, composites with weak interfaces fail at the matrix–filament boundary because the bond is the weakest link in the system. The weak link in a composite will depend on the load system which will be limited to longitudinal tensile loading in this chapter. Four types of interface failure have been identified under these conditions:

(1) Failure under normal tensile stresses generated by Poisson contraction or contraction resulting from plastic deformation.

(2) Failure under shear stresses such as those generated by filament breakage.

(3) Tensile failure of an interaction compound of finite thickness formed around the filament by reaction with the matrix.

(4) Failure of the reinforcement caused by degradation resulting from reaction at the interface.

Theories have been developed for each of these types and each will be reviewed in this section. The first two types of failure are covered by general theories of weak interfaces. The third type of failure relates to the special case where an interaction zone of finite thickness and properties can be identified and will be discussed under the heading of interaction zone theories. The fourth grouping of failure modes have been observed in oxide reinforced composites and will be reviewed under the title of theories of weakening of the reinforcement.

A. General Theories of Weak Interfaces

Tensile loading in the direction of the reinforcement limits the normal stresses applied to the interface to those that result from lateral contractions. But failure of the interface under these conditions is usually a secondary effect. By this is meant that the tensile stress normal to the filament surface reaches the interface strength only after considerable contraction, such as may occur if the filament begins to neck. Jones (1968) and others have observed that the interface in aluminum–stainless steel composites fails when the filaments pull away from the matrix as a result of necking. Similarly, Vennet *et al.* (1969) observed the same sequence of events to occur in composites of brass and tungsten filaments. However, the original failure in both cases began in the filament and its subsequent course did not appear to be influenced by the secondary failures at the interface that resulted from necking. Failure of the interface under normal tensile stresses may be the primary cause of fracture for other types of loading such as transverse tensile loading, but does not seem to be important in longitudinal tensile loading. On the other hand, shear failures at the interface that occur after filament failures do have an important influence on the course of subsequent fracture and will be considered in more detail.

Cooper and Kelly (1968) have divided the mechanical properties of composites into those affected by the tensile strength of the interface σ_i, and those dependent on the shear strength τ_i. For longitudinal tensile loading, they conclude that the tensile strength of the interface is not critical, but the interface shear strength controls the following properties:

(1) Critical or load transfer length (the filament length required for the longitudinal stress in the filament to reach its fracture stress).

(2) Composite fracture under conditions of fiber pullout.

(3) Deformation of the matrix in fracture.

The critical or load transfer length l_c is given by

$$l_c = \sigma_f d / (2\tau_i) \qquad (1)$$

where d is the diameter of the filament, σ_f is the strength of the filament, and τ_i is the shear strength of the interface. This length decreases as the interface strength increases and reaches a minimum value when the shear strength of the interface equals the shear strength of the matrix (this neglects any difference in shear area for interface failure and for failure in

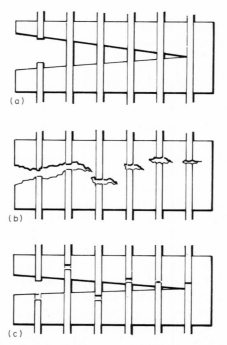

FIG. 1. Failure modes of a composite under longitudinal tensile loading.

the matrix a small distance away from the interface). The load transfer length influences the fracture of composites with noncontinuous filaments in regard to filament pullout, fracture energy, and fracture path. The noncontinuous filaments may arise from filament fractures or be present in the original composite.

A composite with strong interfaces and uniform properties throughout fibers and matrix will fracture on a plane normal to the direction of the applied stress and the fracture surface will be smooth. If the filaments are nonuniform due to weak points or discontinuities, then the crack will be deflected so as to link the weak points. This deflection will require the crack either to pass through additional matrix (strong interface condition) or to pass along the interface. The maximum length of pullout of a filament is governed by the critical length as discussed above. On the other hand, if the failure strain of the matrix is relatively less than that of the filaments, then the matrix will fail first. Figure 1 shows some of these failure types diagrammatically. The failure mode of a composite with low failure strain matrix is shown in Fig. 1a and this behavior is found in aluminum–stainless steel composites according to the work of Jones and Olster (1970). Filaments with low failure strain, such as boron, behave as shown in Fig. 1b. High interface strength has been assumed in case (b) because shearing of the matrix is assumed to be the method by which the cracks link up. In case (c), the filament has low failure strain but the low interface strength allows the crack in the matrix to be propagated with little deflection because the filaments pull out of the matrix easily. Poorly bonded aluminum–boron may behave as shown in Fig. 1c. A low failure strain matrix has been assumed in case (c) but the net result would be the same if the matrix had a higher strain-to-fracture, as shown in case (b), except that the head of the matrix crack will lag further behind the filament cracks as the matrix ductility increases.

In Figs. 1b and 1c, it has been assumed that the filaments are of nonuniform strength so that weak points on the filament removed from the general fracture plane will fail preferentially in spite of the additional work of fracture required for filament pullout. The distribution of these weak points or defects along the filament (length–strength effect) and their relative intensities control the failure of filaments and have an important influence on fracture and fracture energy. In the extreme case, the length–strength effect must include filament breaks leading to discontinuous filament composites. Although Rosen (1964) and others have treated the case of fiber strength characterized by a Weibull distribution, the assumption of a strong interface was made. Cooper and Kelly (1968) have treated the simple case of filaments with uniform properties but containing weak points. If these weakened points have strength σ^* in filaments of uniform strength σ, then the filament will break at a distance y from the fracture plane if

$$y < [(\sigma - \sigma^*)/\sigma]\,(l_c/2) \qquad (2)$$

where l_c is the critical length and is a function of the interface strength (Eq. 1). When y is greater than the distance given by Eq. (2), the interface will be strong enough to reduce the filament stress below the reduced strength σ^* at this distance from the fracture plane. The critical spacing of weak points along the filaments is half this value, or

$$d_{\text{crit}} = [(\sigma - \sigma^*)/\sigma]l_c \qquad (3)$$

If the actual spacing along the filament $d < d_{\text{crit}}$, all filaments break at these weak points. If the spacing $d > d_{\text{crit}}$, then that fraction of the filaments with $d \leq d_{\text{crit}}$ fails out of the plane of the crack.

Cooper and Kelly (1968) developed simple equations for the effective reinforcing strength of the fibers, the average pullout length and the mean work of fracture per fiber. The equations for the mean work of fracture

$\bar{w}$ per fiber were

$$d > d_{\text{crit}}\bar{w} = \pi r \tau' d^2/12 \tag{4}$$

$$d < d_{\text{crit}}\bar{w} = \frac{\pi r \tau' d^2}{12}\left[\left(\frac{\sigma - \sigma^*}{\sigma}\right)l_c\right]^3 \tag{5}$$

The modified interfacial shear strength τ' was used to represent the change in shear strength following yield, or the sliding friction shear stress.

Although the analysis developed by Cooper and Kelly (1968) is idealized, the equations bring out the importance of the statistical distribution of strength of the filaments. Without defects in the filaments; that is, $\sigma^* = \sigma$, the work of fracture for the fibers is zero, but increases to reach its maximum value when σ^* is zero (that is, for discontinuous filaments) and when the critical length $l_c = d$. They point out that under these conditions, the work of fracture per fiber reduces to the value developed by Cottrell (1964) for the pullout fracture toughness of composites reinforced with fibers of length l_c.

B. Interaction Zone Theories

The principal interaction zone theory was summarized in the Introduction and Review because of its importance in composite technology. The theory made it possible to consider reactive (Class III) systems as candidates for engineering composite materials. Since the introduction of this theory for the Class III composites, it has been extended to cover pseudo-Class I composites in a qualitative manner.

In general, the interaction zone theories predict a plateau in composite

TABLE I

STRAIN AT FRACTURE OF COMPONENTS OF COMPOSITES

Material	Modulus E (10^6 psi)	Strength S (psi)	Strain at fracture ($\times 10^6$)
Boron	60	260,000	6000
Silicon Carbide	70	350,000	5000
TiB$_2$	77	193,000	2500
TiSi$_2$	37.7	168,000	4460
TiC	66	197,000	3000

strength and fracture strain until a critical amount of reaction is exceeded. Loss of strength and fracture strain begin to occur at a rapid rate at this critical amount of reaction but continue at a decreasing rate until these parameters merge into a second, lower plateau at a second critical amount of reaction. Although the two critical amounts of reaction depend on other factors such as properties of the filament and properties of the matrix, the strain-to-fracture at the second lower plateau was predicted to be dependent solely on the properties of the interaction zone compound. Only three systems have been studied in detail, and excellent agreement between predicted and observed strain-to-fracture have been obtained in the two cases where the necessary data were available.

1. Theory for Class III Systems

Class III systems are those where reactions begin immediately on contact of filament and matrix and a uniform reaction zone begins to grow by a diffusion controlled process. Examples are: titanium–boron, titanium–silicon carbide, and titanium–alumina.

The theoretical strength of a filament is approximately equal to $E_F/10$ where E_F is the elastic modulus of the filament. The observed strengths S_F result from defects in the filaments with stress concentration factors K_F given by

$$K_F = E_F/(10 \, S_F) \tag{6}$$

This factor will be between 10 and 20 for typical boron filaments corresponding to strengths between 600 and 300 ksi. When such defects occur on the surface, they will be deep, sharp-ended cracks so that the metallic matrix will not be forced into the crack during bonding. No reaction product will form in the crack so that its stress concentration effect will be essentially unchanged. On the other hand, evidence presented by Wawner (1967) shows that the most severe defects in boron are frequently internal around the tungsten core. In either case, the condition will be satisfied that the defects in the filament will be unaffected by the surface reaction.

Metcalfe (1967) proposed that the surface layer formed by reaction will create a new population of defects. These defects arise because the reaction zone cracks at a strain ϵ_R given by

$$\epsilon_R = S_R/E_R \tag{7}$$

where S_R is the strength, and E_R is the elastic modulus of the interaction zone material. Typical values for these strains are given in Table I based on data collected by Samsonov (1964) for massive materials. A further assumption in this analysis is that the interaction zone will contain growth

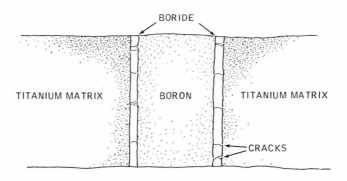

FIG. 2. Cracks in boride layer on boron [from Metcalfe, (1967)].

defects that make its strength similar to that of massive material. Once cracks form in the interaction layer, the situation will be as shown in Fig. 2. These cracks provide a second population of defects. Their stress concentration effect K_R will be given by a term of the type

$$K_R = B(x/r)^{1/2} \tag{8}$$

where x = depth of crack and r = root radius of crack.

The value of B will depend on the stress distribution around the crack, and particularly on the degree of support provided by the matrix. For the appropriate value of the constant B, the value of K_R will increase with increase of x, and hence with the thickness of the reaction layer. A first critical thickness will be reached when K_R becomes equal to K_F, and has the value

$$(x_{\text{crit}})_I = [E_F/(10B \, S_F)]^2 r \tag{9}$$

The cracks in the interaction zone control failure for thicknesses greater than this amount. The failure strain of the filaments in the composite is given by

$$\epsilon_F = (1/10B) (r/x)^{1/2} \tag{10}$$

This equation predicts a parabolic relationship between failure strain and thickness up to the failure strain of the interaction zone. The latter constitutes a second, lower plateau of failure strain starting at a second critical thickness that can be obtained by substituting Eq. (7) into Eq. (10). This gives

$$(x_{\text{crit}})_{II} = [E_R/(10B \, S_R)]^2 r \tag{11}$$

The general form of the relationship between reaction thickness x and strain to fracture, ϵ_F, is shown in Fig. 3 for the titanium–boron system. Metcalfe (1967) assumed that $B = 1$ and $r = 3$ Å (minimum cell size of

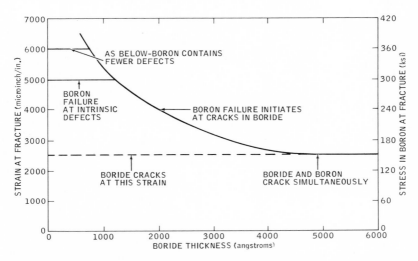

FIG. 3. Failure mechanisms in Ti–B composites [from Metcalfe, (1967)].

TiB$_2$ phase). The assumptions are arbitrary, and the experimental work to be presented later requires some modification of these assumed values (e.g., a root radius of 10 to 15 Å) to obtain agreement with the observed value for the first critical thickness. Figure 3 shows that the first critical thickness is smaller for stronger boron filaments, and that the cracks in the boride layer do not affect the fracture of the filaments or of the composite when the boride is thinner than this critical amount.

Metcalfe (1967) felt that the assumption of crack statics was the most serious weakness in the theory. It was assumed that cracks form at the strain predicted from Eq. (7) but that the filament would not crack until the strain reached the value predicted by Eq. (10). However, the elastic energy release when the crack forms in the reaction zone is likely to propagate into the filament immediately, particularly as the second critical thickness is approached. Hence, the strain at fracture might be expected to fall at a faster rate than given by Eq. (10) and the second critical thickness might be expected to be less than given by Eq. (11). Both effects were found experimentally in subsequent work.

2. *Theory for Pseudo-Class I Systems*

The difference between a Class III (reactive) system and a pseudo-Class I system is that the reaction occurs uniformly in Class III but is initiated only at sites of oxide film breakdown in the case of pseudo-Class I systems. As long as breakdown of the films has not occurred, the composite

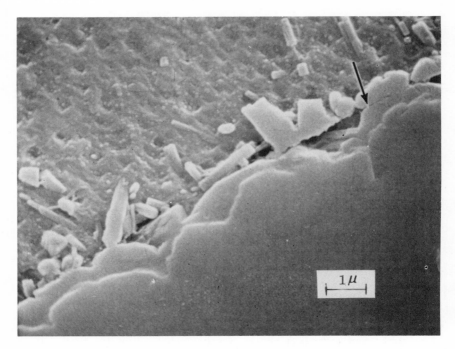

Fig. 4. Vertical section showing AlB_2 interaction phase in a specimen heat-treated for 150 hr at 940°F; the arrow shows growth of AlB_2 through the oxide film at the original interface.

acts as a Class I (nonreactive and insoluble) system. Breakdown occurs in a very irregular manner and reaction is nonuniform. Figure 3 in Chapter 3 depicted some stages in this breakdown in the system aluminum–boron. Pattnaik and Lawley (1971) and Jones (1968) have observed a similar nonuniform reaction in the case of aluminum–stainless steel, and the first authors use the term "corncob" to describe the highly irregular attack.

Results to be presented later show that the relationship of longitudinal tensile strength to amount of reaction is very similar to that predicted for Class III systems (Fig. 3). Yet, the theory developed for Class III systems cannot be applied to pseudo-Class I systems because the interaction zone is not of uniform thickness. Figure 4 shows a specimen of Al(6061)–boron after 12 hr at 940°F, and provides evidence for the irregular growth of the AlB_2 reaction product through points of breakdown of the oxide film. Growth appears to have occurred on both sides of this film, but the original film interface is preserved. The implied approximate equality of diffusion rates of aluminum and boron is supported also by the absence of Kirkendall

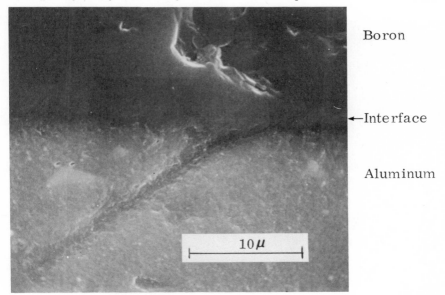

Boron

←Interface

Aluminum

10 μ

FIG. 5. Aluminum–boron interface in a specimen heat-treated for 0.5 hr at 940°F (T-0 temper).

voids. The holes that do form are at the interface and are readily explained by the contraction of 20% that occurs on formation of AlB_2.

Although the theory for Class III systems appears to be inapplicable because no critical reaction thicknesses can be identified, metallographic examination shows that cracking of the AlB_2 is the cause of degradation of strength. Figures 5 and 6 compare notch tensile specimens of Al(6061)–25% boron given 0.5 and 150 hr at 940°F before testing. The specimens were prepared with the filament axes at right angles to the direction of viewing (parallel to the plane of the page). In the specimen heat treated for 0.5 hr, there is no evidence of AlB_2 formation at the interface which is smooth with no evident stress concentrators that could lead to premature filament failure (Fig. 5). There is a recessed region in the boron shown in the top photomicrograph and a diagonal groove in the aluminum, but these features are believed to have been introduced during metallographic preparation.

In contrast with the smooth narrow interface in the 0.5 hr specimen, the interface in the specimen heat-treated for 150 hr before testing is very pitted and irregular due to filament–matrix interaction (Fig. 6). The AlB_2 on the boron side of the original interface remains on the filaments while the AlB_2 on the aluminum side appears to be in part broken-off and em-

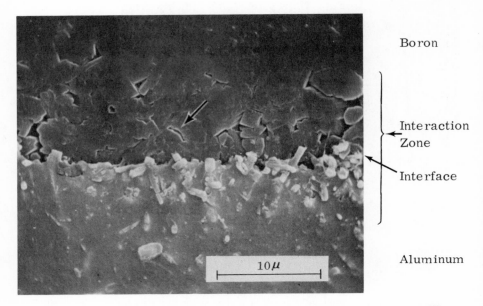

Boron

Interaction
Zone

Interface

Aluminum

FIG. 6. Interface in aluminum–boron specimen heat-treated for 150 hr at 940°F.

bedded in the matrix. The interaction remaining on the filaments near the interface has a rough, granular morphology that yields a very irregular filament surface. As a result, there are many surface features that are probably stress concentrators, and that could certainly account for the reduced tensile strength of the filaments and composite. An example of one of these features is shown adjacent to the arrows.

However, the irregular cracks and notches formed by interaction cannot explain the characteristic relationship between interaction and strength on failure strain (Fig. 3). A satisfactory theory must explain the critical amounts of interaction for both initiation and completion of degradation, as well as the reproducible strain-to-fracture of fully degraded aluminum–boron composites. In the absence of a relationship to the thickness of the interaction zone (characteristic of Class III systems), a tentative theory is proposed based on the critical area of interaction. Justification for this criterion for failure is based on the effect of stress on idealized sections of reacted filaments in a pseudo-Class I system, as shown in Fig. 7. Whereas the uniform interaction zone in a Class III system is strained by tensile load to the same extent as the filament, the interaction product in the pseudo-Class I system can relax at each end by distorting in the manner shown. This relaxation will be large for crystals with a small base attached to the filament, and decreases as the area of contact increases. The critical

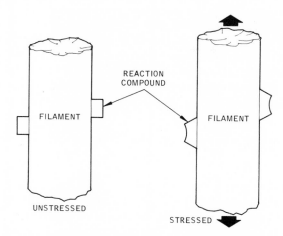

FIG. 7. Behavior of individual crystals of reaction compound under stress.

thicknesses expressed by Eqs. (9) and (11) are exceeded by a large amount before this area becomes large enough to reduce the relaxation to the point where cracking occurs.

In conclusion, it appears that the properties of the interaction compound control fracture of the composite, but a critical amount of interaction is needed before this control is apparent. In the case of Class III systems, this amount is the reaction zone thickness, but in pseudo-Class I systems some other criterion is needed. It is suggested that this is the area over which growth has occurred.

3. Matrix and Filament Effects

The interaction zone theories are based on elastic stress concentrations generated at ends of cracks in the interaction zone. Yielding of the matrix or filament will change the stress situation very markedly.

The support provided by the matrix at the matrix end of the crack in the interaction zone will depend upon the elastic modulus and proportional limit of the matrix. Plastic flow of the matrix will remove this support and increase the stress concentration effect at the filament end of the crack. These effects are not included in the basic theory, except through choice of the constant B in Eq. (8). The influence of matrix support has been investigated experimentally, however, and results will be presented later.

Plastic flow of the filament at the end of the crack will change the stress condition markedly. Pattnaik and Lawley (1971) and Jones (1968) have observed this behavior in aluminum–stainless steel composites. A series of fairly regularly spaced cracks in the interaction compound were noted with a marked concentration of slip lines in the steel filament running from the end of each crack. Figure 5 in the Introduction to this volume showed an example of this type of behavior. Cracking of the interaction zone does not affect composite strength adversely when the stress concentration effects are attenuated by plastic deformation in the filament.

C. Theories of Weakening of the Reinforcement by Surface Damage

An interface interaction can weaken the filaments in a composite by introducing surface defects directly into the virgin filament. These surface defects on the filament can then be the source of premature filament and composite failure. Theories based upon this premise are the subject of this discussion. In contrast, the interaction zone theories discussed above are based upon failure of the unweakened filaments caused by cracks in the brittle interaction zone. The difference in the two approaches is shown schematically in Fig. 8.

Weakening of the filaments can also result from surface damage introduced during filament handling or consolidation of the composite. In both cases, the strength of the filaments is dependent upon the severity of the defects according to the Griffith theory or one of its modifications and the population and distribution of defects. Care in handling glass and oxide filaments to avoid this type of damage is a recognized requirement.

Noone et al. (1969) discussed both of these problems for alumina filament composites. In their work they made use of coatings to minimize mechanical surface damage. Their conclusions are that the function of the coatings are prevention of surface damage of the filament due to mechanical and chemical interaction, and promotion of wetting and bonding.

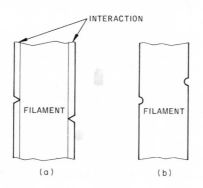

FIG. 8. Comparison of filament weakening induced by (a) the interaction zone and (b) filament damage.

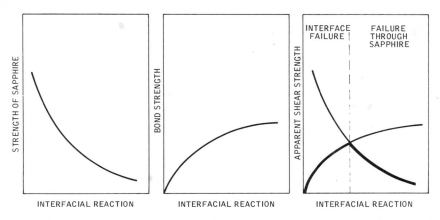

Fɪɢ. 9. Schematic of effects of interfacial reaction on nickel–sapphire joint strength [from Sutton and Feingold, (1966)].

Previous work by Sutton and Feingold (1966) had shown that these requirements tended to oppose each other. Based on their studies of the effect of 1% of various alloying elements in nickel on bonding to alumina (sapphire) plaques, they concluded that surface damage and bonding were related to the amount of interface reaction. Figure 9 shows schematically the competing effects of interface interaction on filament degradation and improved bonding. (The bond strength was measured in shear so that the results are not strictly applicable to a theory for longitudinal tensile testing.) This model may require some modification in detail because the observed shear strengths were nearly the same with pure nickel and with nickel–1% chromium although the amounts of reaction were 100 Å and 50,000 Å, respectively. This suggests a rapid initial rise in the bond strength curve to a plateau in Fig. 9 rather than the gradual and continuous rise indicated by Sutton and Feingold (1966).

III. Effect of Reaction on Strength of Class III Systems

A Class III system was defined earlier as one where reaction between reinforcement and matrix leads to the formation of a layer of compound or compounds. For composites fabricated by diffusion bonding, there will be a short delay time in initiation of this reaction caused by the time to dissipate oxide films on the surface of each constituent. In contrast, pseudo-Class I systems are believed to have quite stable oxide films that break down to permit reaction only after extended holding times at elevated temperatures. The almost instantaneous dissipation of films in Class III systems leads to very uniform reaction zones, whereas the sporadic break-down of the films in pseudo-Class I systems leads to very irregular attack and reaction zones. This difference in the form of the reaction zone affects the way the longitudinal tensile strength is degraded by the reaction.

A. Effect of Reaction on Strength of Titanium–Boron Composites

1. Preparation, Examination, and Heat Treatment of Specimens

Titanium–boron tape containing 30 boron filaments in two layers was made by continuous diffusion bonding using the method described by Schmitz and Metcalfe (1968). Composites were fabricated with two un-alloyed titanium matrices (Ti–40A and Ti–75A) and from three lots of boron.

Klein *et al.* (1969) determined the interaction in as-fabricated tape by several methods that were described in detail in Chapter 3. Briefly, taper sections were used to measure thin interaction zones. Although the reaction layer was complex, the thickness of the titanium diboride zone was used to provide a measure of the degree of interaction.

Heat treatments to change the thickness of reaction product were based on the reaction kinetics described in Chapter 3. The largest series of tests were made after heat treatments at 1600°F but to check that these results were representative of specimens treated at lower temperatures, additional specimens were heat treated at 1000, 1150, and 1400°F. The thickness of the diboride layer was calculated in most cases because prior work had shown that good agreement between observed and calculated values was obtained.

Table II shows that stress relief or conditioning treatments may be very beneficial to increase of strength and to decrease of data scatter. A standard treatment of 30 min at 1200°F was used for all composites to relieve residual stresses without additional growth of the interaction zone.

2. Tensile Properties of Heat-Treated Composites

Table III presents results for the Ti40A–25% boron composite heat treated at 1600°F for various lengths of time. The Ti–40A matrix has a yield strength of 53,000 psi and so there is considerable strengthening of the composite by the boron filaments. Triplicate composite specimens were tested with very little spread in the values for each set of data. The tensile strength is reported in ksi and normalized with respect to the as-fabricated strength (S_0) for each of the times. The strength reaches a lower plateau

TABLE II

EFFECT OF CONDITIONING TREATMENT ON TENSILE PROPERTIES OF Ti40A–B COMPOSITES

| Lot number | Boron (Vol %) | Treatment | Ultimate tensile strength | | | | Strain (μin./in.) | | Elastic modulus[a] (10^6 psi) |
			Obs. (ksi)	Corr.[a] (ksi)	Coeff. of variation (%)		Proportional limit	Failure	
I	28	As fabricated	119	111	8.8		2300	5200	26.6
I	28.4	120 min @ 1000°F	115	107	7.4		2950	5100	26.2
II	25.8	10 min @ 1200°F	112	110	1.5		2250	5710	25.7
II	25.5	30 min @ 1200°F	125	123	0.7		2250	6950	25.4
II	25.4	60 min @ 1200°F	120	119	3.9		2000	6280	27.6

[a] Corrected to 25 vol % boron.

TABLE III

Tensile Properties of Ti40A–25B Composites
Heat-Treated at 1600°F

Time at 1600°F	Diboride thickness (Å)	Ultimate tensile strength		Strain-to-fracture filaments (μin./in.)
		(ksi)	(S_t/S_o)	
0	500	125	1	6950
10 min	4050	121	0.97	5680
0.5 hr	7000	81.5	0.65	2700
1 hr	9900	69.0	0.55	2500
1.5 hr	12,100	71.6	0.57	2800
10 hr	99,000	63.0	0.50	2000

after 0.5 hr at 1600°F, but the strain to fracture of the filaments appears to attain a constant value somewhat before the 0.5 hr period. The average value of the lower limit for fracture strain is 2500 μin. per inch for diboride thicknesses of 7000 Å and greater. This value is that predicted by Metcalfe (1967) based on the properties of titanium diboride, see Table I.

Figure 10 compares the calculated stress-strain curve from the rule-of-mixtures with the stress–strain curves for as-fabricated specimens and for

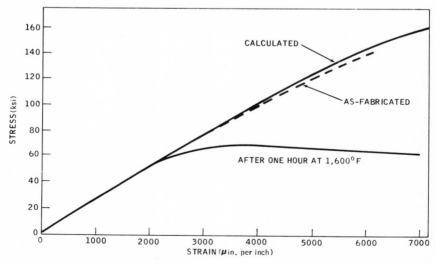

Fig. 10. Stress–strain curves for Ti(40A)–25B composites compared with calculated curve.

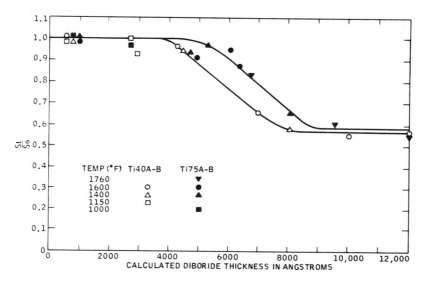

Fig. 11. Effect of diboride thickness on strength of Ti(40A) and Ti(75A) matrix composites with boron.

a specimen heated 1 hr at 1600°F. The as-fabricated specimens follow the calculated curve closely up to a strain of 4000 μin. per inch with gradual departure above the strain as filament breaks begin to occur. The heat-treated specimens have an unchanged modulus but depart abruptly from the calculated curve at 2500 μin. per inch, indicating almost complete loss of the reinforcing effect due to fragmentation of the filaments. The average values of strain-to-fracture of the filaments included in Table III were determined from the average of three individual stress–strain curves such as the one shown in Fig. 10.

Figure 11 includes collected data for heat treatments at lower temperatures and longer times, as well as for the stronger Ti–75A matrix (80,000 psi yield strength at room temperature). The strength values are normalized with respect to the strength before annealing, and are plotted against the calculated thickness of diboride based on the reaction kinetics that were presented in Chapter 3. The plotted data points are the average of 1 to 4 measurements with the majority based on two or three determinations. Although the results relate to specimens annealed over a wide range of temperatures and times, no difference was found in the behavior.

Several studies have been made of the Ti75A–25% boron composites. Schmitz and Metcalfe (1968), Klein *et al.* (1969), and Schmitz *et al.* (1970) have contributed results. One reason for this intensive study was to settle the question of the effect of matrix strength on the tolerance of the system for reaction at the interface. Figure 11 includes data points from each of the sources cited. Each data point is generally the average of two or more determinations. Four batches of Ti75A–boron tape are represented here made from filaments with average strengths of 407 to 494 ksi. The average strength of these batches of tape were 139,000, 143,000, 142,000, and 143,000 psi, and show the consistency of quality of the material used in these studies. Correspondingly, good agreement between results in a group was obtained after any heat treatment.

One set of the specimens tested with the stronger Ti–75A matrix was heat treated for 1500 hr at 1000°F without any significant degradation in tensile strength. Therefore, the Ti75A–B composite is stable for at least this time at 1000°F and probably for much longer since the diboride thickness is only 2800 Å. In addition, the results of this study showed that the stronger Ti–75A matrix is more tolerant of interaction than the weaker Ti–40A matrix. This point was checked by careful examination of one set of specimens that showed no decrease in tensile strength after heat treatment to grow a calculated 5300 Å of diboride (1400°F for 4.3 hr). Measurement of the actual diboride thickness in this set of specimens was 5900 Å. In contrast, the tensile strength of composites made with the weaker Ti–40A matrix begins to diminish at about 4000 Å (Fig. 11).

The results of these studies with Ti–40A and Ti–75A matrices are summarized below:

(1) Boron filament composites do not degrade in 1200 hr at 1150°F (Ti–40A matrix), in 1560 hr at 1000°F (Ti–75A matrix), or in 4.3 hr at 1400°F (Ti–75A matrix). This is an important finding in view of tensile test data on extracted boron filaments showing degradation.

(2) Degradation of composites containing 25% boron commences when the amount of reaction reaches 4000 Å in the case of a Ti–40A matrix and 5500 Å in the case of Ti–75A matrix.

(3) Complete degradation of room temperature tensile strength occurs at a smaller amount of reaction for the weaker matrix than for the stronger matrix (8000 Å versus 9000 Å).

(4) The stress–strain curves show that filament failure in completely degraded composites occurs at a strain of 2500 μin./in.

3. Discussion of the Metcalfe Theory of Weak Interface

The theory presented by Metcalfe (1967) predicted that transition from filament-controlled failure to interface-controlled failure would begin at a diboride thickness of 1000 Å, according to Eq. (9) for the first critical thickness. Completion of the transition would occur at a second critical thickness given by Eq. (11) at a thickness of 5000 Å. It was pointed out that the theory included assumptions of an elastic continuum at the diboride–boron interface, of a crack root radius equal to the cell size of titanium diboride (3 Å), and of an open-ended crack in the diboride with no support from the matrix. [The constant B in Eqs. (8)–(11) was assumed to be unity.] Also, the assumption of crack statics as the second critical thickness was approached was recognized to be an over-simplification.

The first critical thickness has been shown to vary from 4000 Å for Ti–40A to 5500 Å for the stronger Ti–75A matrix. This result suggests that the assumption of open-ended cracks was not valid and that the stress concentration exerted on the boron is reduced by the matrix support at this end of the crack (see Fig. 2). This result will be discussed further under matrix and filament effects on interface-controlled failure.

Comparison of Eqs. (9) and (11) shows that the theory predicts a ratio of the first and second critical thicknesses to be:

$$x_{\mathrm{crit}}\mathrm{II}/x_{\mathrm{crit}}\mathrm{I} = (E_\mathrm{R}S_\mathrm{F}/E_\mathrm{F}S_\mathrm{R})^2 \tag{12}$$

This ratio is independent of the constant B and root radius, r, so that two of the sources of uncertainty in the theory should be eliminated. The ratio equals 5 for typical values of materials properties, in agreement with the values of 5000 and 1000 Å shown in Fig. 3. This ratio should remain constant for other values of the first critical thickness, but the observed ratios are 2 for Ti–40A and 1.7 for Ti–75A. Metcalfe (1967) felt that the assumption of crack statics was "the most serious source of error, particularly as the second critical thickness of 5000 Å is approached when the elastic energy release is more likely to propagate the diboride crack through the boron." Allowance for this factor will reduce the value of the second critical thickness relative to the first critical thickness.

In spite of these modifications necessary to bring the theory in line with experimental observations, it has provided the stimulus to explore composite systems with reactive interfaces and has indicated the possibilities of control of reaction within safe limits in such systems.

4. Matrix and Filament Effects in Interface Failure

Filaments extracted from titanium–boron composites have been found to have an average strength of approximately 150,000 psi according to Klein *et al.* (1969). This corresponds to a strain-to-fracture of 2500 μin./in. in agreement with control by the titanium diboride layer formed during composite fabrication. The critical thickness of diboride in the absence

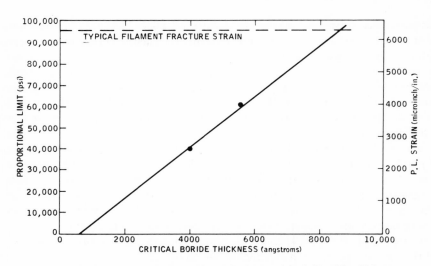

FIG. 12. Effect of matrix proportional limit on critical diboride thickness.

of a matrix is probably less than 1000 Å because the thickness has been established to be 500 to 1500 Å in as-fabricated tape. Figure 12 presents a plot of proportional limit of matrix against critical thickness of diboride for no matrix, Ti–40A (40,000 psi) and Ti–75A (60,000 psi). The simple relationship expresses the contribution of matrix support in reducing the deleterious effects of cracks in the titanium diboride layer. A proportional limit of 90,000 psi in unalloyed titanium would occur at a strain of 6000 μin./in. This is approaching the failure strain of typical boron filaments so that a matrix with a proportional limit greater than this will not be useful in extending the tolerance of the composite for diboride. Figure 12 suggests that up to 8000 Å of diboride could be tolerated in a composite with a titanium matrix that remains elastic up to the failure strains of the boron filaments. This conclusion has not been tested at the present time but continuing work in the area of titanium matrix composites will permit evaluation in the near future.

Preservation of the intrinsic strength of the reinforcement is assumed by Metcalfe (1967) in his theory of the weak interface in a Class III system. Failure at a lower stress occurs only as a result of an increase in the stress concentration factor created by cracks in the interaction zone of greater than critical thickness acting on filaments of undiminished strength.

On the other hand, some measurements of the strength of boron single filaments at elevated temperatures have shown a decrease commencing at temperatures considerably below the safe exposure times and temperatures noted in Fig. 11. These early measurements of the elevated temperature strength of boron filaments were performed in air and show clearly that a marked drop in strength occurred below 1000°F (Wawner, 1967; Vettri and Galasso, 1968). Observation of gross attack on boron filaments by molten boric oxide (melting point 850°F) revealed that one possible cause of strength loss is a surface reaction with the air. Subsequent work was performed in argon atmospheres, but the precautions used to exclude oxygen were not adequate in most of these cases (Herring, 1966). In contrast, oxygen access to a boron filament embedded in a titanium matrix is negligible so that good ground existed to question whether many of these strength data were applicable to the conditions of filaments in a composite. Work was initiated in the author's laboratory by Rose (1966) to measure the tensile and creep strengths of boron filaments in hard vacuum (better than 10^{-5} torr). A plot of the tensile strength and modulus versus temperature up to 2200°F was presented in a subsequent paper by Metcalfe and Schmitz (1969) and is reproduced in Fig. 13. These strengths were derived from short time tensile tests with 5 min holding at temperature prior to the test. The slight rise in strength from room temperature to 1000°F was interpreted to mean that the ductile–brittle transition was located at approximately this temperature. In agreement with this interpretation was the finding by Rose (1966) that plastic deformation occurred in tensile

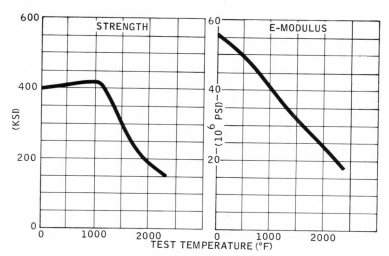

FIG. 13. Effect of temperature on strength and modulus of boron filaments [from Metcalfe and Schmitz, (1969)].

tests at 1000°F prior to failure, but not at lower temperatures. The effect at 1000°F of time on the strength was studied by creep tests on filaments as reported by Metcalfe and Schmitz (1969). This work showed that plastic deformation at fracture was typically 0.2% for stresses in the range 200 to 300 ksi, but also that close correlation could be made between creep tests of filaments in hard vacuum and creep tests of titanium–boron composites continued for at least 100 hr at 1000°F. Further, a tensile test at room temperature of a composite after a 100 hr creep test showed unchanged (actually slightly increased) strength. Therefore, the elevated temperature strength of boron filaments is retained to at least 1000°F if interaction of filaments with the atmosphere is prevented.

These results show that care must be exercised to ensure simulation of conditions within composites in tests to relate single filament, matrix, and composite results. It is concluded that the matrix–filament combination forming the composite is best evaluated as a unit, but that studies of the separate members must be checked thoroughly to assure simulation of the conditions within composites.

Another conclusion is that boron filaments undergo no irreversible loss of properties on exposure for very long periods (in excess of 1000 hr) at temperatures up to 1150°F. Hence, titanium–boron composites will be stable and can be considered for service up to at least this temperature. Similarly, the interface is stable under these exposures. Based on a limit of 8000 Å for a strong titanium alloy matrix, and on the reaction kinetics data presented in Chapter 3, it can be predicted that no loss of strength will be found due to interface degradation until exposure times are of the order of 10^5 to 10^6 hr at this temperature.

Instability in boron filaments occurs at higher temperatures than 1150°F according to Wawner (1967). Wawner cites the results of Gillespie showing grain growth of the beta rhombohedral boron in times as short as 15 min at 1750°F. Wawner also discusses similar results by Lipsitt who made the additional observation that the structural changes could occur at 1300 to 1400°F in the presence of atmospheres of HCl and Cl_2. Hot stage metallography on titanium–boron composites in the author's laboratory (Hutting, 1969) showed that growth occurred at the exposed boron surface in 1 hr at 1550°F. This growth was accompanied by an expansion in volume. Removal of the modified surface showed no subsurface structural changes, in agreement with predictions from the Clausius–Clapeyron equation for a reaction involving an expansion in volume. Containment within a composite will eliminate traces of catalytic elements such as chlorine and inhibit volume changes, so that the temperature range of stability for filaments should be increased when they are embedded within a titanium matrix.

B. Effect of Titanium–Silicon Carbide Reaction on Longitudinal Strength of Composites

The titanium–silicon carbide interface occurs with both silicon carbide and with silicon carbide-coated boron filaments. Fewer data are available on these systems than are available for titanium–boron, but both theory and experiment indicate that the interface-controlled tensile properties follow similar relationships for either filament. The only systematic study of the relationship of interface and strength is that by Klein, *et al.* (1969) on Ti40A–25% Borsic.

Borsic filament with an average strength of 393 ksi and a coefficient of variation of 20.6% (standard deviation 81 ksi) was used to manufacture tape by the method used for titanium–boron tape. The coefficient of variation was greater for this composite after conditioning (5.8%) than for the Ti–B composites. The thickness of the reaction layer was estimated from tape sections using the method described earlier for titanium–boron in Chapter 3.

1. Tensile Properties of Heat-Treated Composites

Specimens (in triplicate) were heat-treated at 1600°F for different times to provide controlled amounts of interface interaction before tensile testing. The results are presented in Table IV. The strength shows a 7% loss after reaction has generated a reaction zone of 4900 Å, increasing to 10 to 15% loss for reaction zones of 12,000 and 14,700 Å. The strength increase

TABLE IV

Tensile Properties of Ti40A–25% Borsic Composites Heat-Treated at 1600°F

Time at 1600°F	Reaction thickness (Å)	Ultimate tensile strength (ksi)	(S_t/S_o)	Elastic modulus (10^6 psi)	Strain-to-fracture filaments (μin./in.)	Reaction product
0	500	105	1	25.8	5550	Silicide
10 min	4900	97.3	0.93	23.6	5300	Silicide
1 hr	12,000	89.7	0.85	23.4	4300	Silicide
1.5 hr	14,700	94.5	0.90	23.5	4400	Silicide
10 hr	38,000	98.3	0.94	29.2	3500	Silicide and some boride
100 hr	85,000	63.3	0.60	32.5	2500	Silicide and much boride

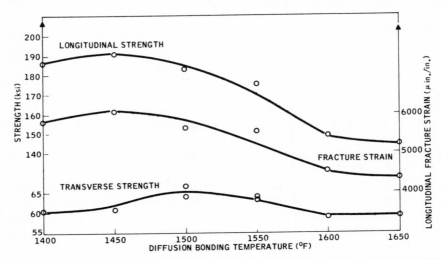

Fig. 14. Effect of diffusion bonding temperature on properties of Ti–6Al–4V/Borsic composite [from Collins, (1972)].

after 10 hr at 1600°F was unexpected. However, examination of the strain-to-fracture of the filaments shows a pattern consistent with that of titanium–boron and with the theory of weak interfaces. The strain-to-fracture begins to decrease when the reaction zone exceeds 4900 Å in thickness, in approximate agreement with titanium–boron, and reaches a plateau at 4300 to 4400 μin./in. for reaction zones in the range of 12,000 to 14,700 Å. This result is in agreement with the value of 4500 μin./in. predicted by Metcalfe (1967) for titanium silicide-controlled failure. The last two results for large amounts of reaction show continued decrease in the strain-to-fracture. Reference to the reaction kinetics data (Chapter 3) shows that diboride has begun to form at 38,000 Å and considerable diboride must be present for a reaction zone of 85,000 Å. The strain-to-fracture of the filaments decreases to 2500 μin./in. when considerable diboride is present, implying control of the fracture process is the same as that in fully degraded titanium–boron composites. In view of the reduced strain-to-fracture of these heavily-reacted specimens, the high strength of the specimens heat-treated for 10 hr at 1600°F is unexpected. The elastic moduli of both the 10 and 100 hr specimens were significantly higher than those of the specimens heated shorter times. An explanation may be that the carbon from the silicon carbide has saturated the matrix while the silicon has formed titanium silicides as shown by Ratliff and Powell (1970) and discussed in Chapter 3. In this regard, the reaction products can vary markedly depending on volume percentage of filaments, composition of matrix, silicon-to-carbon ratio in filament and other factors. The strain-to-fracture of fully-degraded composites might be expected to vary depending on the nature of the reaction product, in agreement with which fracture strains of 3000 to 3400 μin./in. have been reported. Hardening of the titanium matrix by solution of carbon is reported by Schmitz *et al.* (1970).

2. Other Studies of the Titanium–Silicon Carbide Reactions in Composites

Recently, some detailed studies of the processing of Ti–6Al–4V with 45 to 50 vol % of the 5.7 mil diam B–SiC filaments have been made (Collins, 1972). Correlation with the structure of the interface has not been completed, but increasing amounts of reaction at the interface were present as a result of hot pressing at a series of increasing temperatures. The time of pressing was held constant at 30 min and pressures were varied to obtain well-bonded composites at each temperature. Figure 14 shows the results for an average of four specimens, plus some repeat tests for the transverse strength. Although no characterization of the interface was performed for this series, the data are included because the strain-to-fracture of degraded material pressed at 1600 and 1650°F equals that predicted for a titanium–silicon carbide interface.

C. Effect of Reaction on Strength of Titanium–Alumina Composites

Tressler and Moore (1971) have studied the kinetics of reaction of alumina with Ti–40A and Ti–6Al–4V and concluded that useful composites might be made in this system. Their kinetics data were discussed in detail in Chapter 3. The general conclusion was that the principal reaction product was the phase Ti_3Al. Hardening of the matrix outside of this zone and stabilization of alpha phase suggested that most of the oxygen from the reacted alumina was dissolved in the matrix. The reaction rate constants were somewhat greater than in the systems Ti–B or Ti–Borsic, so that the standard hot pressing conditions of 15 min at 1500°F under 14,000 psi would lead to the formation of a minimum of 6000 Å (calculation based on the kinetics published by Tressler and Moore, 1971) neglecting the heating and cooling periods. In spite of this amount of reaction, a peak strength of 125,000 psi for a Ti–6Al–4V/22% alumina composite was obtained with an elastic modulus of 27 million psi and a strain-to-fracture of 5600 μin./in. The modulus is close to the ROM value of 26 million psi. The strength is below that expected for a strain-to-fracture of 5600 μin./in., but the authors point out that the calculated stress in the filament of

270,000 psi is close to the average strength of the original filament of 300,000 psi. In view of the problems encountered with filament misalignment and lack of complete consolidation, the results were believed to indicate that degradation due to interaction was not occurring.

The following explanation is proposed for these results. The intermetallic phase, Ti_3Al, is believed to possess some ductility so that its strain-to-fracture will exceed that of the filaments (5000 μin./in. at an average strength of 300,000 psi). Similarly, the strain-to-fracture of the oxygen-containing titanium matrix will be higher than that of the filaments. Hence, the filaments will be first to fail and will control fracture. In this respect, the behavior differs from that of the other titanium matrix composites examined where the phases TiB_2 and Ti_5Si_3 are the first to fracture.

Further discussion of these composites will be presented in Chapter 8 on oxide-reinforced composites.

IV. Effect of Interface on Strength of Pseudo-Class I Systems

Pseudo-Class I systems are those where preservation of an oxide film at the interface causes the composite to appear to behave as a Class I system (insoluble and nonreactive), but subsequent breakdown of the film allows the true interface behavior to be apparent. This breakdown may permit reaction, as in a Class III system such as aluminum–boron, or may allow solution to occur as in a Class II system. Aluminum–silicon carbide may be an example of the latter type, but more work is needed to establish the form of this system on a firmer basis.

Preservation of the oxide film throughout the processing cycle is essential if pseudo-Class I behavior is desired. This subject will be discussed first, followed by the effect of postfabrication changes in the interface on the longitudinal tensile strength.

A. Effect of Processing

No systematic studies have been made that relate the processing cycle to the interface and to the mechanical properties. Therefore, the following discussion is somewhat speculative.

Early attempts to prepare aluminum–boron composites by molten metal infiltration were completely unsuccessful. Camahort (1968) has reviewed some of this early work and shows examples of the rapid degradation of boron filaments that occurs in the presence of molten aluminum. Rapid attack takes place on the filament with growth of angular crystals of aluminum diboride. In contrast, no apparent attack occurs when aluminum–boron is made by hot pressing at temperatures 200°F lower, although the duration of hot pressing is always much longer than the few seconds needed for molten metal infiltration. These observations led Metcalfe (1968) to suggest that oxide films were preserved at the interface in the case of solid state bonding, but were removed by erosion in the case of molten metal infiltration.

There is a general concensus among producers of aluminum matrix composites that optimum processing conditions exist. When two of the three variables temperature, pressure, and time are held constant, the tensile strength first increases as the third variable is increased, then passes through a maximum, and finally decreases. This experience is consistent with the model based on interfacial oxide films. The increase of tensile strength is associated with decrease of porosity and improvement of the oxide bonds between matrix and filaments. The decrease in tensile strength with higher pressures or temperatures, or at longer times results from general breakdown of the oxide bond and excessive reaction. The optimum represents a balance between completion of bonding and the onset of local reaction at breaks in the film. The mechanism of film breakdown may be by spheroidization at higher temperatures or longer times, or by mechanical rupture by shear at higher pressures. However, the existence of an optimum in processing parameters will result in a marked variation in the composition and structure of the interface. This variation will be from point-to-point within a composite, as well as from hot-pressed part to hot-pressed part in view of the difficulties of exact control of surface conditions, processing cycles, and all of the other variables that control the interface condition.

Further discussion of this model does not appear to be warranted until work has been done to relate interface condition to mechanical properties and to the processing cycle. But, the existence of major variations in the initial interface condition, even for composites with the same strength, must be recognized as a factor to be considered in studies of the relationship between interface and the tensile properties of this class of composites.

B. *Effect of Interface on Strength of Aluminum–Boron Composites*

Loss of strength of aluminum–boron composites has been found to result from continuation of the processing beyond the optimum parameters described earlier. In addition, other studies have shown that heat treatments without applied pressure (as in fabrication by diffusion bonding) will

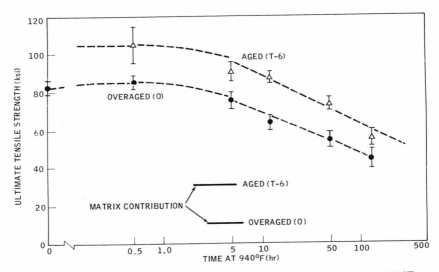

F_{IG}. 15. Strength of Al(6061)–25% boron composites after exposure at 940°F.

cause degradation. The most extensive study was that by Stuhrke (1969) in which specimens of Al(6061)–35 vol % boron were heat treated at 450, 700, and 1000°F for times up to 5000 hr. Figure 8 in Chapter 3 presents his results and shows that an incubation period appears to precede the start of degradation at 450 and 700°F, but must be less than the time of the shortest exposure (1 hr) at 1000°F. Stuhrke did not study the interface, but suggested that the degradation resulted either from uncoupling between filament and matrix so that multiaxial stresses do not develop, or from interaction between boron and aluminum leading to a reduction of the failure strain of the filaments.

Klein and Metcalfe (1971) used Stuhrke's method to vary the strength but characterized the interface by several techniques including: optical microscopy of normal and taper sections; scanning electron microscopy of normal sections; interface isolation followed by examination by transmission electron microscopy, electron diffraction and X-ray diffraction; and by transmission electron microscopy of thin sections prepared by cathodic etching. The interface isolation technique is a powerful method based on solution of both the aluminum and boron by appropriate chemicals to allow the interface to float free in the solution. It is discussed in more detail in Chapter 3 where examples are given showing typical interfaces.

Two composite materials were used to study the relationships between interface and longitudinal tensile properties. Both had the 6061 alloy as matrix, but one contained 25 vol % of 4 mil boron filaments, and the other contained 45 vol % of 5.6 mil boron filaments. The matrix was fully-annealed in most cases by treatment for 2 hr at 800°F followed by slow cool in 6 hr to 500°F. The final treatment was 7 hr at 350°F. The latter is not standard in the full anneal treatment (O condition) but was added so that the internal residual stress condition should be close to that of the T-6 condition where the 350°F, 7 hr cycle is the standard aging treatment. Solution treatment for 6061 alloy requires a short time at 1000°F followed by water quenching, but this was not used for lower temperatures of heat treatment applied to modify the interface. For example, heat treatment at 940°F was followed by water quenching from this temperature because increasing the temperature to the standard solution temperature of 1000°F would have had an unknown effect on the condition of the interface generated by the 940°F treatment. Aging was performed at 350°F for 7 hr in all cases.

Figure 15 compares the strength of the Al–25B in the O and T-6 conditions after various times of treatment at 940°F. The spread of data for the three tests at each temperature are small in most cases. The most striking feature of the curve for the over-aged condition (O) is that the strength remains constant for a period exceeding 1 hr at 940°F before degradation begins. No data could be obtained for exposures below the solution treatment time of 0.5 hr at 940°F for the T-6 matrix, but, by comparison with

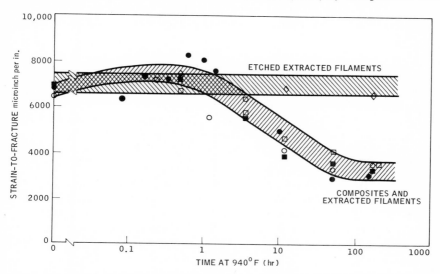

FIG. 16. Strain-to-fracture of Al(6061)–boron composites and filaments after exposure at 940°F (●, Al–45B, O condition; ○, Al–25B, O condition; □, Al–25B, T-6 condition; ■, Al–25B, extracted filaments; ◊, Al–25B, etched extracted filaments).

the curve for the over-aged condition (O), it seems reasonable that the strength remains constant. A plot of the strength normalized with respect to this plateau strength caused the two curves to become coincident. This result supports an explanation based on filament degradation rather than that advanced by Stuhrke (1969) involving decoupling of matrix and filament, because the latter would be expected to be sensitive to the matrix strength.

Examination of the tensile tests results for this composite after various anneals at 940°F showed no trends for the proportional limit or the elastic modulus. The strain-to-fracture did show several interesting effects as depicted in Fig. 16. The apparent plateau in strength was found to coincide with a small but definite increase in the strain-to-fracture from 6400 to between 7000 and 7400 μin./in. for the matrix in the O condition. Of equal theoretical interest was the appearance of a lower plateau in strain-to-fracture at a value of 3300 μin./in. This was more striking because the strength data presented in Fig. 15 did not appear to have reached the minimum values. A second composite, Al(6061)–45B, was exposed to heating cycles at 940°F and tested in tension. The results for strain-to-fracture are included in Fig. 16 and show the three effects noted previously, that is: an incubation period before degradation, a rise in the strain-to-fracture during the incubation period before degradation, and a lower plateau when degradation was complete. An increase in strength was associated with the increase in strain-to-fracture, whereas the strength was essentially constant

TABLE V

STRAIN-TO-FRACTURE OF COMPOSITE AND FILAMENTS AFTER
HEAT TREATMENT OF Al(6061)–25% BORON AT 940°F[a]

Time at 940°F (hr)	Composite (0 condition)		Extracted filaments (fracture strain, 10^{-6})	
	Strength (ksi)	Fracture strain (10^{-6})	As extracted	After AlB$_2$ removal
0	83.3	6400	7100	—
0.5	86.3	7400	7300	7400
5.0	75.7	6500	5500	—
12.0	63.7	4200	4000	6900
50.0	55.2	3200	3500	—
165.0	45.5	3400	3200	6600

[a] The elastic modulus of boron is assumed to be 58 × 10^6 psi.

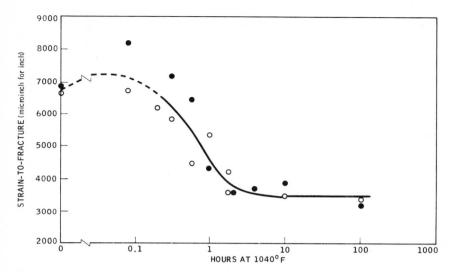

FIG. 17. Strain-to-fracture of Al(6061)–boron composites after exposure at 1040°F (O, Al–25B; ●, Al–45B).

in the case of the 25% boron composite. The 45% boron composites increased from a tensile strength of 183,000 psi in the as-received condition to approximately 215,000 psi for heating times between 10 and 40 min at 940°F. In contrast, the 25% boron composites increased from 83,000 psi (as-received) to 86,000 psi after 30 min at 940°F.

A critical test was devised to test the theory of filament degradation resulting from surface or interface reaction. Filaments were tested after extraction and after removal of any aluminum diboride reaction product by red fuming nitric acid. The strain-to-fracture results are collected in Table V and plotted in Fig. 16. The results for extracted filaments show complete agreement with the three principal effects found for composites. In addition, all of the results fall within a narrow band throughout the transition from nondegraded to degraded conditions. Perhaps the best evidence for the origin of filament degradation is the restoration of the strength and strain-to-fracture of the filaments by etching away the reacted surface layer. Although the points might be linked by a curve showing a slight maximum, restoration to 93% of the original strength by acid etching after extraction from composites exposed for 165 hr at 940°F, shows that a surface reaction is the primary cause of degradation.

Figure 17 shows that the characteristics of the 25 and 45 vol % composites are the same after exposure at 1040°F. There is greater difference between the two composites in the incubation period, but good agreement once degradation is well underway. The strain-to-fracture of degraded composites reaches the almost constant value of 3300 μin./in. for both composites. Superimposition of the curves for 940 and 1040°F is obtained by a shift in the time scale of 15-fold; that is, the reaction rates are speeded up by a factor of 15 when the temperature is increased from 940 to 1040°F.

The Al(6061)–45B composite was heat-treated at 1000°F to study interface reaction and tested in the T-6 condition. Average strengths of 243,000 psi and average strain-to-fracture of 9200 μin./in. were obtained after 5 to 15 min at 1000°F. However, the form of the curve closely resembled those presented in Figs. 16 and 17. The reaction rates were four times those found at 940°F. The lower strength plateau was at a strain of 3200 μin./in.

The peak in strain-to-fracture on strength found in Al–45B composites after short heating times at 940, 1000, and 1040°F is associated with a similar degree of breakdown of the interface and the formation of crystals of AlB$_2$ bridging the original oxide interface. Although the interfacial area of breakdown is quite small, it appears that some interaction is beneficial to the longitudinal tensile strength.

C. *Effect of Reaction in Aluminum–Stainless Steel Composites*

The effect of reaction in a composite with a ductile reinforcement differs in many respects from the behavior found in composites with nonplastic filaments. Unfortunately, no systematic studies have been published of the relationship between interface condition and tensile strength in this system. Studies have been made of other mechanical properties such as fatigue, flexure, and transverse strength and will be included in this section in order to develop an overall picture of the interface.

1. *Effect of Processing on Various Mechanical Properties*

There appears to be good agreement that optimum pressing conditions must be used to achieve the maximum tensile strength in this system, but detailed studies of the development of such conditions have not been published. Davis (1967) and Pinnel and Lawley (1970, 1971) describe their methods of preparation of these composites to achieve optimum properties. The latter concluded that their optimum composites were devoid of significant chemical interaction at the matrix–filament interface. Pattnaik and Lawley (1971) showed that reaction would occur on reheating to form a product consisting mainly of Fe$_2$Al$_5$ phase.

Baker (1966) performed one of the first studies of the relationship between processing conditions and a mechanical property, in this case fatigue strength. He proposed that the compound formed on excessive reaction was Fe_2Al_5, but prepared some composites at low enough temperatures to be free of reaction as judged by metallography. He concluded that the best fatigue performance was found in slightly-reacted composites.

Although these results provide clear-cut evidence for the existence of optimum processing conditions to maximize fatigue strength, the question arises whether this optimum will be the same for the longitudinal tensile property under examination in this chapter. Davis (1967) processed at temperatures near to that found optimum by Baker (1966). Pinnel and Lawley (1970, 1971) pressed in the optimum temperature zone for maximum fatigue strength, but obtained composites essentially free of reaction. The general conclusion is that the processing conditions chosen by all contributors to this field are similar, but it is probable that the interface condition was not precisely the same in each study.

Evidence that the optimum processing conditions are not the same for tension and fatigue loading has been obtained in another aluminum matrix composite system. Using the same test procedures that Baker (1966) used on aluminum–stainless steel, Baker and Cratchley (1964) found that the weak bonds formed in aluminum–silica hot-pressed at 450°C (842°F) were poor in fatigue and failed by the same mechanism as the aluminum–stainless steel. Again, in agreement with the steel-reinforced composite, it was found that the strong bonds formed by hot pressing at 550°C (1022°F) were able to deflect fatigue cracks away from the interface. The cracks were observed to travel in the matrix at some distance from the interface showing that the matrix of pure aluminum was weaker than the bond. On the other hand, Cratchley and Baker (1964) showed that the tensile properties were superior when the composites were pressed at 450°C (842°F) rather than 550°C (1022°F). These results suggest that the degree of bond necessary to transmit shear loads through the matrix in the tensile test is much less than that required to resist propagation of fatigue cracks. But this conclusion for aluminum–silica composites may not be directly applicable to aluminum–stainless steel composites. Baker *et al.* (1966) point out that the reaction between aluminum and silica will generate alumina and silicon. On the basis of the interface types discussed in earlier chapters in the book, it appears that aluminum–silica is of Class III (reactive) in which alumina (or an aluminosilicate) forms in contact with the silica with silicon atoms diffusing through the reaction zone to dissolve in the matrix. Strengthening of the matrix by this means may explain the shift of fatigue cracks away from the interface. Similarly, the high elastic modulus alumina (60 million psi) on low modulus silica (10 million psi) may behave in the same manner as TiB_2 on boron.

Recent work by Paton and Lockhart (1971) provides evidence for an optimum processing temperature to maximize another property, transverse tensile strength. These authors found that the transverse strength of plasma-sprayed aluminum–20% stainless steel reached a peak strength after hot pressing at a temperature near 520°C (986°F). The optimum temperature for transverse strength is in agreement with the optimum noted by Baker (1966) based on fatigue strength.

2. Characteristics of the Interface

Baker (1966) observed that local reaction was characteristic of the interface in aluminum–stainless steel. Further, he noted that the behavior of the bonds under fatigue testing was not consistent so that he introduced the term "mixed bond strength" to describe the interface characteristic. Fatigue cracks propagating along the interfacial region appeared to be deflected at a point of higher bond strength. In other specimens, the path of the fatigue crack was along the interface of unreacted fibers but away from the interface of reacted fibers. In combination with his observations of local breakdown at a number of points on the surface of the steel wire, this leads to the model of a pseudo-Class I type composite. An alumina film on the aluminum matrix delays reaction until breakdown occurs locally and permits the iron aluminide (Fe_2Al_5) to form. Further evidence for a pseudo-Class I type behavior is provided by the observations of Pattnaik and Lawley (1971), discussed in Chapter 3 (see Fig. 10). Baker (1966) and Jones (1968) also noted sporadic breakdown of the interface in this system. Jones (1968) reports that although the bond between wires and matrix appears to be largely mechanical, an interfacial third phase has been noted around some filaments. The mechanical bond interpretation was based on a clean separation from the matrix on necking of the wires in a tensile test, but fits equally well the oxide bond formed initially in a pseudo-Class I system.

3. Effect of Reaction on Longitudinal Tensile Properties

Pattnaik and Lawley (1971) extracted wires from as-fabricated and heat-treated aluminum–stainless steel composites. Some interface reaction was noted over most of the area of the wire but the original wire drawing marks were retained on wires extracted from as-fabricated composites. After reaction at 550°C (1022°F), the 0.006 in. diam wires had a diameter of 0.007 in. when extracted from the composite and a marked corncob appearance. In both cases, the extent of necking of the stainless steel wires was approximately the same if tested alone or in the composite. The center of each wire showed areas of void coalescence characteristic of ductile failure. However, treatment at 625°C (1157°F) increased the wire diameter to 0.008 in., so that the core of unaffected steel was quite small. Both the strength and ductility of such wires were markedly reduced.

Although the 550°C (1022°F) treatment causes very drastic change in the appearance of the composites, so that the "interface" occupies a significant portion of the volume of the composite, Pattnaik and Lawley (1971) found that the strain-to-fracture remained unchanged. This result shows that cracking of the iron aluminide layer prior to fracture has little effect on the overall ductility. Jones (1968) has shown that slip lines originate in the stainless steel from the roots of these cracks, but the localized deformation was observed to develop into deformation bands running entirely across the wire until deformation becomes general and necking takes place. Figure 5 of Chapter 1 showed an example from the work of Jones (1968) of the cracking of the intermetallic phase followed by slip in the wire. On the other hand, cracking of the matrix appeared to result from these cracks in the intermetallic compound.

V. Effect of Interface on Strength of Class II Systems

The copper–tungsten system is one of the few systems where the matrix and filament are essentially insoluble but wet each other. Tensile tests of copper–tungsten composites have been analyzed to show that the tungsten filaments contribute in full measure to the strength of the composite according to McDanels *et al.* (1965). However, when a copper alloy matrix was used, such that the mutual solubility of matrix and filament was increased to make the system of the Class II category, the filament contribution to strength was reduced. For example, a copper–5% cobalt matrix caused progressive recrystallization of the tungsten as a result of mutual solution effects at the interface. Petrasek and Weeton (1963) show that the ductile, necking failure of tungsten filaments in a pure copper matrix at a stress of approximately 330,000 psi is changed to a brittle fracture in a copper–5% cobalt matrix. The contribution to strength decreased as the filament became progressively recrystallized; after 50% of the cross section was recrystallized, the original contribution of 330,000 psi had decreased to nearly 200,000 psi with no evidence of plastic flow preceding fracture.

This system provides an illustration of a composite where a minor change at the interface has caused a marked change in the intrinsic strength of the reinforcement. The minor change is primarily transfer of the surface active element cobalt into the tungsten filament, and is a result of the properties of this element. Other copper alloys that form a Class II system with tungsten, such as copper–chromium, do not have this drastic effect on the reinforcement.

VI. Relationship of Interface Condition and Longitudinal Strength

Growth of interest in the study of interfaces has accompanied the advance from model systems to those with matrices of the three important structural metals: aluminum, titanium, and the ferrous metals. These metals tend to be considerably more reactive than matrices of silver and copper used to study model systems such as $Ag–Al_2O_3$ and Cu–W. However, the results reported in this chapter show that some reactivity may be desirable to attain the overall mechanical properties desired. Several examples have been noted of cases where some reaction at the interface gave the otpimum condition. Baker (1966) showed that the best fatigue performance of aluminum–stainless steel was found in slightly-reacted composites, and Baker and Cratchley (1964) found the same result for the best fatigue strength of aluminum–silica. Forrest and Christian (1970) compared Al–B, Al–SiC-coated B, and Al–BN-coated B, and concluded that the best all-round properties were obtained with the most reactive of these systems, Al–B. This choice may reflect other factors such as method of manufacture, optimization of processing, and so on, but the fact remains that the systems developed to reduce reaction by coating the boron with SiC or BN were not found to be superior to the more reactive Al–B system. In confirmation of this conclusion, Klein and Metcalfe (1971) found that the longitudinal failure strain and strength of Al–B composites reach peaks after a small amount of reaction has occurred.

Longitudinal tensile properties place less demand on the strength of the bond than most other properties. The work of Baker and Cratchley (1964) on fatigue of $Al–SiO_2$ showed that much stronger bonds were necessary than those determined to be optimum for the tensile properties by Cratchley and Baker (1964). This problem of bond optimization arises principally in pseudo-Class I systems. Much additional work will be needed before these systems can be fully understood.

The systems selected for detailed discussion in the earlier sections of this chapter were those where adequate information was available to warrant inclusion. The factors controlling fracture were quite different in many cases. Table VI represents an attempt to summarize the information

TABLE VI

SUMMARY OF TENSILE FAILURE MODES IN CLASS III AND PSEUDO-CLASS I SYSTEMS

| System | Class | Principal compound | Elastic modulus[a] | Reaction zone | | | Probable failure control factor |
				Failure strain[b]	Filament failure strain[b]	Critical reaction thickness	
Ti–B	III	TiB_2	77	2500	6000+	Up to 8000 Å	Thickness of reaction zone
Ti–SiC(B)	III	Ti_5Si_3	37	4500	6000+	Up to 8000 Å	Thickness of reaction zone
Ti–Al₂O₃	III	Ti_3Al	—	High	5000+	No critical thickness	Filaments fail before reaction zone
Al–B	Ps. I	AlB_2	—	3300	6000+	—	Minimum area of reaction required to control fracture
Al–SiO₂	(III)	Al_2O_3	60	Low	50,000	Not known, may be small	Not known
Al–S. Steel	Ps. I	Fe_2Al_5	—	Low	20,000+	No critical thickness	Filaments yield to eliminate effect of reaction zone
Al–C	Ps. I	Al_4C_3	—	Low	5000	Not known, may be small	Not known, probably Al–B type
Al–SiC(B)	(Ps. I)	None	—	—	6000+	Not applicable	Filaments
Ti–Be	III	Ti beryllide	—	Low	High	No critical thickness	Filaments yield to eliminate any effect of reaction zone

[a] Elastic modulus in million psi.
[b] Strain in microinch per inch.

relating reaction product, failure strain of reaction product and filament, critical reaction extent (if any), and probable factor controlling fracture.

Although the class of the composite system (e.g., pseudo-Class I or Class III) is very important in terms of the longitudinal tensile strength, another important factor is whether the filament possesses sufficient ductility to offset the formation of a brittle reaction compound. The interaction compound controls failure only in the case of brittle (elastic) filaments. Class III systems of this type are represented by Ti–B that forms a continuous reaction zone of uniform thickness and limited strain-to-fracture. It cracks before the filament, and its subsequent effect depends on the thickness of the reaction zone. Titanium reinforced with boron or silicon carbide surface filaments belongs to this class, although variations in the reaction product with conditions may make the failure strain variable for silicon carbide surface filaments. Pseudo-Class I systems are represented by Al–B that forms isolated patches of reaction product of limited strain-to-fracture. The thickness exceeds the critical value very rapidly, but the limited area of contact with the filament attenuates the strain so that fracture is believed to be controlled by area of contact rather than thickness. Aluminum–boron and possibly aluminum–carbon belong to this class. Aluminum–silica may belong to either class.

It is likely that titanium–alumina belongs to a special grouping because the principal product of Ti_3Al is believed to possess a small degree of ductility. Unlike the system discussed above, the reaction product will not crack prior to filament failure in such a case.

In composites with ductile filaments, the stress concentrations generated by rupture of the reaction product are attenuated by plastic flow of the filament. Aluminum–steel and titanium–beryllium are examples of this grouping of composites. It is believed that the iron aluminide and titanium beryllide are very brittle and develop a series of closely spaced cracks, but plastic flow in the filaments rapidly reduces the stress concentrations caused by these very short cracks.

The remaining composite in Table VI is the aluminum–silicon carbide surface filament combination. Although it is believed that no compound forms in this system, this has not been established definitely.

References

Baker, A. A. (1966). *Appl. Mater. Res.* **5**, 143–153.
Baker, A. A., and Cratchley, D. (1964). *Appl. Mater. Res.* **3**, 215–222.
Baker, A. A., Cratchley, D., and Mason, J. E. (1966). *J. Mater. Sci.* **1**, 229–237.

Camahort, J. L. (1968). *J. Comp. Mater.* **2** (1), 104–112.

Collins, B. R. (1972). Private communication of the results of Dr. I. Toth on Contract F33615-71-C-1044.

Cooper, G. A., and Kelly, A. (1968). "Interfaces in Composites." Amer. Soc. Test. Mater., Philadelphia, Pennsylvania.

Cottrell, A. H. (1964). *Proc. Roy. Soc. (London)* **282A,** 2.

Cratchley, D., and Baker, A. A. (1964). *Metallurgia* **69,** 153.

Davis, L. W. (1967). *Metal Progr.* **91,** 105–114.

Forest, J. D., and Christian, J. L. (1970). *Metals Eng. Quart.* **10** (1), 1–6.

Herring, H. W. (1966). NASA Tech. Note TD D-3202.

Hutting, R. (1969). Unpublished work in Solar Res. Lab.

Jones, R. C. (1968). "Deformation of Wire Reinforced Metal Matrix Composites," STP 438, pp. 183–217. Amer. Soc. Test. Mater., Philadelphia, Pennsylvania.

Jones, R. C., and Olster, E. F. (1970). "Toughening Mechanisms in Fiber Reinforced Metal Matrix Composites." Res. Rep. R70-75. Massachusetts Inst. of Technol., Cambridge, Massachusetts.

Klein, M. J., and Metcalfe, A. G. (1971). "Effect of Interfaces in Metal Matrix Composites on Mechanical Properties." Tech. Rep. AFML-TR-71-189.

Klein, M. J., Reid, M. L., and Metcalfe, A. G. (1969). "Compatibility Studies for Viable Titanium Matrix Composites." AFML-TR-69-242.

McDanels, D. L., Jech, R. W., and Weeton, J. W. (1965). *Trans. AIME* **233,** 636.

Metcalfe, A. G. (1967). *J. Comp. Mater.* **1,** 356–365.

Metcalfe, A. G. (1968). Unpublished work.

Metcalfe, A. G., and Schmitz, G. K. (1969). *Trans ASME J. Eng. Power* **91,** 297–303.

Noone, M. J., Feingold, E., and Sutton, W. H. (1969). The importance of coatings in the preparation of Al_2O_3 filament/metal matrix composites, *Amer. Soc. Test. Mater. Symp. Interfaces Composites, San Francisco,* Publ. as ASTM STP 452.

Paton, W., and Lockhart, A. (1971). NEL Rep. 475, Nat. Eng. Lab., Glasgow, Scotland.

Pattnaik, A., and Lawley, A. (1971). Tech. Rep. No. 6 on Contract N00014-67A 0406-001.

Petrasek, D. W., and Weeton, J. W. (1964). *Trans. AIME* **230,** 977–990.

Pinnel, M. R., and Lawley, A. (1971). *Metall. Trans.* **1,** 1415.

Pinnel, M. R., and Lawley, A. (1970). *Metall. Trans.* **1,** 1337.

Ratliff, J. L., and Powell, G. W. (1970). "Research on Diffusion in Multi-Phase Systems: Reaction Diffusion in the Ti–SiC and Ti–6Al–4V/SiC Systems." AFML-TR-70-242.

Rose, F. K. (1966). Unpublished work in Solar Res. Lab.

Rosen, B. W. (1964). Mechanics of composite strengthening *in* "Fiber Composite Materials," Chap. 3. Amer. Soc. Metals, Metals Park, Ohio.

Samsonov, G. V. (1964). "Handbook of High Temperature Materials," No. 2. Plenum Press, New York.

Schmitz, G. K., and Metcalfe, A. G. (1968). "Development of Continuous Filament Reinforced Metal Tape." AFML-TR-68-41.

Schmitz, G. K., Klein, M. J., Reid, M. L., and Metcalfe, A. G. (1970). "Compatibility Studies for Viable Titanium Matrix Composites." AFML-TR-70-237.

Stuhrke, W. F. (1969). *AIME Symp. Metal Matrix Composites, Pittsburgh, Pennsylvania.*

Sutton, W. H., and Feingold, E. (1966). *Mater. Sci. Res.* **3,** 577–611.

Tressler, R. E., and Moore, T. L. (1971). *Metals Eng. Quart.* **11** (1), 16–22.
Vennett, R. M., Wolf, S. M., and Levitt, A. P. (1969). *Metall. Trans.* **1,** 1569–1575.
Vettri, R., and Galasso, F. (1968). *Nature (London)* **220,** 781.
Wawner, F. E., Jr. (1967). Boron filaments *in* "Modern Composite Materials" (L. J. Broutman and R. H. Krock, eds.), Chapter 10, pp. 244–269. Addison-Wesley, Reading, Massachusetts.

5

Effect of the Filament–Matrix Interface on Off-Axis Tensile Strength

MARK J. KLEIN

Solar Division of International Harvester Company
San Diego, California

The effect of the interface on the tensile strength of composites with filaments aligned parallel to the direction of loading was discussed in the previous chapter. This chapter is concerned with the effects of the interface on the tensile strength of composites where the loading direction is at an angle to the filament axis. Only continuous unidirectional filaments and uniaxial loading are considered because of the lack of experimental data and interface theory for the more complex filament arrangements and stress states. Although tensile data for off-axis loading of many composites have been determined, few studies have considered the effects of the interface on tensile strength. The ideal conditions considered in this chapter are less complex than those likely to be encountered in practical applications. Nevertheless, consideration of off-axis loading adds an additional degree of complexity to the study of the effect of interfaces on mechanical properties and is a first step before considering more practical situations.

The theoretical approaches that have been used to analyze off-axis tensile strengths will be summarized first. This will provide a background for the subsequent discussion of real systems and the presentation of experimental results where interface effects have been considered.

I. Theoretical Analyses of Off-Axis Strength

There are several different ways that interfaces can be taken into account in off-axis strength analyses. First, it can be assumed that the interface is strong enough to transmit off-axis loads between filaments and matrix to the point of composite failure. For this condition, it is assumed that the interface is strong and does not fail. Therefore, theories of this type can be termed "theories for strong interfaces" in analogy with the terminology used in the previous chapter on longitudinal tensile strength. A second appraoch that can be used is to consider interface failure as well as all other failure mechanisms only indirectly. Theories of this type can be catagorized as "phenomenological" since interface failure is taken into account only through the experimental strength measurements of the composite that are required in the proposed solutions. For this type of analysis, the experimental strengths must agree with theory at the zero and 90° orientations because of the way the solutions are formulated. The predicted strengths at intermediate angles are dependent upon experimental measurements of composite longitudinal, transverse and shear strengths, regardless of failure mechanism. The most complex theoretical approach is to consider the properties of the interface and to predict both interface failure as well as the effect of the interface on off-axis composite tensile strength. This is a much more complex situation and has only been attempted for the transverse orientation (filament axis at 90° to the loading direction). This type of analysis can be termed "theory for weak interfaces" since interface failure is assumed to be at least in part responsible for composite failure.

Various experimental data have been cited as evidence for the validity of these different theoretical approaches. Theory for strong interfaces is expected to be in reasonable agreement with experimental strength data for composites with strong filament–matrix bonds that do not embrittle the composite. The predicted strengths are, in effect, the upper limits to composite strength. When failure occurs at the interface because of poor bonding or formation of a weak interaction product, the off-axis strength is reduced, and theory for weak interfaces will apply. In this regard, the lower limit for off-axis strength is approached for the case of little or no bond at the interface. The upper and lower limits for off-axis strength will be discussed more fully later in this chapter.

Only a brief review will be given of the phenomenological and strong interface theories since these theories have been discussed elsewhere (Tsai, 1968) and detailed presentations are available in the referenced literature. In addition, these theories as they are formulated do not really deal with the effect of the interface other than to assume perfect load transfer through it or to ignore the problems of interface failure. Emphasis will be placed upon the limited theory that deals with the effects of imperfect interfaces on off-axis strength and the limiting value of composite strength for strong and weak interfaces. This approach will lead to a better understanding of the effect of imperfect interfaces on off-axis strength and is required to identify the problems that must be solved to attain maximum off-axis strength for many real systems.

A. Theory for Strong Interfaces

Kelly and Davies (1965) and Stowell and Liu (1961) have formulated equations for composite strength as a function of loading angle based upon three failure modes. The three types of tensile failure and the equations describing the variation in composite strength σ_c, with the angle between the loading direction and the filament axis θ, are listed below:

(1) Longitudinal filament fracture (low angles)

$$\sigma_c = \sigma_l/\cos^2\theta \tag{1}$$

(2) Matrix shear failure (intermediate angles)

$$\sigma_c = \tau_m/\sin\theta\cos\theta \tag{2}$$

(3) Matrix tensile failure (high angles)

$$\sigma_c = \sigma_m/\sin^2\theta \tag{3}$$

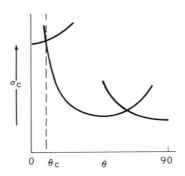

FIG. 1. Variation in composite strength σ_c with loading angle θ.

Here σ_l is the longitudinal failure strength of the composite (i.e., $\theta = 0°$), τ_m is the matrix shear strength, and σ_m is the matrix tensile strength after allowance for matrix constraints. A similar set of three equations can be formulated for failure at an ultimate strain rather than stress (Tsai, 1968). The general shape of the $\sigma_c - \theta$ profile predicted by the failure stress and strain theories is shown in Fig. 1.

An important prediction of these theories and one that would distinguish them from some others, is the existence of a critical angle θ_c. The critical angle is the angle at which the composite begins to lose strength and is, therefore, an important parameter to define. However, the existence of a critical angle has not been established so that no check has been possible on a theoretical basis.

Interface failure is not considered in this approach to off-axis strength and cannot be introduced merely by broadening the definitions of matrix shear (τ_m) and tensile strength (σ_m) as suggested by Cratchely *et al.* (1968) and Petrasek *et al.* (1967). For example, if τ_m and σ_m are taken to be the interface strengths in shear and tension, respectively, the off-axis composite strength would approach zero as the interface shear and tensile strengths approach zero [Eqs. (2) and (3)]. This cannot be the case because the matrix forms a continuous network between filaments that supports load and provides a lower limit to off-axis strength.

Petrasek *et al.* (1967) have modified the Stowell and Liu (1961) and Kelly and Davies (1965) approaches so that the composite strength will not increase for off-axis loading at small angles of misalignment. In their modification, Eq. (1) is $\sigma_c = \sigma_l$ rather than $\sigma_c = \sigma_l/\cos^2 \theta$, and the composite strength is constant as θ increases until shear failure occurs.

A model based upon composite failure induced by flaws in composites was used to derive the composite strength as a function of loading angle by Lauraitis (1971). The flaws were assumed to be cracks oriented parallel to the filament axis, and the critical crack length was determined from experi-

mental data. Excellent agreement between theory and experimental results was obtained for glass fiber-reinforced epoxy for loading angles from
zero to 90°. Lauraitis also found her theory to be almost coincident with
the theory formulated by Azzi and Tsai (1965) when comparison was
made with her experimental results although the basis for the two theories
is entirely different. The latter theory, which will be discussed in the next
section, is based upon a maximum energy of distortion criterion.

B. *Phenomenological Theories*

Azzi and Tsai adopted the Hill yield criterion to predict the strength of
composites (laminates as well as unidirectional filaments) for any combination of combined stresses. To obtain the predicted uniaxial $\sigma_c - \theta$ profile
for the composite, the longitudinal and transverse composite strengths as
well as the strength in shear, must be determined. This theory, in contrast with those discussed earlier, is phenomenological and thus is independent of failure mechanism. The equation given by these authors to
describe the $\sigma_c - \theta$ profile is

$$\sigma_c = \sigma_l / [\alpha^2 \sin^4 \theta + (\beta^2 - 1) \sin^2 \theta \cos^2 \theta \cos^4 \theta]^{1/2} \tag{4}$$

where $\alpha = \sigma_l/\sigma_t$, $\beta = \sigma_l/\tau$, σ_t is the transverse composite strength, and τ is
the composite strength in shear on a plane of anisotropic symmetry. The
shapes of the $\sigma_c - \theta$ profiles that can be derived from this equation are
shown schematically in Fig. 2. Azzi and Tsai found good agreement between experiment and theory for composites of glass filaments in resin,
but no information on the failure mechanism was given.

This equation must agree with experimental data at zero and 90° be-

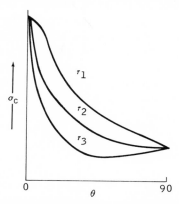

FIG. 2. Variation in composite strength with loading angle for different values of
shear strength τ.

cause of the way the equation is formulated. For fixed values of σ_l and σ_t the shape of the $\sigma_c - \theta$ profile is dependent upon the value of τ and can show a minimum, at intermediate angles, for low values of τ as indicated in Fig. 2. At midpoint in the $\sigma_c - \theta$ profile, that is, where $\theta = 45°$, $\sigma_c = 2\sigma_t\tau[(\tau^2 + \sigma_t^2)^{-1/2}]$. Therefore, the strength of the composite when tested at an angle of 45° is independent of σ_l, but is dependent upon both τ and σ_t. There is no critical angle in the Azzi and Tsai theory, but the initial slope of the $\sigma_c - \theta$ profile at low angles is determined primarily by the relative values of σ_l and τ. The effect of the interface on off-axis strength is not taken into account but would influence composite strength through the values of σ_l, σ_t, and τ which are required to solve Eq. (4).

Tsai and Wu (1971) have reviewed strength theories for anisotropic materials and have developed their own general theory. This is also a phenomenological theory but has the freedom of operation derived from the use of strength tensors. The basic assumption in their theory is that failure occurs when

$$F_i\sigma_i + F_{ij}\sigma_i\sigma_j \geq 1 \tag{5}$$

where i, j, k equal 1, 2, . . . , 6 and F_i and F_{ij} are second- and fourth-ranked strength tensors. A further requirement is

$$F_{ii}F_{jj} \geq F_{ij}^2 \tag{6}$$

Equation (5) describes the failure surface and Eq. (6) limits the magnitude of the interaction terms and ensures that the failure surface will intercept each stress axis.

Solution of these equations to determine the $\sigma_c - \theta$ profile requires considerable experimental data: longitudinal, transverse, and shear strengths of the composite in both compression and tension. The theory is independent of failure mechanism, and the effect of the interface on the off-axis tensile strength can only be reflected indirectly through the experimental results at zero and 90°, and the shape of the curve at midpoint in the profile is largely determined by the composite shear strength and the magnitude of the interaction term F_{ij}. The authors found good agreement with experimental results for the uniaxial off-axis strength of graphite epoxy, but the theory has not been checked further using other composites or more complex stress states.

C. Theory for Weak Interfaces

Analysis of the effect of interfaces on the composite off-axis strength has only been attempted for the transverse orientation ($\theta = 90°$). Although this is the simplest case to consider, present theories still do not take into account the full complexity of the actual conditions of deformation but rather are approximations. An important approach to the problem, however, is the establishment of upper and lower limits for transverse strength and the possible extension of these limits to other off-axis angles. The establishment of the upper limit to transverse strength requires, of course, a strong interface. However, this phase of the discussion will be included here because it is considered in conjunction with the lower limit and is a starting point for further development of theories for weak interfaces.

Several approaches have been pursued to establish upper and lower limits for transverse composite strength. For the maximum transverse strength (upper limit), the interface would have to be strong enough so that it would not fail before the matrix (including matrix failure induced by filament splitting); for the minimum transverse strength (lower limit), the interface would have to be weak enough so that it would fail before failure would occur in the matrix of a composite in which filaments were not bonded.

Chen and Lin (1969) used the finite element method and a maximum distortion energy criterion to establish these limits. In their analysis, they considered the cases for perfect bonding (strong interface) and for complete debonding (no interface bond) for two filament arrays, square and hexagonal (close packed). Their results show finite values of transverse strength for no interface bond and maximum filament packing, i.e., 78.5 vol % for square and 90.6 vol % for hexagonal. However, at this point, filaments are touching and for infinitely thick composites, the amount of matrix holding the composite together would approach zero, and the composite could have no strength. Predicted upper and lower limits for transverse strength were compared with experimental results for Al(6061)–B and for Al(2024)–stainless steel composites for volume fractions to 50% reinforcement. Good agreement between experiment and theory was obtained for Al(6061)–B composites for the case of perfect bonding for the square array. Agreement for the Al(2024)–stainless steel composite, taken to be unbonded, was not as good but was closest for the hexagonal array, although a random array appears to have been used. Failure apparently occurred at the interface for the aluminum–stainless steel composites, but no mention was made of the type of failure in the Al(6061)–B composite specimens (e.g., matrix, filament splitting, or interface failure). In this regard, careful examination of transverse failure modes in many Al(6061)–B composite specimens tested in transverse tension almost always show two or three different failure modes (Klein and Metcalfe, 1971).

Lin *et al.* (1971) did additional work on these composites to show the effect of matrix heat treatment and composite fabrication conditions on transverse strength. They found agreement between experimental results and theory and a dependence of interface bond upon the surface condition of the filaments.

Ebert *et al.* (1971) applied the finite element method to predict the elastic microstress distribution within composites subjected to uniaxial and biaxial loading. Their analysis predicted yielding at midpoint between filaments in an Al–50B composite subjected to uniaxial transverse tension for a square filament array. The effect of filament–matrix interaction and diffusion in Ni–Cu composites on tensile properties was predicted for longitudinal but not for off-axis orientations. The latter is discussed in more detail in Chapter 2.

Adams (1970) also applied the finite element analysis utilizing the tangent modulus method to define the stress state in a composite subjected to transverse tension. The stress in the composite was analyzed beyond the elastic limit to the point of first component failure for square and rectangular filament arrays. Matrix failure was assumed to occur when the stress within the composite reached the bulk matrix strength. First matrix failure for Al–34B with filaments in a rectangular array was predicted between close-spaced filaments for an applied transverse stress of 24.6 ksi, a stress well below the bulk matrix stress of over 33 ksi. However, failure was determined experimentally to occur in this composite by filament splitting. The stresses at the interface and within the filaments were determined, but failure by these modes could not be predicted because the strengths of the interface and the filaments in transverse tension were not known. The analysis is being developed further to allow for crack propagation and final composite failure. In theory, failure at the interface or by other failure modes could be predicted by this method if the mechanical properties of the interface, matrix, and filaments were known.

Paton and Lockhart (1971) used a somewhat different approach to establish limits for the transverse tensile strength of aluminum–stainless steel composites. They applied an analysis by Drucker (1964) to determine the upper limit for transverse strength based upon a plastic constraint factor. Drucker's analysis was derived for different packing densities for a close-packed array of rigid inclusions in a rigid plastic matrix. Paton and Lockhart's analysis indicated that the maximum transverse composite strength would be about equal to the matrix strength until the volume fraction of filaments exceeds about 35%. For greater volume fractions, the transverse stregnth would exceed the bulk matrix strength and would approach twice this value at about 65 vol % filaments. The latter has not been shown experimentally and is in contrast with the predictions of Chen

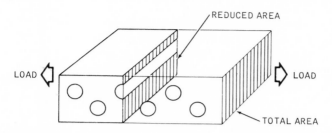

FIG. 3. Sketch showing reduced and total areas for transverse tensile loading.

and Lin who calculate maximum transverse strengths equal to or less than the bulk matrix strength, depending upon the volume fraction of filaments and their array.

Paton and Lockhart derived values for the lower limit of transverse strength by assuming that the filaments are replaced by holes (Fig. 3) and that the composite transverse strength varies with volume fraction of filaments according to the reduced cross section. Their results show finite values of transverse strength for volume fractions at which the filaments touch. However, for thick composites (or where filaments are at the composite surface), the transverse strength for maximum packing must approach zero as the load-bearing cross section approaches zero.

The reduction in cross section and hence the lower limit for transverse strength is also dependent upon the filament array (Fig. 4). Patton and Lockhart compared their predicted lower limit for the hexagonal array with the lower limit derived by Chen and Lin for the same filament array. The authors reported close agreement for the two approaches. However, the lower limit using the geometric approach is dependent upon the direction of applied stress with respect to the filament array and Patton and Lockhart did not take this into account in their comparison. As will be shown later, agreement between the geometric and Chen and Lin analyses

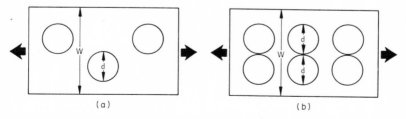

FIG. 4. Influence of filament placement on effective width of transverse tensile specimens. Width (W) reduced by (a) one and (b) two filament diameters (d).

Mark J. Klein

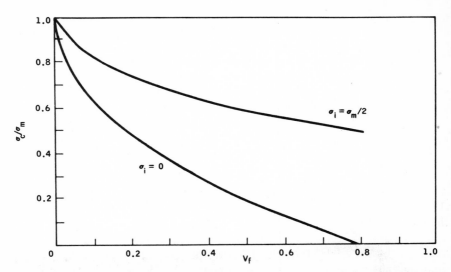

FIG. 5. Variation of normalized transverse strength (σ_c/σ_m) with fiber volume fraction (V_f) for different interface strengths [from Cooper and Kelly, (1968)].

for the hexagonal array is not good when the direction of applied stress is the same.

Cooper and Kelly (1968) analyzed the effect of interface strength on transverse composite strength and proposed upper and lower limits for weak and strong interfaces. The lower limit was also taken to be the strength of the matrix in which holes replace filaments. If matrix constraint and stress concentrations are attenuated by plastic flow, this condition would be approached when the interface is very weak or when the filaments have a low transverse strength (low resistance to splitting). As described earlier, the strength is reduced because the load-bearing area is decreased by filament or interface failure. The upper limit in the Cooper and Kelly analysis (the case for strong interfaces) was taken to be the strength of the matrix for conditions of plane strain. On this basis, the maximum transverse composite strength was approximated at 1.15 times the bulk (unconstrained) matrix strength.

The Cooper and Kelly analysis is shown graphically in Fig. 5. In this figure, the ratio of transverse composite strength to matrix strength σ_c/σ_m, is shown as a function of the volume fraction of filaments V_f, for a random filament array for two conditions of interface strength σ_i: where the interface has no strength ($\sigma_i = 0$), and where the interface is half as strong as the matrix ($\sigma_i = \sigma_m/2$). In this analysis, σ_i is a complex term that denotes the average tensile strength at the interface necessary to cause the filaments to part from the matrix under transverse loading and, therefore, can be a function of volume fraction and filament placement through changing stress concentrations. This analysis is an important step in depicting the general relationship between the interface strength and the transverse strength, but there are several points that should be considered further.

The interface strength σ_i, was defined by Cooper and Kelly as the tensile stress necessary to part the filament from its interface in transverse loading, but this definition of σ_i should be broadened to include the stress σ_f to fracture (split) the filaments, if $\sigma_f < \sigma_i$, since filament splitting is a common mode of transverse failure. As discussed in Chapter 2, the tensile stress at the interface under transverse loading will, in general, be different from the stress applied to the composite specimen. Therefore, it is probably better to define σ_i as the transverse stress applied to the composite that is required to induce interface failure or filament splitting within the composite for a specific test condition. In addition, in the Cooper and Kelly analysis, σ_c was derived by assuming that the matrix and interface failure stresses are additive according to the rule-of-mixtures for an equal strain condition, whereas the actual stress state will be much more complex. Furthermore, it should be emphasized that the line indicated as $\sigma_i = 0$ in Fig. 5 need not define a condition where no stress is required to part the interface and filament, as indicated by Cooper and Kelly. It actually defines the strength of a composite where either the interface or transverse filament strength is less than the strength of the matrix in which filaments are unbonded or for the Cooper and Kelly analysis, in which holes replace

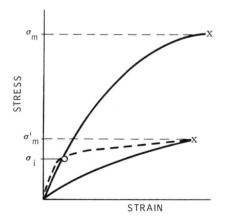

Fig. 6. Schematic transverse tensile test showing bulk matrix strength (σ_m), matrix strength in which holes replace filaments (σ_m'), and a composite stress–strain curve (dashed lines) in which the filaments or interface strengths (σ_i) are less than (σ_m').

filaments. This point can be shown by reference to Fig. 6. Here, σ_m is the strength of the matrix or the strength of the composite in which neither the filaments nor the interface break (the upper limit of transverse strength) and σ_m' is the lower limit to transverse strength defined as the strength of the matrix in which the filaments are unbonded or are replaced by holes. As indicated by the dotted curve in Fig. 6, composites in which σ_i is less than σ_m' will always have a strength of σ_m'. For the geometric approach, the strength is independent of the value of σ_i, where $\sigma_i < \sigma_m'$. The point is that σ_i can have a finite value, but this will not be revealed by composites whose strength lie along the line indicated by $\sigma_i = 0$ (Fig. 5) in the Cooper and Kelly analysis. However, σ_i will affect the composite transverse strength when its value is greater than σ_m' but less than σ_m. When σ_i exceeds σ_m, the matrix will fail and the transverse strength of the composite will equal the matrix strength with allowance made for stress concentrations and matrix constraint.

In brief, $\sigma_c = f(\sigma_i)$ when $\sigma_m > \sigma_i > \sigma_m'$. Using Cooper and Kelly's definition of σ_m' based on a random filament distribution, the composite is strengthened by filaments when $\sigma_i > \sigma_m [1 - (4V_f/\pi)^{1/2}]$ and this will occur when $\sigma_c/\sigma_m > [1 - (4V_f/\pi)^{1/2}]$. When $\sigma_c/\sigma_m = 1$ then $\sigma_i \geq \sigma_m$ and

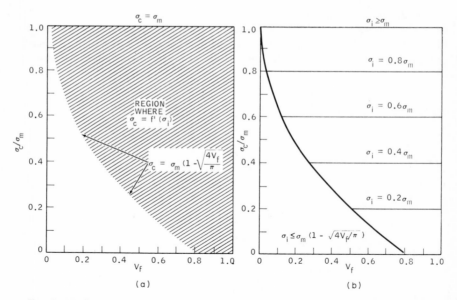

FIG. 7. Variation of composite–matrix strength ratio (σ_c/σ_m) with volume fraction of filaments (V_f) for transverse tensile tests. (a) Region where interface strength or filament strength (σ_i) affects composite strength. (b) Relationship between σ_c/σ_m and σ_i.

the transverse tests show only that the interface (and filament transverse strength) exceed that of the matrix. These results are summarized in Fig. **7**.

Equal strains in filaments and matrix are assumed in analyses of the stress state in composites tested with fibers aligned parallel to the applied load. With fibers aligned perpendicular to the applied load, the conditions as failure is approached may be closer to those for equal stress than the equal strain conditions assumed by Cooper and Kelly. Therefore, as a first approximation, the equal stress condition should also be considered even though the true state of affairs is much more complex. In this regard, analyses of mechanical interactions under transverse loading, presented in Chapter **2**, show maximum stresses at the interface on the order of that applied to the composite with only slight dependence upon filament array and volume fraction. Assumption of an equal stress condition is equivalent to defining σ_i as the transverse stress applied to the composite that is required to induce interface failure. On this basis, in the region where $\sigma_c = f(\sigma_i)$ in Fig. **7a**, the failure stress of the composite will be equal to σ_i and it will be independent of V_f. Thus, $\sigma_c/\sigma_m = \sigma_i/\sigma_m$ and the results of the analysis are as shown in Fig. **7b**.

Thus far, the effect of filament geometry on the lower limit for transverse composite strength has not been considered quantitatively. The Cooper and Kelly analysis is based upon a random arrangement of parallel filaments with the lower limit to transverse strength determined by a com-

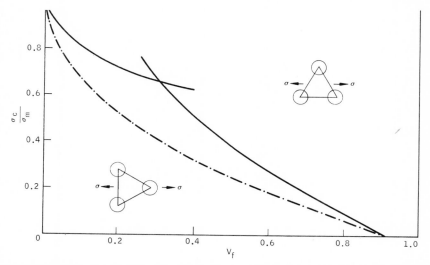

FIG. 8. Effect of stress direction on the lower limit for transverse strength for close-packed filament array.

posite in which filaments are replaced by holes. A simple failure criterion is
assumed in which only tensile failure occurs when the stress on the minimum cross section equals the matrix strength.

The effect of filament geometry on the lower limit can be shown by considering two basic arrangements of parallel fibers: the close packed and the
orthogonal arrays. Consider a close-packed hexagonal array of equally-
spaced filaments in which the unit cells form equilateral triangles, as indicated in Fig. 8. Analysis of this configuration for the stress applied parallel
to the base of the triangle yields the lower limit for transverse composite
strength shown by the upper curve in Fig. 8. There are two equations
governing the variation of σ_c/σ_m with volume fraction and their intersec-
tion occurs at $V_f \simeq 0.30$. The transverse strength decreases more rapidly
with increasing V_f when the volume fraction exceeds about 0.30, and
reaches zero when the filaments touch at $V_f = 0.906$. When the direction
of applied stress is as indicated in the bottom sketch in Fig. 8, the lower
limit is much reduced (lower curve). A similar analysis yields the lower
limits for transverse strength for the square array. The results for two
directions of applied stress for the square array are shown in Fig. 9. It is
interesting to note that the transverse strength for the square array shown
by the lower curve in Fig. 9 is the same as that for the random array, i.e.,
$\sigma_c/\sigma_m = 1 - (4V_f/\pi)^{1/2}$.

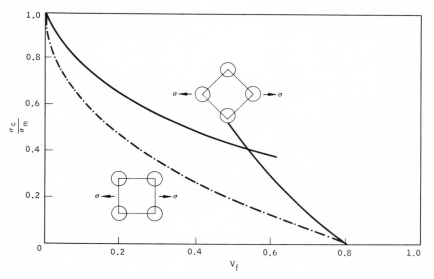

FIG. 9. Effect of stress direction on the lower limit for transverse strength for the
square filament array.

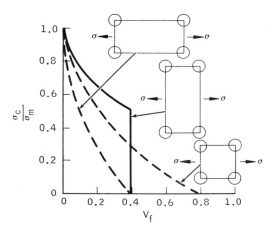

FIG. 10. Effect of filament geometry on transverse strength for orthogonal filament arrays.

The direction of applied stress with respect to the orientation of the filaments would certainly have a large effect on the lower limit for transverse strength according to the geometric analysis. For example, at 50 vol % reinforcement, σ_c/σ_m for the square array in Fig. 9 is about 0.44 for the orientation shown in the upper sketch but only about 0.20 for the orientation shown in the lower sketch. The strength differences, as a result of the change in orientation, would be over 100%. In contrast, the profiles for the minimum lower limits for the close-packed and square arrays (lower curves in Figs. 8 and 9) and the profiles for the maximum lower limits for the close-packed and square arrays (upper curves in Figs. 8 and 9) are in much closer agreement. Figure 10 shows some further orthogonal arrangements and their effect on the lower limits for transverse strength.

According to the geometric analysis, the dependence of the lower limit for transverse strength upon filament placement is large and, therefore, must be considered to distinguish between the effects of the interface and filament placement on transverse strength.

The geometric and the Chen and Lin analyses are compared for two composites for the square and hexagonal close-packed arrays in Figs. 11 and 12. The direction of applied stress with respect to the filament array is indicated in the sketch in each figure. Agreement for the lower limit to transverse strength is close for the square array, but there is considerable difference in the projected strengths at intermediate volume fractions for the hexagonal array. These figures illustrate a previously discussed difficulty with the Chen and Lin analysis which is a finite transverse strength

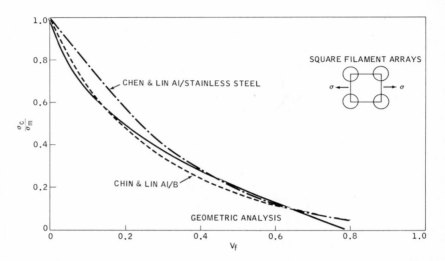

FIG. 11. Comparison of Chen and Lin (1969) and geometric analysis for minimum transverse strength for the square array.

at maximum packing density where filaments touch even though the filaments are unbonded. There are experimental data for these composites to compare the two approaches but most of the data are for random filament placement. As indicated previously, this is equivalent to the square array for the geometric approach. However, it will be difficult to distinguish

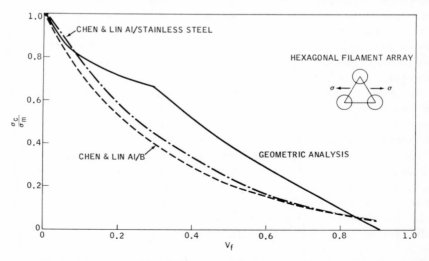

FIG. 12. Comparison of Chen and Lin (1969) and geometric analysis for minimum transverse strength for the hexagonal array.

between the two approaches for the square array with the stress applied as shown in Fig. 11 because of the close agreement of the curves. Experimental data defining the lower limit to transverse strength are not available to compare the predicted curves for the hexagonal array. Such a comparison of theory and experiment would require careful filament placement because of the sensitivity of the lower limit to filament placement.

In summation, theoretical predictions of composite off-axis strength have been made for composites in which the interface does not fail and in which interface failure is taken into account indirectly. Less progress has been made on the more complex problem of directly predicting the effect of the interface on off-axis strength, although the finite element analysis (Adams, 1970) offers insight into the problem. Reasonable upper and lower limits for composite transverse strength have been predicted by theories that consider the interface directly, but these limits have not been satisfactorily verified experimentally. More exact treatment defining the quantitative effects of the interface on off-axis strength is a very difficult problem that has not been completely resolved. One family of problems arises from the practical difficulty of measuring meaningful interface properties and another from the difficulty in determining how to use the measured properties to predict composite strength. These are complex, practical and theoretical problems. However, a start can be made by measuring off-axis composite tensile properties and examining the fractured specimens to determine the role the interface plays in tensile failure. The results of a number of studies of this type are presented in the following sections.

II. Off-Axis Failure of Composites

The presentation and discussion in this section will be limited to experimental results showing the various types of failure mode and failure strengths. Results that show only off-axis strength without consideration of failure mechanism will not be discussed because these results do not enhance our knowledge of the interface. However, the experimental results presented will include those showing not only interface failure but also other competitive failure mechanisms since the interface does not act independently.

A. *Columbium–Tungsten Composite*

Brentall *et al.* (1969) and Klein *et al.* (1970) studied failure modes in a Cb(Alloy)–W composite at room temperature and at 2200°F. This compo-

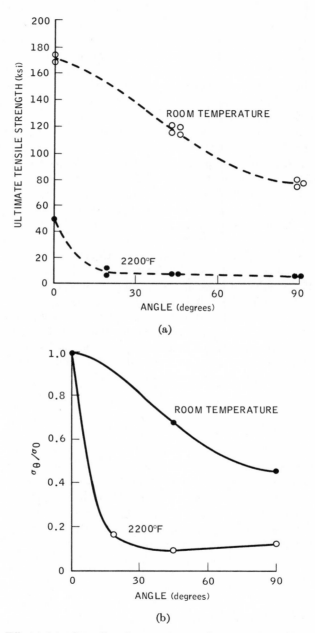

(a)

(b)

Fig. 13. Effect of loading direction on the tensile strength of Cb(alloy)–24W at room temperature and at 2200°F. (a) Variation in tensile strength with loading angle; (b) Normalized to unity for 0° loading angle. [From Brentnall *et al.* (1970).]

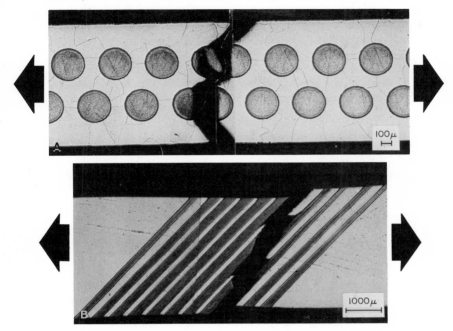

FIG. 14. Tensile specimens of Cb(alloy)–W tested at room temperature. (A) Filament splitting in transverse tensile specimen; (B) complex failure in 45° specimen. [From Klein *et al.* (1970).]

site is designed for high temperature use in an oxidizing environment and is composed of an oxidation resistant, columbium base matrix with tungsten filaments. Since filaments and matrix are soluble, but no reaction product is formed, the composite is of the Class II type. The off-axis strength of this composite was measured at room temperature and at 2200°F to determine the effect of temperature on the failure mode and on strength. Figure 13a shows the variation in tensile strength with angle between the loading direction and the filament axis. At 2200°F the composite is relatively much more sensitive to the loading direction than at room temperature. This is shown more clearly in Fig. 13b where the off-axis strength is normalized to unity for 0° loading.

The failure modes at room temperature for the 90° and 45° orientations are shown in Fig. 14. The fracture path for the 90° orientation (Fig. 14A) is through the filaments (filament splitting) rather than through the matrix or interface. Therefore, transverse strength is independent of the interface strength. However, the fracture path for the 45° specimen (Fig. 14B) is, in part, along the interface, but also goes through the filaments and the matrix. Therefore, fracture cannot be described in a simple way for this orientation.

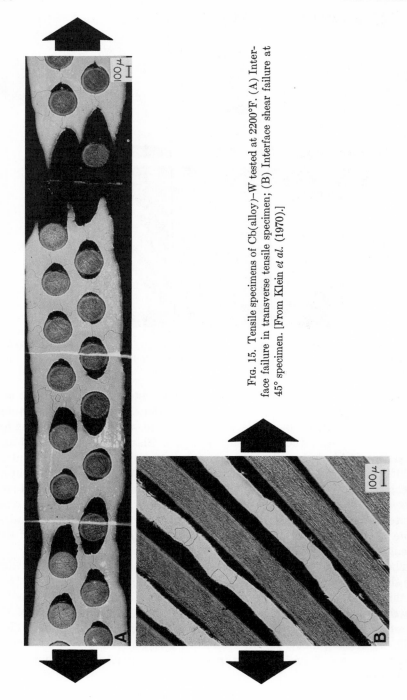

FIG. 15. Tensile specimens of Cb(alloy)–W tested at 2200°F. (A) Interface failure in transverse tensile specimen; (B) Interface shear failure at 45° specimen. [From Klein *et al.* (1970).]

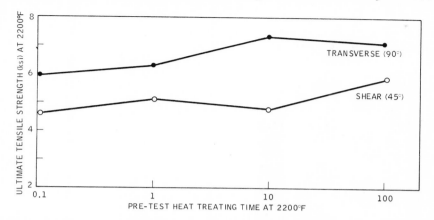

Fɪɢ. 16. Effect of pretest heat treatments at 2200°F on strength of Cb(alloy)–24W at 2200°F for different filament orientations. [From Klein *et al.* (1970).]

The fracture path at 2200°F for the 90° and 45° orientations are shown in Fig. 15. Failure is at the interface for both orientations and, therefore, these off-axis strengths are limited by interface strength. An increase in the strength of the interface should result in an increase in composite strength and a shift in failure from the interface to the matrix or to the filaments. One way to strengthen the interface in this composite is to heat-treat the composite to promote filament–matrix interdiffusion. Therefore, a series of specimens were given pretest heat treatments at 2200°F before tensile testing at this temperature. The change in composite strength as a result of the pretest heat treatments is shown in Fig. 16. After pretest heat treatments of 100 hr at 2200°F, the 45° strength has increased from 4.6 to 6.3 ksi and the transverse strength has increased from 5.8 to 7.2 ksi.

The fracture path in the transverse specimen heat treated for 100 hr at 2200°F before testing at 2200°F is shown in Fig. 17. The pretest heat treatment has strengthened the matrix and the interface by tungsten enrichment from the filaments. This has reduced matrix failure strain and shifted the fracture path away from the interface and, as a result, the composite strength at 2200°F has increased. Further increase in composite strength appears to be limited by filament splitting, or interface failure resulting from diffusion-induced porosity. Although not shown here, failure by the latter mode was also observed in some of these transverse tensile specimens that were given pretest heat treatments.

Failure typical of the 45° specimens given the 100 hr pretest treatments at 2200°F and then tested at 2200°F is shown in Fig. 17B. The 45° specimen has again failed at the interface. Although both the unheat-treated specimen (Fig. 15B) and the specimen given the 100 hr pretest treatment

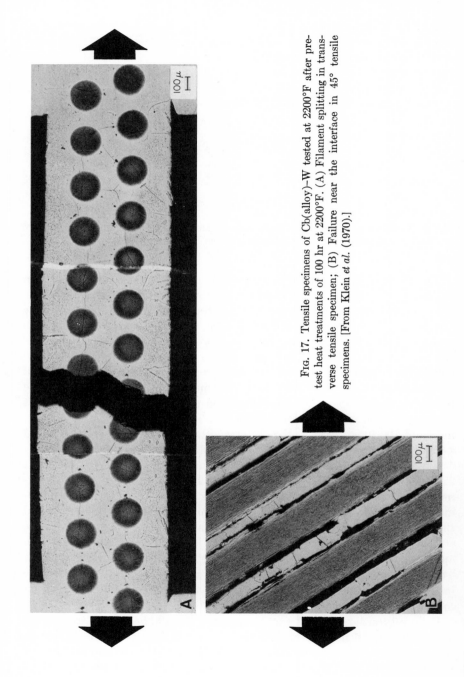

FIG. 17. Tensile specimens of Cb(alloy)–W tested at 2200°F after pretest heat treatments of 100 hr at 2200°F. (A) Filament splitting in transverse tensile specimen; (B) Failure near the interface in 45° tensile specimens. [From Klein *et al.* (1970).]

fail at the interface, the reason for the interface failures are probably different. The initial bond is not adequate to withstand the transverse tensile loading but improves on heat treatment. However, after long heat-treating times, the interface is weakened by diffusion-induced porosity. This porosity may have contributed to parting of the filament and matrix near the interface. Although the interface porosity is located in the diffusion zone outside the position of the original interface, it may be regarded as interface-related failure. Despite the weakening effect of the porosity, the overall interface strength must be greater after the pretest heat treatment because the strength of the composite is increased.

In summation, the Cb(Alloy)–W composite shows a number of different failure modes, including interface failure, that affect off-axis composite strength. At room temperature, transverse strength is limited by filament splitting rather than by interface failure, while at 45°, failure is complex but occurs, in part, along the interface. When the composite is tested at 2200°F, both transverse and 45° strengths are limited by interface failure. If the interface is strengthened by pretest heat treatments, strength is then limited by filament splitting or by failure near the interface at diffusion-induced porosity. However, when the interface is strengthened by the pre-test heat treatments, the off-axis strength of the composite is increased.

B. *Titanium–Boron and Titanium–Borsic† Composites*

Both the titanium–boron (Ti–B) and the titanium–Borsic (Ti–Borsic) composites are of the Class III type since reaction products form at the filament–matrix interface. The effect of the interface interaction on the strength of these composites was studied at Solar for composites with 4 mil filaments (Schmitz *et al.*, 1970). In this study, off-axis strength measurements were made before and after heat treatments, and the change in strength was correlated with the fracture mode.

In the as-fabricated condition, the Ti–B composite had a titanium diboride reaction product of 500 to 1500 Å, whereas no reaction product was detected in the Ti–Borsic composite. The interaction at the interface induced by heat treating the composites for 90 min at 1600°F is shown in Fig. 18. This heat-treatment resulted in the growth of the TiB_2 interaction product in the Ti–B composite to a thickness of about 1.2 μ and the formation of porosity at the interface. Porosity in this system results from a volume contraction upon formation of TiB_2 and from unequal filament–matrix diffusion. There was also an interaction between the SiC coating and the titanium in the Ti–Borsic composite which resulted in the forma-

---

† The trade name of Hamilton Standard Division of United Aircraft Corporation for boron filaments coated with a thin layer of silicon carbide.

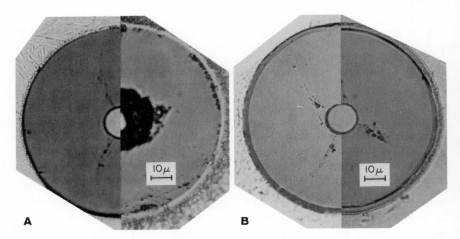

Fig. 18. Comparison of as-bonded and heat-treated structures for (A) Ti–B and (B) Ti–Borsic. As-bonded structure shown at left of each photograph and heat-treated structure at right (90 min at 1600°F).

tion and growth of several intermediate phases to a total thickness of about 1.5 μ. However, there was no evidence of porosity in this composite.

1. Titanium–Boron (Ti–B)

The variation in tensile strength with loading angle for Ti(75A)–25B is shown in Fig. 19. The tensile strength of the composite decreases from 143

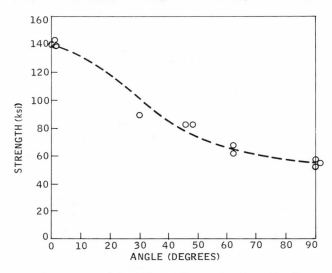

Fig. 19. Tensile strength of Ti–75A/B as a function of loading angle.

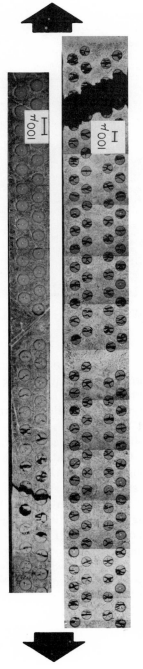

Fig. 20. Cross sections of transverse tensile specimens typical of Ti–75A/B after fracture. Single thickness tape and double thickness tape.

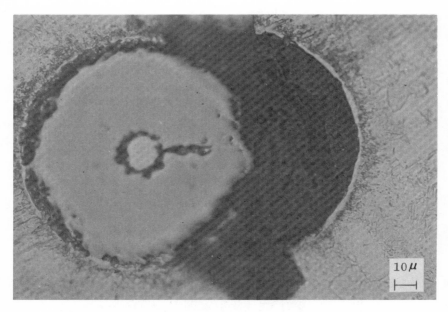

Fig. 21. Fracture in transverse Ti–40A/B specimen heat-treated 90 min at 1600°F before testing.

ksi for the 0° orientation to about 60 ksi for the 90° orientation. The strength decrease is very gradual with increasing angle because of the relatively high matrix strength (∼100 ksi).

Examination of these specimens after failure showed that filaments in the 30° and 45° specimens had separated from the matrix at the interface and had also failed in tension. The dominant failure mode for the 60° and 90° orientations was filament splitting rather than interface failure and, therefore, the interface strength exceeds the transverse filament strength for the conditions of these tests. Filament splitting in transverse specimens is shown in Fig. 20. Both the two- and four-layer specimens have about the same transverse tensile strength, but the distribution of fractured filaments is different in the two types of specimen. Filament splitting in the thicker specimen is distributed throughout the specimen gage length. When a filament splits, its share of the load is shifted to the matrix and the matrix can sustain the extra load if it strain hardens sufficiently. Splitting of a filament in the thinner specimens reduces the cross section to such an extent that the matrix is unable to compensate for the reduced section by strain hardening before the composite fails. For the thicker specimen, the cross section is reduced proportionately less by a split filament and matrix

TABLE I

TRANSVERSE TENSILE STRENGTH FOR TI(40A)–25B COMPOSITE SPECIMENS

Treatment	Number of tests	Diboride thickness (Å)	Tensile strength (ksi)	Type of failure
As-fabricated	4	~1,000	49.5	Filament splitting
90 min/1600°F	2	12,000	48.4	Interface

strain hardening is able to take up the load shed by the split filament before the specimen fails at this point.

Transverse specimens of Ti–25B will fail at the interface, however, if they are heat-treated before testing. The structure of a transverse specimen pulled to failure after a pretest heat treatment of 90 min at 1600°F is shown in Fig. 21. The transverse specimens fail at the interface in a region that is weakened by porosity. Interface porosity is believed to be the result of a volume contraction when TiB_2 forms from its reactants but could also be in part caused by unequal filament–matrix diffusion. The diboride layer (white phase) remains attached to the titanium surface adjacent to the mating filament. Although the failure mechanism changes with increased

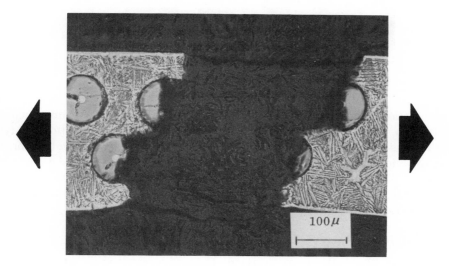

FIG. 22. Section of Ti–Borsic transverse specimen in as-bonded condition showing fracture through Borsic filaments.

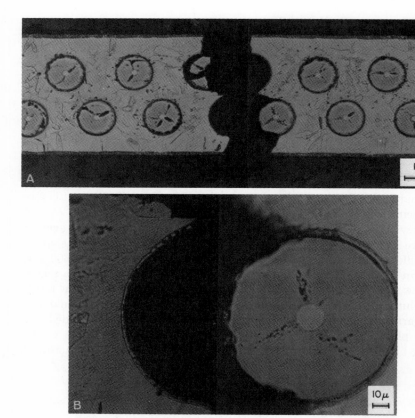

Fɪɢ. 23. Fracture in transverse Ti–Borsic specimen heat-treated 90 min at 1600°F before testing.

interaction zone thickness, this change does not affect the transverse strength. The strengths of Ti–B specimens with different interaction zone thicknesses are listed in Table I. The almost constant strength values for both failure modes suggests that the transverse strength is at the lower limit. This would be the case where the stress to cause interface failure or filament splitting is less than the minimum strength for an unbonded composite.

2. Titanium–Borsic (Ti–Borsic)

The fracture path in the Ti–Borsic composite tested in the transverse direction in the as-fabricated condition is also through the filaments rather

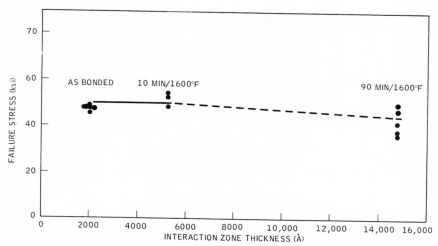

FIG. 24. Effect of interaction on transverse strength of Ti–40A/Borsic; exposure times are shown above the datum points.

than through the interface (Fig. 22). Therefore, the interface strength in this composite also exceeds the transverse filament strength. After the 90 min heat treatment at 1600°F, however, the fracture path is also shifted to the interface. As shown in Fig. 23, the fracture path appears to be within the interaction zone or between the silicon carbide coating and the interaction zone. Part of the interaction layer remains on the titanium surrounding the filaments. However, as shown in Fig. 24, there is little change in transverse strength with an increase in interaction zone thickness induced by the heat treatments. Again the results suggest that the transverse strength is near the lower limit and thus is unaffected by change in failure mechanism (i.e., filament splitting to interface failure). Enrichment of the matrix with carbon from the SiC coating is a complicating factor that may increase matrix strength and contribute to the scatter in strength values shown in Fig. 24.

C. Aluminum–Boron (Al–B) and Aluminum–Borsic (Al–Borsic) Composites

There have been a number of studies in which the off-axis tensile properties of Al–B and Al–Borsic composites were determined, but only a few considered the effect of the interface on off-axis strength. Al–B composites may be initially nonreactive (Class I) because of oxide barriers between

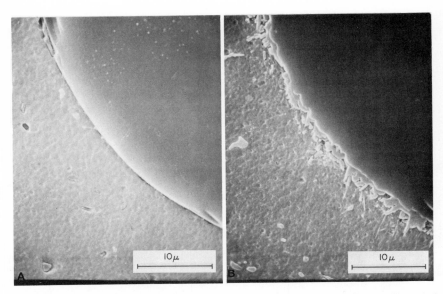

Fig. 25. Interface interaction of Al(6061)–B composites (Klein and Metcalfe, 1971). (A) After 0.5 hr at 940°F; (B) after 50 hr at 940°F.

TABLE II

Effect of 1000°F Heat Treatments[a] on Tensile Properties and Fracture of Al(6061)–45B for T-6 Temper (5.6 Mil Filaments)

Time at 1000°F	Failure strain (μin./in.)	Ultimate tensile strength (ksi)	Fracture type (%)		
			Filament splitting	Matrix	Interface
0	1900	20.2	15	15	70
5 min	2300	34.8	25	50	25
10 min	1770	36.8	25	60	15
15 min	2300	36.7	10	50	40
30 min	1100	33.5	20	60	20
1 hr	2700	31.2	25	40	35
2 hr	5200	20.0	40	30	30
5 hr	4600	17.1	75	—	25
10 hr	3900	20.0	15	—	85
150 hr	3100	19.0	20	—	80

[a] Zero time specimens were heat treated for 7 hr at 350°F. All other specimens were first heat treated at 1000°F, quenched in water and aged 7 hr at 350°F.

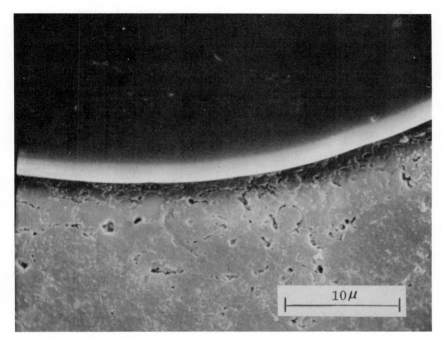

Fɪɢ. 26. Borsic filament in Al(6061) matrix after 160 hr heat treatment at 1000°F; the band on the rim of the filament is the SiC coating [from Klein and Metcalfe, (1971)].

the filaments and the matrix, but after heat treatment this system readily becomes reactive with formation of AlB_2 at the interface. This interface reaction in Al(6061)–B composite induced by heat treatment is shown in Fig. 25. In contrast, Al–Borsic composites appear to be nonreactive and insoluble. Although an aluminum carbide might be expected to form from phase considerations, no reaction product has been identified. The interface in Al(6061)–Borsic composites after a 160 hr heat treatment at 1000°F is shown in Fig. 26. There is no evident formation of a reaction product or an interdiffusion zone at the SiC–Al interface, but there appears to be some porosity remaining in the plasma-sprayed matrix.

1. Aluminum–Boron (Al–B)

The effect of the interface on transverse tensile strength of an Al(6061)– B composite (5.6 mil filaments) was studied by Klein and Metcalfe (1971). The condition of the interface was varied by pretest heat treatments at 1000°F, and then the matrix was placed in the T-6 temper by quenching the composite specimens in water and aging at 350°F. The transverse strength and fracture modes are listed in Table II (in general, the average of triplicate specimens). Three types of transverse failure are considered: filament splitting, failure at the filament–matrix interface or interaction zone, and complete matrix failure. Some matrix failure must occur in all composite specimens since the matrix forms a continuous load-bearing network between filaments. However, the matrix failure listed in this table refers to failure occurring near the filaments where the fracture path preferentially-selected was through the matrix rather than through the interface or filament. The percentages of the fracture types listed in the table are only approximate. They are qualitative estimates based upon examination of the fracture surface using stereographic and scanning electron microscopes.

In general, the transverse tensile strength decreases with an increase in pretest heat-treating time at 1000°F, while the failure strain shows a tendency to increase. The strength for the first specimen listed in Table II (zero-time specimen) is low because the matrix was not solution-annealed. Mixed types of failures were observed in all of the specimens, indicating a wide range of interface strength and resistance to filament splitting. This behavior may, in part, result from the gradual breakdown of an oxide barrier between the filaments and the matrix. Although the percentages listed under fracture types are qualitative rather than quantitative, the results are sufficient to show the following general trends. When failure is predominantly through the matrix, the transverse strength is high and when interface failure and filament splitting predominate, the transverse strength is low. Filament splitting occurs in all specimens but is usually not the major failure mode.

The transverse strength is shown as a function of pretest heat-treating time in Fig. 27. Strengths are above 30 ksi for heat-treating times of 1 hr or less and decrease to a plateau of about 19 ksi for longer pretest heat-treating times. As indicated in this figure, 19 ksi is well above the minimum matrix contribution for the T-6 temper based upon the lower limit for transverse strength (geometric or Chen and Lin analysis for square array). Therefore, the interface is probably contributing to transverse strength even when the interaction zone is quite extensive.

The scatter in strength values are greater for specimens given the shorter heat treatments and greatest for the two shortest heat treatments, 5 and 10 min. The larger scatter for the shorter heat treatments may be caused by experimental difficulties in controlling the heat-treating time and temperature for short times, or incomplete solution annealing of the matrix prior to age hardening. Although the average strength of the specimens

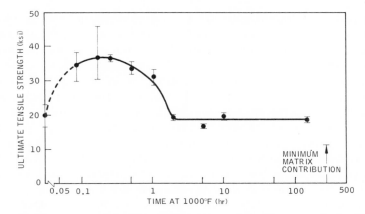

FIG. 27. Effect of pretest heat treatments at 1000°F on transverse tensile strength of Al(6061)–45B for the T-6 temper.

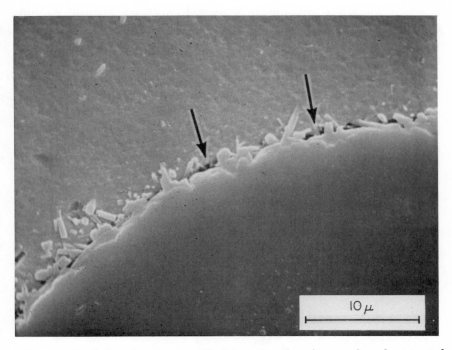

FIG. 28. Vertical section showing AlB_2 interaction phase in a specimen heat-treated for 150 hr at 940°F; the arrows designate regions where interface gaps may exist because of a volume contraction upon formation of AlB_2.

heat-treated for 10 min is about 37 ksi, one specimen in this series had a
transverse strength of 46 ksi. The strength of the matrix (bonded matrix
without filaments) heat-treated at 1000°F for 30 min followed by quenching and aging is 44 ksi.

The results of these tests indicate that the interface plays a major role
in determining the transverse strength. Failure of the zero time specimen
is predominantly at the interface. For this condition, the interface is probably weak because of incomplete filament–matrix bonding. For short
heat-treating times (1 hr or less), the first reaction at the interface increases interface strength and causes failure to be shifted to the matrix.
Specimens heat-treated for 2 and 5 hr have less matrix failure, but more
filament splitting. The reason for the increased filament splitting in these
specimens is not known. However, when heat treatments are extended beyond 5 hr, there is a tendency for failure to be shifted again to the interface,
and the transverse strength is low. This suggests that there is a limit to the
amount of the interaction at the interface that can be tolerated before the
interface and the composite are weakened. Although filament splitting
may contribute in an unpredictable way to the stress-time profile shown in
Fig. 27, the general trend of higher transverse strength associated with
matrix failure and lower transverse strength associated with interface failure seems to be valid.

One reason that the transverse strength is reduced and apparent inter-

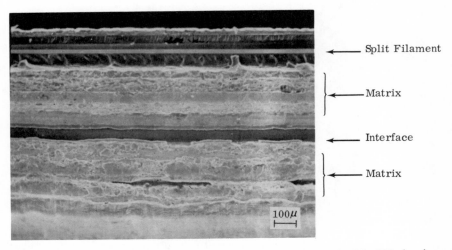

Fig. 29. Fracture surface of transverse tensile specimen of Al(6061)–45B showing
failure at split filament, interface, and matrix; specimen was heat-treated 1 hr at 1000°F
(T-6 temper) and had a transverse tensile strength of 33 ksi.

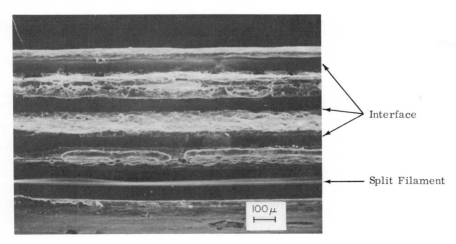

FIG. 30. Fracture surface of transverse tensile specimen of Al(6061)–45B showing failure at split filament and at interface; specimen was heat-treated for 150 hr at 1000°F (T-6 temper) and had a transverse strength of 18 ksi.

face failure is increased with increased AlB_2 formation may be the 20% volume contraction that occurs when this compound forms. This volume contraction may give rise to internal stresses as well as porosity at the interface that would lower transverse strength. Figure 28 shows structural features (indicated by arrows) believed to be gaps at the interface arising from this source.

Some of the structures observed at the fracture surfaces of the transverse tensile specimens are shown in Figs. 29 and 30. Figure 29 shows the different types of fracture detected in one of the specimens heat treated for 1 hr at 1000°F and having a transverse strength of 33 ksi. This figure shows the fractured end of the specimen viewed at a tilt angle of about 25°. The primary failure mode in this specimen was matrix failure, but some filament splitting and interface failure were observed, as indicated in this figure. The fracture surface of a specimen that was given a pretest heat treatment of 150 hr at 1000°F is shown in Fig. 30. This specimen, with a transverse strength of 18 ksi, shows some filament splitting but more interface failure.

It is of interest to note that there is a correlation between the change in longitudinal and transverse strengths with heat treatments that increase the interface interaction. This correlation for specimens from the same Al–B panel is shown in Fig. 31. The results show that optimization of both longitudinal and transverse tensile strengths occurs at the same combina-

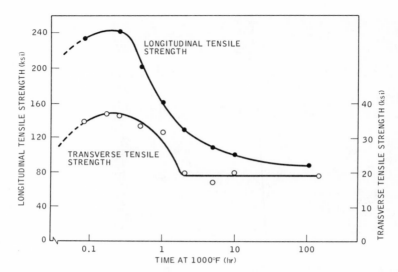

Fɪɢ. 31. Comparison of the longitudinal and transverse tensile strengths for Al(6061)–45B for the T-6 temper.

tion of time and temperature and, therefore, for the same interface condition. Since some interface interaction has taken place in these specimens, the initial reaction at the interface may be beneficial or at least may not degrade longitudinal and transverse tensile strengths. However, this point will be difficult to establish for the T-6 temper since incomplete solution annealing of the matrix for specimens given the short heat-treating times may contribute to their somewhat lower tensile strengths.

The effect of the interface on 30° and 90° off-axis strengths of Al(6061)–40B was studied at Midwest Research Institute (Swanson and Hancock, 1971), but the off-axis strength appeared to be limited by filament splitting rather than by interface failure. No interface failure was reported in this study.

2. Aluminum–Borsic (Al–Borsic)

The role of the interface in determining the transverse strength of Al(6061)–25Borsic composites was studied by Klein *et al.* (1972). These composites were made from hot-pressed, plasma arc-sprayed tapes containing 5.7 mil diam filaments. Composite specimens were given pretest heat treatments at 1000°F to change the condition of the interface and then the specimens were quenched in water and aged at 350°F (T-6 temper) or held for 2 hr at 800°F and slowly cooled to 350°F and held there for 7 hr

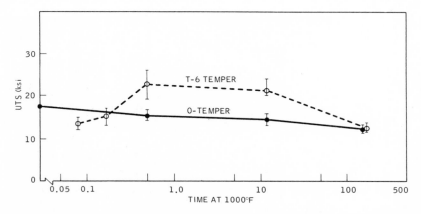

FIG. 32. Effect of heat treatments at 1000°F on the transverse strength of Al(6061)–25 Borsic (5.7 mil filaments).

(O temper). The 350°F heat treatment served to relieve residual stresses as well as to age the quenched specimens.

The results of these tests are shown in Fig. 32. The specimens given the O-temper after the interface heat treatment at 1000°F show a slight decrease in transverse strength with increased heat-treating time at 1000°F while the specimens given the T-6 temper show maximum values for intermediate heat-treating times. Fracture surfaces were examined in all of these specimens and the dominant failure modes were matrix failure and matrix failure combined with filament splitting. Matrix failure occurred in part by delamination of the plasma-sprayed material from the matrix foil. In this regard, the plasma-sprayed aluminum may be weaker than foil because of incomplete matrix consolidation. A minor amount of interface failure was detected in all of these specimens, but there was no evident relationship between the amount of interface failure and the duration of the pretest heat treatments. The low strengths for the short heat-treating times for the T-6 temper are probably caused by incomplete solution annealing and the low strength for the longest heat-treating time (160 hr) is caused by increased filament splitting although the reason for the increase in filament splitting is not known. In general, the change in the condition of the interface induced by heat treatments did not affect transverse strength for these tests because the filaments split or the matrix failed before sufficient load could be applied to induce interface failure.

Prewo and McCarthy (1972) tested Al(6061)–Borsic composites with more completely consolidated plasma-sprayed matrices and with filaments that were more resistant to splitting. Al(6061)–50Borsic panels were

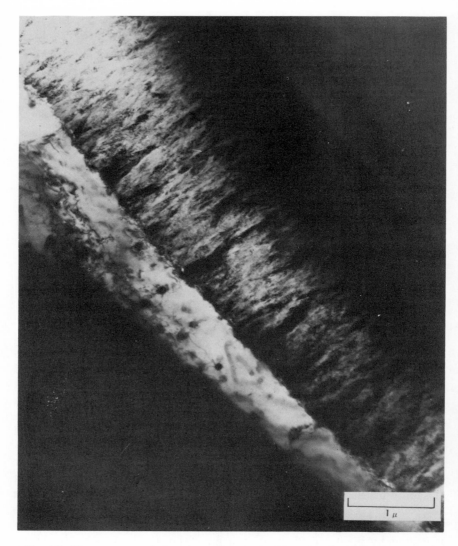

FIG. 33. Electron transmission photomicrograph of interface in Al–Borsic composite [from Prewo and McCarthy, (1972)].

fabricated by press-bonding layers of plasma-sprayed composite tape and were heat-treated to age-harden the matrix. The authors reported that composites made with 4 mil diam filaments were not as strong in transverse tension as composites made with 5.7 mil diam filaments. The transverse strength of the composites made with the smaller diameter filaments was about 22 ksi and was limited by filament splitting rather than by interface failure. The transverse strength of the composites made with the larger diameter filaments was about 36 ksi, and failure occurred primarily in the matrix with only a small amount of interface failure and filament splitting. This high transverse strength was found to be contingent upon full consolidation of plasma-sprayed tapes to ensure a sound filament–matrix bond and matrix consolidation.

The interface in the composite specimens was carefully examined by Prewo and McCarthy using scanning and transmission electron microscopy. The silicon carbide coating and the aluminum matrix were found to be in intimate contact, but there was no evidence of chemical interaction. The high interface strength was attributed to a sound filament–matrix bond established during consolidation of the composite tape. The structure of the interface revealed by transmission electron microscopy is shown in Fig. 33.

D. Other Composite Systems

The effect of the interface in Al–20% stainless steel composites on transverse strength was studied at the National Engineering Laboratory in the United Kingdom (Paton and Lockhart, 1971). Composite specimens were fabricated by hot pressing plasma-sprayed layers of composite at temperatures ranging from 750 to 1110°F. The resulting composites were tested at room temperature to determine the effect of processing temperature on transverse strength. The room-temperature transverse strength increased with increased processing temperature to 972°F as the bond strength increased. The increase in transverse strength was greater than could be accounted for by the increase in matrix strength due to sintering. For higher processing temperatures, the transverse strength decreased, and this was attributed to formation of a weak, brittle phase (Fe_2Al_5) at the interface where fracture was initiated.

The effect of interface bond on transverse strength of $NiCr–Al_2O_3$ composites was studied by Mehan and Harris (1971). The results of their work indicate that a filament–matrix bond was not achieved despite pretest heat treatments and coating of the filaments prior to hot pressing. In this system, the transverse strength for volume fractions to about 40% was reported to be that of the matrix in which holes replace the filaments in the composite, or in which there is no interface bond according to the Chen and Lin analysis. The comparison is presumably based upon the square filament array (Fig. 11A) where both analyses are in approximate agreement.

III. Discussion and Summary

In this presentation, theories predicting off-axis tensile strength have been divided into three different classifications according to the way that the interface is taken into account. Theories for strong interfaces are based upon an interface that does not fail before the composite fails. Phenomenological theories consider interface failure indirectly through mechanical property values required in the proposed solutions. Theories for weak interfaces take interface failure into account directly, but this class of theories is the most complex and least developed of the three. At present there are no theories that have been developed sufficiently to permit prediction of composite off-axis strength from the mechanical properties of the interface, filaments, and matrix. Theories for weak interfaces have been limited to the transverse orientation and, for the most part, have been restricted to establishment of upper and lower limits for transverse strength. Where the interface does not fail, upper limits for transverse strength approaching or exceeding the bulk matrix strength have been predicted. Where the interface is very weak, lower limits for transverse strength have been proposed for two cases: where the filaments are not bonded and where the filaments are replaced by holes. On the basis of limited comparison, both approaches yield about the same results for the square filament array but predict different results for the close-packed array.

The effect of the interface on off-axis strength has been studied experimentally in a number of composite systems. In addition to interface failure, the results show that other competitive failure modes also limit strength. Most of these studies are for the transverse orientation where the interface can have a large effect on composite strength and are restricted to relating the fracture mode to strength. However, even for the transverse orientation, comparison of experiment and theory can only be qualitative because theory is not sufficiently developed and experimental results are incomplete with respect to the mechanical properties of the composite and its components.

Work done thus far on the Ti–B, Ti–Borsic and NiCr–Al_2O_3 composites indicates that these systems are near the lower limit for transverse strength. Ti–B and Ti–Borsic in the as-bonded condition have low transverse strengths because of filament splitting, and heat treatments that promote the interface interaction cause the failure mode to shift to interface failure without a significant change in strength. The NiCr–Al$_2$O$_3$ composite has low transverse strength in the as-bonded condition because of both filament splitting and a very weak interface bond.

In well-bonded Al–B composites the interface does not appear to limit transverse composite strength because failure occurs in the matrix or is induced by filament splitting. For certain heat treatments, and for filaments that resist splitting, the transverse strength is well above the lower limit. With extensive interface interaction, however, the transverse strength is greatly reduced and failure is shifted to the interface.

The transverse strength of fully consolidated plasma-sprayed Al–Borsic composites also does not appear to be limited by interface failure. Where filament splitting is minimized and the matrix is fully consolidated, the transverse strength is well above the lower strength limit. The transverse strength is greatly reduced, however, by filament splitting and incomplete matrix consolidation.

Formation of extensive interaction products in the reactive composites, aluminum–stainless steel and Al–B, decrease transverse strength. However, limited interface interaction may not degrade transverse strength and may increase it. For example, the room temperature transverse strength of aluminum–stainless steel is increased with increased bonding temperatures to the first formation of Fe$_2$Al$_5$ at 972°F. For higher bonding temperatures, the transverse strength decreases with extensive formation of Fe$_2$Al$_5$. Similarly, when as-fabricated Al(6061)–B is heat-treated at 1000°F and then given the T-6 temper, the transverse strength does not decrease upon first-formation of AlB$_2$ at the interface but only after some nucleation and growth of AlB$_2$ has taken place. In this regard, the longitudinal strength of Al(6061)–B also shows the same trend. Therefore, there may be bonding conditions that would yield optimum mechanical properties for a controlled amount of interaction at the interface. This is a phase of composite research that warrants further study.

Acknowledgments

The author would like to thank Solar Division of International Harvester Company, for support and permission to publish this paper and Dr. A. G. Metcalfe for review of the manuscript.

The support of the Solar Research Laboratory personnel is appreciated, especially that of Mr. R. Hutting for metallographic preparation of composite specimens and that of Mrs. P. J. Lind for manuscript preparation.

References

Adams, D. F. (1970). *J. Composite Mater.* **4,** 310.

Azzi, V. D., and Tsai, S. W. (1965). *Exp. Mech.* **5,** 283.

Brentnall, W. D., Klein, M. J., and Metcalfe, A. G. (1970). "Tungsten Reinforced Oxidation Resistant Columbium Alloys." First Ann. Rep. on Contract N00019-69-C-0137, Naval Air Syst. Command.

Chen, P. E., and Lin, J. M. (1969). *Mater. Res. Std.* **9,** 29.

Cooper, G. A., and Kelly, A. (1968). "Interfaces in Composites." Amer. Soc. Test. Mater., Philadelphia, Pennsylvania.

Cratchley, D., Baker, A. A., and Jackson, P. W. (1968). "Metal Matrix Composites." Amer. Soc. Test. Mater., Philadelphia, Pennsylvania.

Drucker, D. C. (1964). "Engineering Continium Aspects of High Strength Materials." Tech. Rep. 7, Brown Univ., Providence, Rhode Island.

Ebert, L. J., Claxton, R. J., and Wright, P. K. (1971). "Analytical Approach to Composite Behavior." Tech. Rep. AFML-TR-71-113.

Kelly, A., and Davies, G. J. (1965). *Metall. Rev.* **10,** 1.

Klein, M. J., and Metcalfe, A. G. (1971). "Effect of Interfaces in Metal Matrix Composites on Mechanical Properties." Tech. Rep. AFML-TR-71-189.

Klein, M. J., Metcalfe, A. G., and Domes, R. B. (1970). "Tungsten Reinforced Oxidation Resistant Alloys." Final Rep. on Contract N00019-69-C-0137.

Klein, M. J., Metcalfe, A. G., and Gulden, M. E. (1972). "Effect of Interfaces in Metal Matrix Composites on Mechanical Properties." Tech. Rep. AFML-TR-72-226.

Lauraitis, K. (1971). "Failure Modes and Strength of Angle-Ply Laminates." T&AM Rep. 345, Univ. of Illinois, Urbana, Illinois.

Lin, J. M., Chen. P. E., and DiBenedetto, A. T. (1971). *Polym.-Eng. Sci.* **11,** 344.

Mehan, R. L., and Harris, T. A. (1971). "Stability of Oxides in Metal or Metal Alloy Matrices." Tech. Rep. AFML-TR-71-160.

Paton, W., and Lockhart, A. (1971). "Factors Affecting the Transverse Strength of Fiber-Reinforced Metals." NEL Rep. 475.

Petrasek, D. W., and Signorelli, R. A. (1967). "Factors Affecting Tensile Properties of Discontinuous Fiber Composites." NASA TN D 3886.

Prewo, K. M., and McCarthy, G. (1972). *J. Mater. Sci.* (to be published).

Schmitz, G. K., Klein, M. J., Reid, M. L., and Metcalfe, A. G. (1970). "Compatibility of Viable Titanium Matrix Composites." Tech. Rep. AFML-TR-70-237.

Stowell, E. Z., and Liu, T. S. (1961). *J. Mech. Phys. Solids* **9,** 242.

Swanson, G. D., and Hancock, J. R. (1971). "Effects of Interfaces on the Off-Axis and Transverse Tensile Properties of Boron-Reinforced Aluminum Alloys." Midwest Res. Inst., Contract N00014-70-C-0212.

Tsai, S. W. (1968). "Fundamental Aspects of Fiber Reinforced Plastic Composites." Wiley (Interscience), New York.

Tsai, S. W., and Wu, E. M. (1971). *J. Composite Mater.* **5,** 58.

6

Role of the Interface on Elastic–Plastic Composite Behavior

A. LAWLEY and M. J. KOCZAK

Department of Metallurgical Engineering
Drexel University
Philadelphia, Pennsylvania

Key to an understanding of the elastic and plastic behavior of composite materials is a characterization of the nature and role of the interface or interfacial region between phases. The current status of analytic and experimental studies is reviewed in relation to filamentary composites and those produced by directional solidification. Attention is directed to interface structure and bond integrity, efficiency of load transfer, interface stability, and the effect of mode of loading.

I. Introduction

In light of current and anticipated materials requirements of the Department of Defense, metal matrix composites are of particular interest. The unique combinations of high stiffness (modulus)-to-density and high strength-to-density, coupled with ductility, toughness, and often oxidation–corrosion resistance are compatible with use as aircraft and submarine structural elements, and fan blading in gas or steam turbines. These unique property combinations pose new complexities and constraints in the design process. A first requirement in the utilization of metal-matrix composites is a comprehensive knowledge of mechanical (elastic–plastic) properties. In design for dimensional stability, elastic moduli and microstrain characteristics are of primary concern. Limits on loading and failure criteria mandate a determination of the stress levels for yielding, flow, and fracture. Many in-service conditions involve time-dependent stress or strain so that the composite behavior in creep or fatigue must be known.

For a detailed understanding of the elastic–plastic behavior of composite materials, it is necessary to characterize the role of the interface or interfacial region. The interface provides the means for transfer of stress from the matrix to the reinforcing phase in the composite. Various classifications of the interface are possible. From physical–chemical considerations, the following are identified either individually or in combination (Metcalfe, 1974, Chapter 3): the mechanical bond, the dissolution and wetting bond, the oxide reaction bond, the exchange reaction bond, and mixed bonds. Alternatively, the interface can be considered in relation to the mode of composite fabrication or growth. This provides two major catagories: the interface characteristic of directionally-solidified (*in-situ*) composites, and the interface in filamentary composites (wire- or fiber-reinforced) fabricated by diffusion bonding, liquid metal infiltration, or electroplating techniques. In directionally-solidified composites, the phases are essentially in equilibrium; however, physicochemical instability is possible (Salkind, 1969; Bayles *et al.*, 1967) which leads to spheroidization or coarsening of the structure with little net change in composition or amount of each phase. In comparison, differences in chemical potential at interfaces in filamentary composites provide a driving force for chemical reaction and/or diffusion which can lead to changes in the composition and volume fraction of each phase.

Analytically, it is possible to predict mechanical properties and general dynamic performance of a composite structure from the basic properties of the constituent materials (Kelly and Davies, 1965; Cratchely, 1965; Hill, 1965). This approach has its origins in continuum mechanics and leads to the familiar "rule of mixtures" relationship for a given property in terms of the volume fractions of each phase in the composite. Residual stress, Poisson's ratio differences, fiber geometry, and other composite parameters can be accommodated in these micromechanical treatments. Credibility of the analytic approach is vested primarily in a comparison of predicted and measured moduli and tensile strengths, though recently attention has also been directed to a characterization of micro- and macro-yielding (Pinnel and Lawley, 1970, 1971). In the prediction of composite behavior, it is assumed that an efficient transfer of load is possible across interfaces. Account must be taken of any differences between the expected and actual character or behavior of the interface or interfacial region.

This review attempts a comprehensive evaluation of the current understanding of the role of the interface on elastic–plastic behavior of metal matrix composites. In view of the distinct differences in interface structure and stability for directionally solidified (in situ) and fabricated filamentary composites, both forms are considered. Particular attention is directed to interface structure, interface stability, and to response as a function of mode of loading, i.e., tension, compression, creep, and fatigue. It will be seen that with the advent of scanning electron microscopy, the intelligent use of both transmission electron microscopy and optical metallography, coupled with microprobe work, a clearer understanding of the details and role of the interface is now emerging.

II. Micromechanics of Elastic and Plastic Deformation

Much of the data on composite materials are discussed in light of the various relationships of the rule of mixtures analysis, which provides a good indication of bond integrity and efficiency of load transfer at interfaces. Since a similar approach is adopted in this review, it is appropriate to present pertinent relationships and to list important assumptions. Attention is directed to elastic modulus, microstrain behavior, macroyielding, ultimate strength, and creep.

A. The Response to Tensile Loading

To predict the tensile behavior of continuous metal matrix composites from the additive properties of the individual components, McDanels *et al.* (1965) made the following assumptions:
(1) Isostrain conditions exist in each phase at all stress levels.
(2) The fibers are aligned parallel to the loading axis.

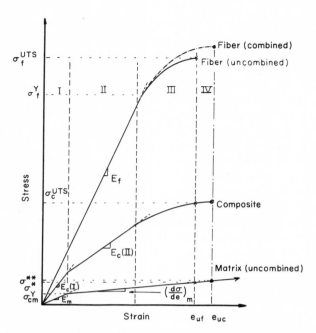

FIG. 1. Schematic representation of the stress–strain behavior of continuous fiber metal matrix composites [from McDaniels *et al.* (1965)].

(3) A "perfect" bond exists between the fiber and matrix.
(4) The properties of the phases are unaltered in the combined state.
(5) No transverse stresses develop as a result of fiber–matrix interaction.
(6) No residual stresses exist in either component prior to loading.

The tensile stress–strain curve of the composite is classified into four stages (Fig. 1), namely:

State I: Fibers and matrix deform elastically.
Stage II: Matrix deforms plastically, fibers elastically.
Stage III: Fibers and matrix deform plastically.
Stage IV: Tensile fracture of the composite.

During stage I, the composite modulus of elasticity is given by

$$E_c(I) = E_f V_f + E_m(1 - V_f) \tag{1}$$

where $E_c(I)$ is the initial modulus of elasticity of the composite during stage I, E_f and E_m the elastic moduli of the uncombined fiber and matrix

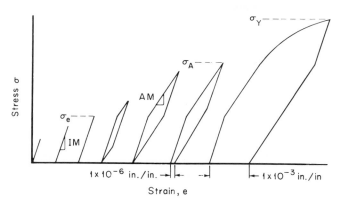

FIG. 2. Schematic representation of microyielding stress–strain parameters where IM is the initial modulus, AM is the anelastic modulus, σ_A the anelastic limit, σ_y the macroyield stress, and σ_e the precision elastic limit.

respectively, and V_f the volume fraction of fibers. For stage II, a change of the modulus occurs and a "secondary modulus of elasticity" is defined by

$$E_c(II) = E_f V_f + (\partial\sigma/\partial e)_m (1 - V_f) \tag{2}$$

where $E_c(II)$ is the secondary modulus of elasticity during stage II, and $(\partial\sigma/\partial e)_m$ the slope of the matrix stress–strain curve in the uncombined state at a given strain in stage II.

Stage II behavior is of practical importance since most service conditions lie in this region. If the composite is loaded into stage II and then unloaded, the fibers are in a state of tension while the matrix is in compression. Stage III only exists if the reinforcement can undergo plastic deformation.

A generalized equation for the state of stress in the composite is

$$\sigma_c = \sigma_f V_f + \sigma_m (1 - V_f) \tag{3}$$

where σ_c, σ_m, and σ_f are the respective stresses in the composite, matrix and fiber at equal strains. The stress σ_c can refer to the precision elastic limit, the microyield stress (Fig. 2), the stress for macroyielding, or the tensile stress of the composite.

B. Creep Analysis

Theoretical considerations of creep behavior in composite materials have been formulated by several investigators. The analysis of McDanels *et al.* (1967) incorporates the rule of mixture in order to predict continuous

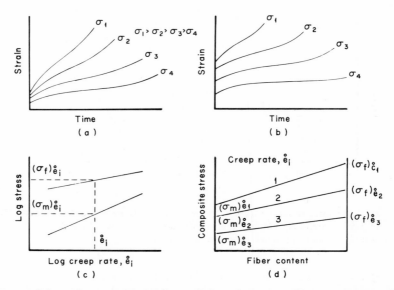

Fig. 3. Schematic representation for prediction of creep rate in continuous fiber metal matrix composites [from McDanels *et al.* (1967)]. (a) Fiber, (b) matrix, (c) log stress–log creep rate plots, (d) composite stress–fiber content plots for several creep rates.

fiber metal matrix composite behavior. The following assumptions are made:

(1) Isostrain conditions exist.
(2) The fibers are aligned parallel to the load axis.
(3) Fiber and matrix are insoluble.
(4) Tensile failure of both fibers and matrix occurs.

Analogous to the micromechanical model of tensile deformation, the creep rate of a composite may be predicted by an exponential form of the rule of mixtures

$$\sigma_c = (\sigma_f)_0 \dot{e}^{\Psi_f} V_f + (\sigma_m)_0 \dot{e}^{\Psi_m} V_m \qquad (4)$$

where c, f, and m refer to the composite, fiber, and matrix, respectively; $\dot{e}$ is the creep rate, σ the stress on one component tested separately, σ_0 the stress required to give a creep rate of 1% per hour, and Ψ the slope of the log stress–log creep rate curve $(\Delta \ln \sigma / \Delta \ln \dot{e})$. A graphical representation of the creep relationships for continuous fiber reinforcement is given in Fig. 3.

C. Compressive Loading

In comprehensive loading, a simple transfer of shear stress across the interface does not occur; rather, this form of loading promotes buckling. The mode of yielding and eventual failure is attributed to buckling of the fiber reinforcement, either "in-phase" (shear mode) or "out-of-phase" (extension mode) as illustrated in Fig. 4.

For a model system of elastic fibers in a perfectly plastic matrix, Dow *et al.* (1966) predict compressive strength as

$$\sigma_c = \{V_f E_f \sigma_y / [3(1 - V_f)]\}^{1/2} \tag{5}$$

where σ_y is the yield strength of the unreinforced matrix. A refinement of the model to take account of work hardening of the matrix was developed

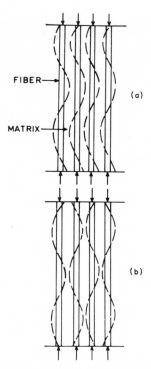

FIBER

MATRIX

(a)

(b)

FIG. 4. Schematic representation of two possible failure modes for continuous fiber metal matrix composites loaded in compression. (a) Shear mode, (b) extension mode.

by Yue *et al.* (1968) the compressive strength is given as

$$\sigma_c = A\left\{\frac{\alpha\theta_m + [(\alpha\theta_m)^2 + \frac{4}{3}V_f(1 - V_f)\tau_0 E_f]^{1/2}}{2(1 - V_f)V_f}\right\} \tag{6}$$

where A is the structure factor varying from 0 to 1 depending on the continuity of the fibers ($A = 1$ for continuous fibers), θ_m the work hardening characteristic, τ_0 the critical shear stress of the matrix, and α the parameter varying from 0 to 1 to account for the amount of plastic deformation.

III. Interfacial Effects on Mechanical Behavior–Structure Relationships

A. Filamentary Composites

Since physical, chemical, and mechanical incompatibilities exist at the interface of filamentary composites, an understanding of the contribution of the interface to composite strength is necessary. Analytic models may simply predict mechanical response, assuming a perfect interface. In

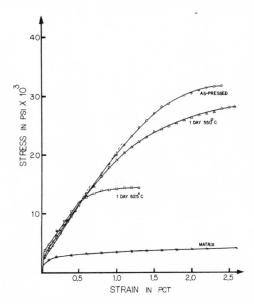

FIG. 5. Representative tensile stress–strain curves for aluminum–stainless steel composites ($V_f = 6.5\%$).

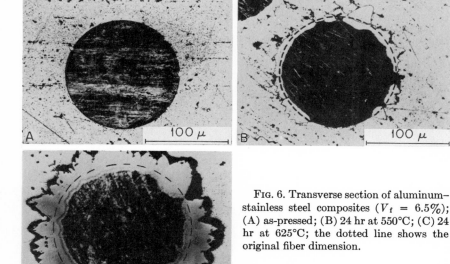

FIG. 6. Transverse section of aluminum–stainless steel composites ($V_f = 6.5\%$); (A) as-pressed; (B) 24 hr at 550°C; (C) 24 hr at 625°C; the dotted line shows the original fiber dimension.

reality, degradation can and often does occur (Metcalfe, 1974, Chapter 3). As a result, the emphasis of the following sections will be directed toward an examination of well-defined filamentary composites. The most detailed work available pertains to aluminum–stainless steel. In addition, boron and tungsten filamentary systems will be considered. When available, a microstructural analysis will serve as a check for the applicability of ideal behavior.

Considering the mechanism of stress transfer in a composite, it is necessary to examine the form of testing (e.g., compression versus tension) as well as the mode of loading. The goal is to assess the efficiency of load transfer and to analyze deviations from rule of mixtures' predictions for tensile, compressive, creep, and fatigue behavior.

1. Tensile Behavior

The bulk mechanical properties of aluminum–stainless steel composites have been examined by a number of investigators (Pinnel and Lawley, 1970, 1971; Baker, 1968; Hancock, 1967; Cratchley, 1963; Gulbransen, 1968; Hancock and Grosskreutz, 1968; Pattnaik, 1972). The role of the interface, as well as substructural properties, as a function of mechanical

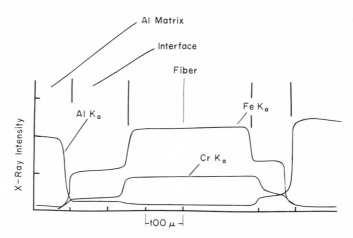

FIG. 7. Microprobe analysis of aluminum–stainless steel composite following degradation at 625°C for 24 hr.

behavior has been considered by Pattnaik (1972). The tensile stress–strain behavior was examined for three conditions: as-pressed, after 24 hr at 550°C, and after 24 hr at 625°C (Fig. 5). With thermal exposure, interfacial growth occurs and an attendant deterioration of mechanical properties results (Figs. 5 and 6). Diffusion produces a brittle intermetallic compound which drastically reduces the load bearing capacity of the interface. Following thermal exposure, microprobe analysis reveals a single phase Fe–Al–Cr ternary intermetallic (Fig. 7) with a discontinuous layer of Kirkendall porosity. In addition, gaps at the interface result from differences in structure and thermal expansion coefficients of the composite phases.

A comparison of rule of mixtures behavior with experimentally-determined composite properties reveals general agreement (Table I). With increased thermal exposure an increase in Young's modulus during stage I occurs, and can be attributed to an increased fiber volume fraction as a result of interfacial growth. With degradation, composite tensile strength will also follow rule of mixtures' behavior providing the strength of the extracted fiber is used (Pattnaik, 1972). Composite deterioration following thermal degradation results from premature cracking of the brittle intermetallic interface reducing its load carrying capacity.

In order to show the validity of the isostrain condition for rule of mixtures behavior, four composite conditions were examined: as-pressed, annealed for 24 hr at 550°C, annealed for 24 hr at 625°C, and an as-pressed pure aluminum "composite" to serve as a standard of reference. Each

TABLE I

TENSILE STRENGTH PROPERTIES OF ALUMINUM–STAINLESS STEEL COMPOSITES[a]

V_{f_0} (%)	V_f (%)	Composite conditions	Mean UTS of extracted fibers (psi)	S^b (psi)	Measured UTS of composites (psi)	UTS of composites ROM[c] (psi)	Young's modulus, Stage I (psi)	Cumulative weakening failure strength (psi)
—	—	As-received fibers	435,000	7000	—	—	—	—
6.5	6.5	As-pressed	455,000	7000	31,620	33,460	11.84×10^6	—
	9.35	Annealed 550°C	230,000	2700	28,000	25,175	11.92×10^6	—
	11.5	Annealed 625°C	76,000 77,200[d]	19,500	14,500	12,310 12,460	12.38×10^6	13,740
19.5	19.6	As-pressed	447,000	8850	88,000	90,810	13.50×10^6	—
	25.5	Annealed 550°C	227,500	6150	69,000	62,000	13.68×10^6	—
	30.2	Annealed 625°C	89,700 102,000[d]	8750	34,000	27,930 30,230	14.22×10^6	33,650

[a] Pattnaik (1972).
[b] Standard deviation.
[c] Rule of Mixtures.
[d] Average UTS.

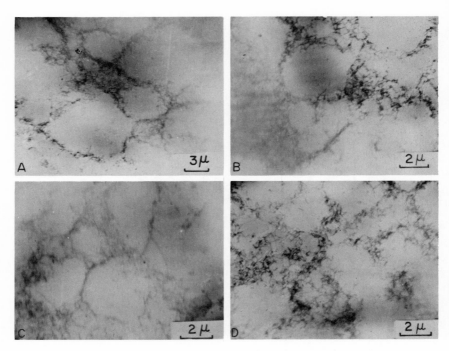

Fig. 8. Deformation matrix substructure as a function of heat treatment, volume fraction, and distance from the interface where $\bar{x}$ is the distance from the interface and l the average cell size. (A) $V_f = 0$, $e \sim 2\%$, $l = 3.28$ μm; (B) as-pressed, $V_f = 6.5\%$, $e \sim 2\%$, $\bar{x} = 90\mu$m, $l = 3.0$ μm; (C) 24 hr at 550°C, $V_f = 6.5\%$, $e \sim 2\%$, $\bar{x}$ 150μm, $l = 2.3\mu$m; (D) 24 hr at 625°C, $V_f = 6.5\%$, $e \sim 1.2\%$, $\bar{x} = 350$ μm, $l = 3.6\mu$m.

composite was strained ($\sim 2\%$) and the dislocation substructure examined. A comparable cell structure develops for the pure aluminum as well as for the reinforced composites (Fig. 8). In addition, the cell structure appears to be independent of volume fraction, verifying the isostrain criterion. A study of the dislocation cell size revealed a substructure relatively independent of distance from the matrix–fiber interface (Fig. 9), indicating a minimum fiber–matrix interaction.

As described earlier, formation of a brittle intermetallic can degrade composite properties. Thermal degradation may be limited by producing a composite with no mutual solubility, or by a coating or plating process which limits interdiffusion between the matrix and fiber. If diffusional degradation would be permitted, interfacial growth of the fiber may result in an assymetric intermetallic as shown earlier for aluminum–stainless steel (Fig. 5). An alternative approach involving the nickel–aluminum system has been suggested by Darroudi et al. (1971) where a complete and

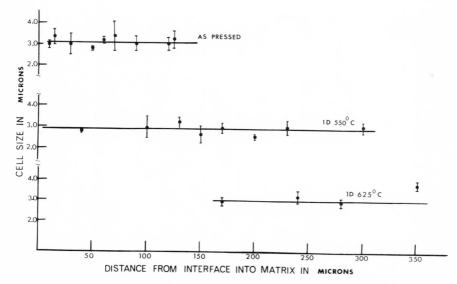

FIG. 9. Dislocation cell size as a function of distance from the matrix–wire interface for as-pressed aluminum–stainless steel composites ($V_f = 6.5\%$).

symmetric reaction of the fiber and matrix occurs. The progressive interfacial growth stages for the nickel–aluminum system are depicted in Fig. 10, revealing uniform interfacial reaction producing a thermodynamically stable composite. In principle, a unidirectional thermodynamically stable, filamentary composite may be produced by solid state diffusion processing.

The study of aluminum–boron and titanium–boron has been rather extensive as a result of its desirable properties and the relative ease of composite fabrication. The tensile behavior of aluminum–Borsic has been well-documented by Kreider *et al.* (1971). In addition, the effect of the interface on tensile behavior has been considered by Metcalfe (1974, Chapter 4), and off-axis tensile strength of filamentary systems has been reviewed by Klein (Metcalfe, 1974, Chapter 5).

Correlation of ideal interface behavior with properties is generally good. Values for the elastic modulus agree with rule of mixtures behavior (Krieder and Marciano, 1969). Composite strengths of aluminum–boron may exceed predicted behavior by 20 to 30%, as a consequence of synergistic effects (Joseph *et al.*, 1968; Stuhrke, 1967). However, the structural analysis has only been considered by a few investigators (Blucher *et al.*, 1970; Swanson and Hancock, 1970; Hancock, 1970). Blucher *et al.* (1970), investigating the interface of aluminum–boron in the as-pressed condition, found no observable interaction at the interface (Fig. 11). Considering aluminum

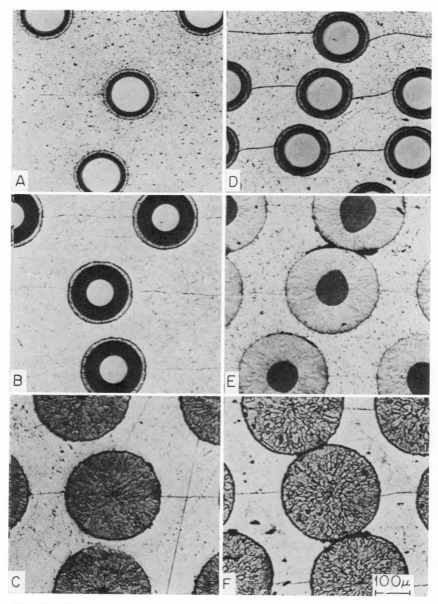

FIG. 10. States of diffusional growth of Al–NiAl₃ composites [from Darroudi *et al.*
(1971)]. Left column: A1–7 vol % Ni; (A) as-pressed, (B) annealed at 600°C for 9×10^3
sec, (C) annealed at 600°C for 2.6×10^5 sec. Right column: A1–13 vol % Ni; (D)
as-pressed, (E) annealed at 600°C for 1.7×10^5 sec, (F) annealed at 600°C for 2.6×10^5
sec.

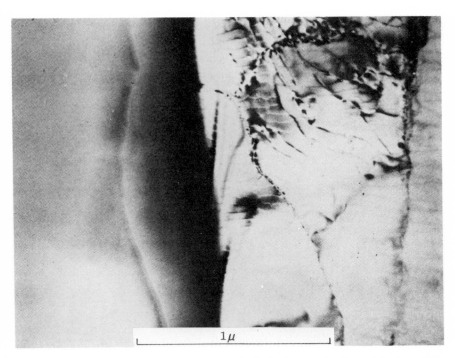

FIG. 11. Transmission electron micrograph of the boron–aluminum interface [from Blucher *et al.* (1970)].

(7075)–boron, Swanson and Hancock (1970) and Hancock (1970) found sound interfaces free of microvoids, although a degree of precipitate segregation occurred at the interface as well as the matrix grain boundaries. Precipitate-free zones, found adjacent to grain boundaries, were not observed at composite interfaces (Swanson and Hancock, 1970). The substructural interface studies of refractory metal and metal carbide systems remain a relatively unexplored area which have not developed as rapidly as the mechanical behavior studies (McDanels *et al.*, 1965, 1967; Ohnysty and Stetson, 1967; Dean, 1967; Petrasek and Weeton, 1964; Petrasek and Signorelli, 1970; Petrasek *et al.*, 1968; Botie *et al.*, 1971).

2. Micromechanical Behavior

An extension of the rule of mixtures for filamentary composites has been considered in the area of micromechanical behavior. The objective of the microstrain approach is a correlation of composite properties with microyielding, complimented by a transmission electron microscopy analysis of

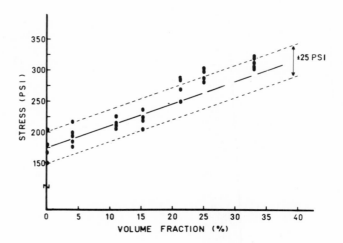

FIG. 12. Precision elastic limit versus volume fraction for aluminum–stainless steel; uniaxial tension [from Pinnel and Lawley (1970)]. Plotted curve represents rule of mixtures behavior as given by Eq. (7).

dislocation behavior (Pinnel and Lawley, 1970; Rutherford, 1966; Hughes and Rutherford, 1969). Briefly, the parameters in composite microstrain behavior may be summarized with the aid of Fig. 2 and defined as follows:

Precision Elastic Limit (σ_e): The stress at which the first deviation in elastic behavior is noted, where the composite stress, σ_c, may be given as

$$\sigma_c = \sigma_e(1 - V_f) + E_f(\sigma_e/E_m)V_f \tag{7}$$

Microyield Stress (σ_{yc}): The first sign of plastic deformation prior to the occurrence of the engineering yield stress can be given by

$$\sigma_{yc} = \sigma_{ym}(1 - V_f) + E_f e_{ym}V_f \tag{8}$$

where σ_{ym} and e_{ym} are the experimentally-determined microyield stress and strain, respectively.

Macroyield stress: At $e_p = 1 \times 10^{-3}$ is defined as

$$\sigma_{0.1c} = \sigma_{0.1m}(1 - V_f) + E_f e_{0.1m}V_f \tag{9}$$

where $\sigma_{0.1m}$, $\sigma_{0.1c}$ and $e_{0.1m}$ are, respectively, the macroyield stress of the matrix and composite and the corresponding strain.

Pinnel and Lawley (1970) examined the micromechanical response of an as-pressed aluminum–stainless steel composite as a function of volume fraction of reinforcement. For tensile loading, the precision elastic limit,

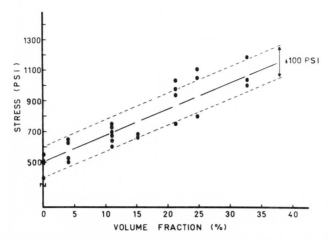

FIG. 13. Microyield stress versus volume fraction for aluminum–stainless steel; uniaxial tension [from Pinnel and Lawley (1970)]. Plotted curve represents rule of mixtures behavior as given by Eq. (8).

the microyield stress, and the 0.1% macroyield stress reveal good agreement with calculated rule of mixtures behavior (Figs. 12, 13, and 14). Structural characterization revealed the dislocation substructure to be independent of volume fraction for a given composite strain, i.e., no significant matrix–fiber interaction, validating rule of mixtures assumptions.

Hughes and Rutherford (1969) and Rutherford (1966) considered a microstrain approach to describe tensile yielding of copper–tungsten. The microyield and macroyield stresses were linearly dependent on the volume fraction of tungsten reinforcement (Fig. 15). In addition, the microyield and frictional stresses were shown to increase with increased prestrain and were qualitatively evaluated in terms of a dislocation model for the copper matrix (Hughes and Rutherford, 1969). A microstrain investigation supplemented by transmission microscopy is particularly useful since it may provide information concerning the role of the interface as a barrier to dislocation motion, or as a source or sink for dislocations.

3. Compressive Behavior

Compressive properties of filamentary composites show marked deviations from rule of mixtures behavior (Pinnel and Lawley, 1970; Krieder and Marciano, 1969), e.g., as-pressed aluminum–stainless steel composites exhibit increases of the elastic limit by a factor of ~2, and by a factor of 5 to 8 for the microyield stress depending on volume fraction. The com-

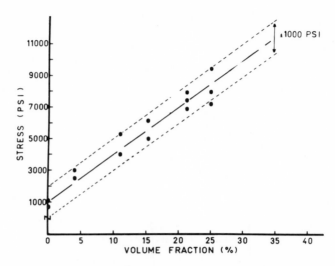

Fig. 14.   Macroyield stress (0.1%) versus volume fraction for aluminum–stainless steel; uniaxial tension [from Pinnel and Lawley (1970)]. Plotted curve represents rule of mixtures behavior as given by Eq. (9).

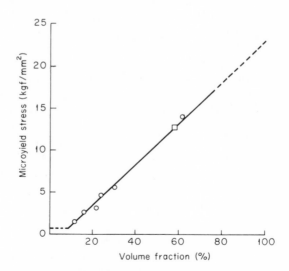

Fig. 15.   Microyield stress versus volume fraction for copper–tungsten filamentary composites [from Hughes and Rutherford (1969)].

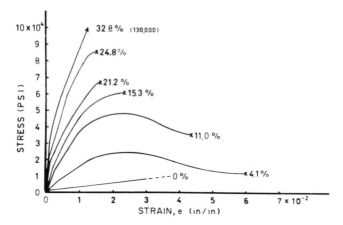

Fig. 16. Compressive stress versus strain behavior of aluminum–stainless steel under uniaxial compression as a function of volume fraction [from Pinnel and Lawley (1970)]. Failure ($\times$) is by buckling in all cases except for pure aluminum.

pressive stress–strain relationships for various volume fractions of aluminum–stainless steel are depicted in Fig. 16. Failure was shown to occur by in-phase or shear buckling (Fig. 4) with no delamination at the interface. Consequently, uniaxial compression of composites has been analyzed theoretically to consider the buckling of fibers surrounded by a restraining medium (Dow *et al.*, 1966; Yue *et al.*, 1968; Rosen, 1965; Schverek, 1965; Lager and June, 1969; Chung and Testa, 1969). The formulation by Dow *et al.* (1966), when modified to incorporate work hardening, provides a good correlation with observed compressive behavior of aluminum–stainless steel (Pinnel and Lawley, 1970; Dow *et al.*, 1966).

Krieder and Marciano, (1969), in the study of tensile and compressive strength of aluminum–Borsic composites found a significant variation of composite strength as a function of loading. At a volume fraction of 50%, the ultimate tensile strength was 160,000 psi while compressive strength reached 297,000 psi (Krieder and Marciano, 1969). The compressive load is primarily supported by the reinforcing fibers as compared with a shear transfer across the interface as in the case of tension. As a result, a composite failure in uniaxial compression constitutes a form of buckling test.

4. Creep Behavior

Creep and stress rupture behavior of filamentary composites has been documented for a number of systems including aluminum–stainless steel

(Pinnel and Lawley, 1970), aluminum–boron (Breinan and Kreider, 1970; Antony and Chang, 1968), magnesium alloy–stainless steel (Wilcox and Clauer, 1969), silver–tungsten (Kelly and Tyson, 1966), and nickel alloy–tungsten composites (Petrasek and Signorelli, 1970; Petrasek et al., 1968). Pinnel and Lawley (1971) investigated ambient temperature tensile creep behavior and associated substructural detail for aluminum–stainless steel for volume fractions in the range 0 to 0.33. Reinforcement was found to markedly increase the creep resistance in agreement with the behavior predicted by McDanels et al. (1967) (Eq. 4). Both matrix and fibers gave an exponential dependence of creep rate on stress with power exponents of 2.7 and 3.3, respectively. From a characterization of the dislocation substructure, it was observed that the density and cellular configurations are uniform throughout the matrix. Similar behavior was previously shown for uniaxial tensile (Pinnel and Lawley, 1970; Pattnaik, 1972) and compressive (Pinnel and Lawley, 1970) loading, where the matrix and fiber experience the same level of deformation.

Wilcox and Clauer (1969), considering a magnesium alloy–stainless steel composite, observed that steady state creep was established in the unreinforced matrix, while logarithmic creep was observed for only the wire and the reinforced matrix. They concluded that the fibers were rate-controlling, in agreement with the analysis of de Silva (1968) and McDanels et al. (1967). Studies in aluminum–boron composites (Antony and Chang, 1968) concurred that the fibers were rate-controlling (de Silva, 1968) and the rule of mixtures analysis of McDanels et al. (1967) was applicable.

5. Fatigue Behavior

The analysis of composite fatigue is dependent to a large extent on the mode of fatigue (Baker, 1968), as well as the composite system and geometry. An examination of tungsten- and steel-reinforced silver composites, for tension–tension fatigue (Morris and Steigerwald, 1967), revealed marked improvements in fatigue strength with increased fiber reinforcement. Furthermore, the silver–tungsten system revealed no surface cracking while the silver–steel composite did. Forsyth et al. (1964) also noted increased fatigue life with additions of steel and tungsten fibers in an aluminum matrix.

Examination of the aluminum–boron system by Hancock (1970), indicates that fatigue cracks initiated at free surfaces with branching occurring at the filament matrix interface. Krieder et al. (1971), considering aluminum–Borsic, demonstrated the ability of the ductile matrix to isolate and blunt crack tips. At the current state of knowledge of composite fatigue, no unified theory exists to provide firm design guidelines for service involving cyclic loading. As a consequence, each composite system must be treated uniquely until an improved understanding of the role of the interface is acquired.

B. Directionally Solidified Composites

A broad spectrum of eutectic alloys have now been examined which exhibit composite behavior. Although diverse structural morphologies are observed, most can be considered as either an aligned rod-like (or whisker) reinforcement in the matrix or an aligned lamellar structure (Fig. 17). The structural form chosen is primarily a function of the volume fraction of second phase which in turn affects the total interfacial area per unit volume. In general, eutectics with more than ~30% second phase exhibit a lamellar morphology (Hunt and Chilton, 1962–1963; Cooksey, 1964). In considering the role of the interface in the deformation of eutectic composites, it is necessary to first examine the crystallographic properties, structure, and stability of such interfaces.

1. Interface and Interphase Crystallography

In lamellar eutectic microstructures there is a tendency toward a preferred relative crystal orientation between the two phases. A complete specification includes the lamellar habit plane, the crystallographic planes in contact at the interface, and mutually parallel directions in these planes. The growth direction is a further variable; during steady state growth, the lamellae generally lie normal to the solid–liquid interface so that the lamellar interface necessarily contains the growth direction.

Fiber or rod-like eutectic composites usually grow with a preferred crystallographic direction parallel to the fiber axis. To date, most of the rod-like reinforcements are faceted in cross section with facets parallel in adjacent fibers. The "absence" of faceting is probably a false conclusion in that high resolution metallography frequently shows that the cross section is not perfectly "round."

Examples of the microstructure and associated crystallography of selected eutectic composites are summarized in Table II. These directionally-solidified composites represent materials having potential in structural applications, systems for which crystallographic relationships are clearly-defined, or those eutectic composites in which the structure–mechanical property relationship has been examined.

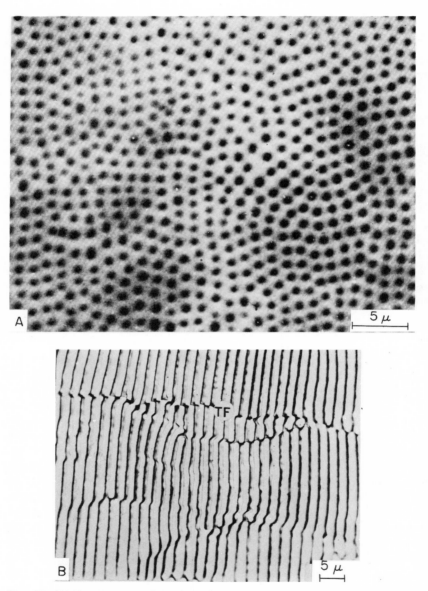

FIG. 17. (A) Transverse section of rod-like eutectic composite Al–Al₃Ni [from Smartt
et al. (1971)]. (B) Transverse section of lamellar Al–CuAl₂ eutectic composite showing
a transverse fault (TF).

TABLE II

THE MICROSTRUCTURE AND CRYSTALLOGRAPHY OF
DIRECTIONALLY-SOLIDIFIED COMPOSITES[a]

Composite	Type of structure	Crystallographic relationships
Ag–Cu	Lamellar	Interface $\|$ $(211)_{Ag}$ $\|$ $(211)_{Cu}$ Growth direction $\|$ $[110]_{Ag}$ $\|$ $[110]_{Cu}$
Al–CuAl$_2$	Lamellar	Interface $\|$ $\{111\}_{Al}$ $\|$ $\{211\}_{CuAl_2}$ $<110>_{Al}$ $\|$ $<120>_{CuAl_2}$
Al–Al$_3$Ni	Rods	Growth direction $\|$ $<010>_{Al_3Ni}$ $\|$ $<110>_{Al}$
Cd–Zn	Lamellar	$(0001)_{Cd}$ $\|$ $(0001)_{Zn}$ $[01\bar{1}0]_{Cd}$ $\|$ $[0110]_{Zn}$ Growth direction $\|$ $[11\bar{2}0]$
Co–CoAl	Lamellar/rods	Growth direction $\|$ $[10\bar{1}]_{CoAl}$ $\|$ $[112]_{Co}$ Interface $\|$ $(101)_{CoAl}$ $\|$ $(111)_{Co}$ $[010]_{CoAl}$ $\|$ $[110]_{Co}$
Fe–Fe$_2$B	Square rods	Not known
Ni–Ni$_3$Nb	Lamellar	Interface $\|$ $(\bar{1}11)_{Ni}$ $\|$ $(010)_{Ni_3Cb}$ Growth direction $\|$ $[110]_{Ni}$ $\|$ $[100]_{Ni_3Nb}$
NiAl–Cr	Rods/lamellar	Interface $\|$ $\{112\}_{NiAl}$ $\|$ $\{112\}_{Cr}$ Growth direction $\|$ $<111>_{NiAl}$ $\|$ $<111>_{Cr}$

[a] Adapted from Hogan *et al.* (1971).

2. Interface Structure

Boundaries between phases in eutectic composites are usually semi-coherent. The mismatch in lattice parameter across the interface can be accommodated by interface dislocation networks. Interface dislocations have been observed in Ni–Cr (Kossowsky, 1970), Cr–NiAl (Walter *et al.*, 1969; Cline *et al.*, 1970), Al–CuAl$_2$ (Pattnaik, 1972; Weatherly, 1968; Davies and Hellawell, 1969; Pattnaik and Lawley, 1971), Al–Al$_2$Ni (Breinan *et al.*, 1972) and the ternary system Al–Cu–Mg (Garmong and Rhodes, 1972). Since glide dislocations can interact with interfacial dislocation networks, the latter will play an important part in plastic deformation; this form of interaction is discussed in a subsequent section.

In Cr–NiAl, Cline *et al.* (1971) and Walter *et al.* (1969) identified dislocation networks composed primarily of a $\langle 100 \rangle$ dislocations at the interface between Cr-rich rods and the NiAl matrix. From the dislocation spac-

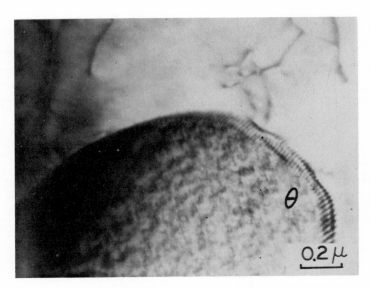

FIG. 18. Interface dislocations at a terminal fault in Al–CuAl₂ (θ).

ing, an interface network energy of ~ 140 ergs cm^{-2} was calculated. Weatherly (1968) and Pattnaik and Lawley (1971) observe a high density of dislocations running parallel to the growth direction at lamellar fault lines in Al–CuAl₂ (Fig. 18). The spacing between these dislocations (~ 220 Å) is of the correct order to relieve misfit between $\{111\}_{Al}$ and $\{211\}_{CuAl_2}$ planes; the energy of the dislocation array is ~ 50 ergs cm^{-2}. In a more detailed study of this system, Davies and Hellawell (1969) examined the dislocation structure at terminal lamellae and at the normal interfaces between phases. All the dislocations present were not identified unambiguously; observed spacings (Weatherly, 1968; Davies and Hellawell, 1969) are consistent with either $a/3 \langle 111 \rangle$ or $a/2 \langle 110 \rangle$.

The ternary eutectic in the Al–Cu–Mg system consists of alternate lamellae of aluminum solid solution and CuMgAl₂ interspersed with rods of CuAl₂. Interfaces between the aluminum-rich and the CuMgAl₂ phases are semicoherent (Garmong and Rhodes, 1972). $a/2 \langle 110 \rangle$ misfit dislocations are imaged only in the aluminum-rich-phase at the interface, with a spacing ~ 300 Å. In this ternary system, and also in the Cr–NiAl eutectic (Cline et al., 1971), it is possible that the misfit dislocations are in fact glide dislocations in the matrix phase. If this is true, then a continuous readjustment of the interfacial dislocation configuration can occur; the interface can then act as a source or sink for matrix glide dislocations.

3. Interface Stability

When Kraft *et al.* (1963) originally reported on the high thermal stability of lamellar Al–CuAl$_2$, it was termed "anomalous." Since 1963, studies of many eutectic composites have served to verify the generality of this phenomenon; recent examples include: Al–Al$_3$Ni (Bayles *et al.*, 1967; Salkind *et al.*, 1967, 1969; Marich, 1970; Smartt *et al.*, 1971; Jaffrey and Chadwick, 1970), Ni–Ni$_3$Nb (Quinn *et al.*, 1969), Ni–Ni$_3$Ti (Sheffler *et al.*, 1969), Co–CoAl (Cline, 1967), Cd–Zn (Soutiere and Kerr, 1969), Nb–Nb$_2$C (Lemkey and Salkind, 1967), and Fe–Cr–Nb (Jaffrey and Marich, 1972). By thermal stability is meant structural invariance at temperatures up to $\sim 0.9T_m$ for times of several hundred hours.

Stability is a consequence of the chemical equilibrium existing between the phases of the composite up to the melting point; the only exceptions to this are phase changes below the eutectic temperature or small changes in solid solubility over a temperature range. However, eutectic composites are characterized by a large total interfacial area. The total interfacial energy per unit volume can be reduced by coarsening of the structure, by the growth of existing lower-energy interfaces at the expense of existing higher-energy interfaces, or by the formation of new lower-energy interfaces at the expense of existing interfaces (Bayles *et al.*, 1967). This is the basis of the physiochemical instability in eutectic composites that becomes apparent at temperatures $\gtrsim 0.9T_m$.

Detailed observations on Al–CuAl$_2$ show that the lamellar structure coarsens above $0.9T_m$ but that the preferred crystallography is maintained (Pattnaik, 1972; Gangloff *et al.*, 1972; Graham and Kraft (1966). Coarsening involves the recession of θ platelets from faulted regions with an associated thickening of adjacent platelets. Three primary conclusions are drawn from these observations namely:

(1) Thermal stability is dependent on fault density since coarsening is initiated at faulted regions in the lamellar structure.

(2) Coarsening rates increase with decreasing interlamellar spacing over the range $\lesssim 2\mu$ to $\sim 12\mu$.

(3) Prolonged high temperature exposure leads to faceting of the CuAl$_2$ phase (Fig. 19), so that the interfaces in the directionally-solidified composite are not necessarily those of minimum energy.

Where data are available on other lamellar structures, these conclusions appear to have general validity. The rod-like Al–Al$_3$Ni structure behaves in a similar manner to Al–CuAl$_2$ in that coarsening and faceting of the Al$_3$Ni occurs at temperatures $\gtrsim 0.9T_m$ (Bayles *et al.*, 1967; Breinan *et al.*, 1972). Cold working (Salkind *et al.*, 1969) or the presence of inclusions (Gangloff

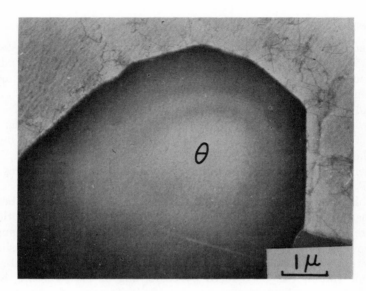

FIG. 19.   Faceting of CuAl$_2$ (θ) after elevated temperature exposure at 530°C for 100 hr.

et al., 1972) enhances coarsening in Al–Al$_3$Ni since either effect perturbs the low-energy interface structure and produces faulting.

4. Interface and Mechanical Behavior

(a) BOND INTREGRITY. Integrity of the bond in directionally-solidified composites is confirmed from mechanical property–microstructure correlations in several lamellar and rod-like structures. The most common form of loading has been tension or compression with less information available on creep and fatigue response. A number of examples will serve to illustrate bond integrity; from these studies it is clear that the sole function of the interface is to act as the vehicle for load transfer from matrix to reinforcement. In composites of small interlamellar or rod-spacing ($\lesssim 5\mu$), other effects due to the presence of the interface can become significant. Mechanical constraints and interaction between slip dislocations and dislocation networks at the semicoherent interfaces are discussed in Section b.

In Al–CuAl$_2$ lamellar composites with interlamellar spacings $\gtrsim 3\mu$, excellent interfacial bond strengths exist (Yue *et al.*, 1968; Pattnaik and Lawley, 1971; Crossman *et al.*, 1969; Lawson and Kerr, 1971). From fracture morphology following tensile or compressive failure, it is concluded that the strength of the Al–CuAl$_2$ bond is equal to or greater than the transverse strength of the CuAl$_2$ platelets. Tensile strength can be predicted from the rule of mixtures provided the *"in situ"* matrix strength and an "effective" volume fraction for the reinforcing phase are used (Pattnaik, 1972; Crossman *et al.*, 1969). The latter is necessary since premature cracking of the CuAl$_2$ platelets occurs in Stage II which lowers the volume fraction of reinforcement through stress concentrations at points of breakage.

Similar conclusions are drawn from tensile (Bayles *et al.*, 1967; Hertzberg *et al.*, 1965) or constant load creep studies (Breinan *et al.*, 1972) on the rod-like Al–Al$_3$Ni composite. The rod–matrix interface bond formed during directional solidification allows for efficient load transfer from matrix to reinforcing phase. As in Al–CuAl$_2$, the rule of mixtures can be used for a prediction of strength if account is taken of the premature failure of the rod-like Al$_3$Ni whiskers (Hertzberg *et al.*, 1965). Typically, this directionally-solidified eutectic composite exhibits a tensile strength at least three times higher than that of the same alloy composition in the as-cast condition.

Cyclic loading would be expected to provide a particularly stringent test of interface strength. Although only limited data are available, the evidence clearly supports bond integrity. In Al–Al$_3$Ni (Salkind *et al.*, 1966), fatigue failure occurs by propagation of one or more cracks along zones of slip in the matrix, with no major delamination. The fatigue behavior of lamellar Ni–Ni$_3$Nb is similar to that of Al–Al$_3$Ni, even though the two eutectic composites are quite different (Hoover and Hertzberg, 1971). In both alloys, high stress–low cycle life is controlled by the fracture resistance of the reinforcing phase whereas fatigue crack propagation in the matrix phase is responsible for low stress–high cycle fatigue resistance. Delamination at interfaces is not a controlling mechanism.

At ambient temperature and at 500°C, the fatigue strength of the rod-like Fe–Fe$_2$B eutectic composite is greater than that of pure iron (de Silva and Chadwick, 1970). Fatigue cracks are initiated at the specimen surface in the matrix, not at matrix–fiber interfaces. de Silva and Chadwick (1970) consider possible dislocation mechanisms in eutectic composites subjected to cyclic loading. If matrix dislocations can glide across the interface and through the reinforcing phase, the reinforcement would be cut into shorter lengths. Alternatively, the pile-up of matrix glide dislocations at the fiber–matrix interface could lead to stress concentrations and fiber breakage. Either mechanism produces a free path for further deformation of the matrix phase. Since neither mechanism operates in Fe–Fe$_2$B the eutectic behaves like a conventional fiber-reinforced metal; compatible deformation must occur between fiber and matrix. This is not the case in the cyclic deformation of Al–Al$_3$Ni or Ni–Ni$_3$Nb.

(b) INTERFACE CONSTRAINTS AND SUBSTRUCTURE IN PLASTIC DEFORMATION. It is a general observation that yield or flow stress increases with decreasing interlamellar spacing λ for values of $\lambda \lesssim 5\mu$. Micromechanical models based on the concept of the rule of mixtures do not account for this effect. In a number of lamellar structures, the stress level varies as $\lambda^{-1/2}$; eutectic alloys conforming to this relationship include Cd–Zn (Shaw, 1967), Ag–Cu, (Cline and Stein, 1969; Cline and Lee, 1970), and NiAl–Cr (Walter and Cline, 1970). Analysis of this phenomenon in terms of dislocation theory provides some interesting possibilities.

Shaw (1967) rationalizes the $\lambda^{-1/2}$ dependence of yielding in terms of a dislocation pile-up model. Qualitatively yielding requires that a slip system break through the lamellar interface. Thereafter, either the interface offers much less resistance to slip or dislocation multiplication occurs with an accompanying yield drop. Cline and Stein (1969), Cline and Lee (1970), and Walter and Cline (1970) propose that lamellar interfaces act as barriers in one of the following ways:

(1) An image force exists by virtue of the shear modulus difference ΔG between adjacent phases. It is calculated that

$$\tau_{\max} = \Delta G/(8\pi) \tag{10}$$

(2) Interaction takes place between slip dislocations and the interface dislocations, in which case

$$\tau_{\max} = Gb/(2\pi D) \tag{11}$$

where D is the spacing of interface dislocations and b is the Burgers vector.

The magnitude of this effect [mechanisms (1) or (2)] is about 100,000 psi in Ag–Cu. Kim and Stoloff (1971) find a strong dependence of flow stress on ordering of the Ag$_3$Mg phase in fine lamellar Ag$_3$Mg–AgMg eutectic composites. It is suggested that ordering increases the resistance to slip across the lamellar interface.

Direct evidence for glide dislocation–interface dislocation network interaction is difficult to obtain. However, the recent observations by Pattnaik and Lawley (1971) on Al–CuAl$_2$ of fine lamellar spacing ($\lesssim 2\mu$) are convincing. Although the deformation substructure in the aluminum-rich phase is complex, the average dislocation density adjacent to the interface is higher than in the central regions of the aluminum-rich lamellae (Fig. 20). In these fine lamellar structures, the in situ yield strength of the aluminum-rich phase is approximately three times higher than in the un-

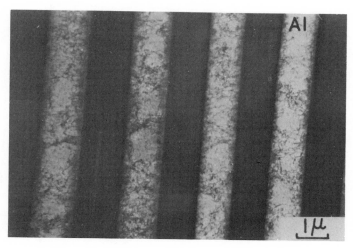

FIG. 20. Deformation substructure in lamellae of Al–CuAl₂ (longitudinal section) [from Pattnaik and Lawley (1971)].

constrained condition. Matrix dislocation–interface dislocation interactions have recently been confirmed in the creep deformation of Al–Al₃Ni (Breinan *et al.*, 1972). Due to this interaction, regions of heavy deformation tend to form as networks or cells in which the cell walls are keyed to the Al₃Ni whiskers.

In the Fe–Fe₂B fibrous eutectic, de Silva and Chadwick (1969) observe enhanced strength characteristics of the matrix phase at fiber spacings ∼4μ. This is the result of a combined effect of the closeness of the fibers and a condition of strain compatibility at the matrix–fiber interface. When the matrix is deforming plastically, flow is more restricted in the vicinity of the interface than in interfiber regions. de Silva and Chadwick make an analogy with the hydrodynamic boundary layer in laminar fluid flow.

Lawson and Kerr (1971) attribute high matrix work-hardening rates in fine-scale Al–Al₂Cu and Al–Al₃Ni to elastic constraint effects of the matrix aluminum phase by the reinforcing phases. The origin of the effect lies in the development of transverse stresses due to differences in Poisson's ratio of the two phases (Lilholt and Kelly, 1969).

(c) ELEVATED TEMPERATURE STRENGTH. In light of the structural stability of directionally-solidified eutectic composites it is expected that strength levels will be maintained at temperatures close to the eutectic melting point. Although the high temperature stability of microstructure has been verified in a number of lamellar and rod-like composites, only a

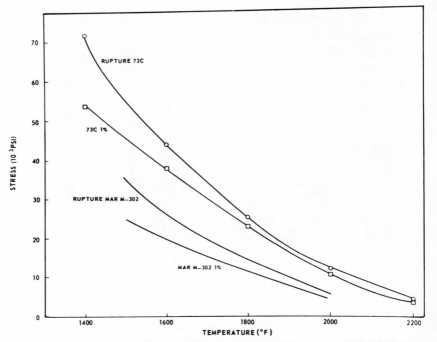

FIG. 21. A comparison of stress and creep rupture (1% creep in 100 hr) behavior for a carbide-reinforced Co–Cr eutectic alloy (73C) and a cobalt-base superalloy (Mar M-302).

limited number of studies have been concerned with the associated strength, i.e., creep, stress rupture, and creep rupture.

Ni–Ni₃Nb exhibits a higher tensile strength than the as-cast alloy over the temperature range 25–1000°C (Quinn *et al.*, 1969). Above ∼1000°C, the composite matches the performance of the best available superalloys (Annarumma and Turpin, 1972). In Nb–Nb₂C, both structure and strength are retained at temperatures close to the eutectic point (Salkind *et al.*, 1970).

Al–Al₃Ni exhibits excellent thermal stability up to ∼97% of the eutectic temperature with no reduction in ambient temperature strength (Bayles *et al.*, 1967). Creep resistance, i.e., 100 hr at 90% of the eutectic temperature is also not impaired (Salkind *et al.*, 1967). Creep resistance is found to increase with decreasing whisker (rod) spacing at $T/T_m \lesssim 0.6$ (Breinan *et al.*, 1972). However, the creep properties are extremely sensitive to structural imperfections; a "dual microstructure" which leads to whisker misalignment drastically reduces creep resistance at these temperatures (Breinan *et al.*, 1972).

The creep studies of Thompson *et al.* (1970) on a carbide (rod)-reinforced Co–Cr eutectic alloy attest to the integrity and high temperature stability of the interface. Creep rupture and stress rupture data are illustrated in Fig. 21. Up to 2000°F the eutectic microstructure is essentially stable, and the material more creep resistant than the conventional cobalt-base super-alloy Mar M-302. The activation energy for creep suggests that the reinforcing carbide phase is the controlling factor, again a reflection on the effectiveness of load transfer across the interface.

IV. Summary

To assess the influence of the interface or interfacial region on mechanical behavior, analytic and experimental data on metal matrix composites have been reviewed. In terms of the application of composites as structural materials, the properties of primary importance are: elastic modulus, microstrain behavior, macroyielding, ultimate strength, and creep and fatigue behavior. The interface is best characterized by its structure, stability, and bond integrity. A simple rule of mixtures approach is useful in the assessment of bond integrity and efficiency of load transfer, provided all the assumptions and limitations of the analysis are taken into consideration.

From a consideration of the behavior of filamentary and directionally-solidified composites, the following conclusions are drawn:

(1) The mechanical response of the interface is a function of the mode of loading since the latter affects the manner in which stress is transferred at the interface.

(2) Characterization of interface and matrix substructure provides a means for evaluating matrix–fiber interactions and possible deviations from isostrain conditions.

(3) Apart from being the vehicle for stress transfer, the interface can give rise to additional strengthening (true synergism) in fine-scale composite microstructures. The origin of this effect may reside in glide dislocation–interface dislocation interactions, image forces associated with the interface, or mechanical constraints on matrix deformation.

(4) The retention of strength at high temperature ($T/T_m \gtrsim 0.9$) in directionally-solidified composites is a result of the interface stability.

(5) Although mechanical property data are available for a spectrum of metal matrix composites, detailed microstructural studies, particularly of the interfacial region are restricted largely to aluminum–stainless steel, aluminum–boron, aluminum–CuAl₂, and aluminum–Al₃Ni.

Acknowledgments

The authors wish to cite the contributions made by A. Pattnaik and M. R. Pinnel, formerly graduate assistants in the Materials Engineering program at Drexel University. Support by the Office of Naval Research, Arlington, Virginia, and The Pennsylvania Science and Engineering Foundation, Harrisburg, Pennsylvania, of the studies on composite materials is gratefully acknowledged.

References

Annarumma, P., and Turpin, M. (1972). *Met. Trans.* **3,** 137.

Antony, K. C., and Chang, W. H. (1968). *Trans. ASM* **61,** 550.

Baker, A. A. (1968). *J. Mater. Sci.* **3,** 412.

Bayles, B. J., Ford, J. A., and Salkind, M. J. (1967). *Trans. TMS-AIME* **239,** 844.

Blucher, I. D., Spencer, W. R., and Stuhrke, W. F. (1970). Transmission and scanning electron microscopy of boron/aluminum interfaces. *Refractory Composites Working Group, 17th, Williamsburg, Virginia, June 1970.*

Breinan, E. M., and Kreider, K. G. (1970). *Met. Trans.* **1,** 93.

Breinan, E. M., Thompson, E. R., McCarthy, G. P., and Herman, W. J. (1972). *Met. Trans.* **3,** 221.

Breinan, E. M., Thompson, E. R., and Tice, W. K. (1972). *Met. Trans.* **3,** 211.

Botie, B., Spencer, W. R., and Stuhrke, W. F. (1971). AFML-TR-71-134, July.

Chung, W., and Testa, R. B. (1969). *J. Compos. Mater.* **3,** 58.

Cline, H. E. (1967). *Trans. TMS-AIME* **239,** 1906.

Cline, H. E., and Lee, D. (1970). *Acta Met.* **18,** 315.

Cline, H. E., and Stein, D. F. (1969). *Trans. TMS-AIME* **245,** 84.

Cline, H. E., Walter, J. L., Koch, E. F., and Osika, L. M. (1971). *Acta Met.* **19,** 405.

Cratchley, D. (1963). *Powder Met.* No. 11, 59.

Cratchley, D. (1965). *Met. Rev.* **10,** 79.

Cooksey, D. (1964). *Phil. Mag.* **10,** 745.

Crossman, F. W., Yue, A. S., and Vidoz, A. E. (1969). *Trans. TMS-AIME* **245,** 397.

Darroudi, T., Vedula, K. M., and Heckel, R. W. (1971). *Met. Trans.* **2,** 325.

Davies, I. G., and Hellawell, A. (1969). *Phil. Mag.* **19,** 1285.

Dean, A. V. (1967). *J. Inst. Metals* **95,** 79.

de Silva, A. R. T. (1968). *J. Mech. Phys. Solids* **16,** 169.

de Silva, A. R. T., and Chadwick, G. A. (1969). *Metal Sci. J.* **3,** 168.

de Silva, A. R. T., and Chadwick, G. A. (1970). *Metal Sci. J.* **4,** 63.

Dow, N. F., Rosen, B. W., and Hashin, Z. (1966). NASA-CR-492, June.

Forsyth, P. J. E., George, R. W., and Ryder, D. A. (1964). *Appl. Mater. Res.* **3,** 223.
Gangloff, R. P., Kraft, R. W., and Wood, J. D. (1972). *Met. Trans.* **3,** 348.
Garmond, G., and Rhodes, C. G. (1972). *Met. Trans.* **3,** 533.
Graham, L. D., and Kraft, R. W. (1966). *Trans. TMS-AIME* **236,** 94.
Gulbransen, L. (1968). *Trans. ASME* **90,** 292.
Hancock, J. R. (1967). *J. Compos. Mat.* **1,** 136.
Hancock, J. R. (1970). NR-031-743 Tech. Rep. No. 2, Sept.
Hancock, J. R., and Grosskreutz, J. C. (1968). "Metal Matrix Composites." ASTM-STP 438, p. 134. Amer. Soc. Test. Mater. Philadelphia, Pennsylvania.
Hertzberg, R. W., Lemkey, F. D., and Ford, J. A. (1965). *Trans. TMS-AIME* **233,** 342.
Hill, R. (1965). *J. Mech. Phys. Solids* **13,** 189.
Hogan, L. M., Kraft, R. W., and Lemkey, F. D. (1971). *Advan. Mater. Res.* **5,** 83.
Hoover, W. R., and Hertzberg, R. W. (1971). *Met. Trans.* **2,** 1289.
Hughes, E. J., and Rutherford, J. L. (1969). "Composite Materials Testing and Design," ASTM STP 440, p. 562. Amer. Soc. Test. Mater. Philadelphia, Pennsylvania.
Hunt, J., and Chilton, J. (1962–1963). *J. Inst. Metals* **91,** 338.
Jaffrey, D., and Chadwick, G. A. (1970). *Met. Trans.* **1,** 3389.
Jaffrey, D., and Marich, S. (1972). *Met. Trans.* **3,** 551.
Joseph, E., Meyers, E. J., and Stuhrke, W. F. (1968). *J. Compos. Mater.* **2,** 56.
Kelly, A., and Davies, G. J. (1965). *Met. Rev.* **10,** 1.
Kelly, A., and Tyson, W. R. (1966). *J. Mech. Phys. Solids* **14,** 177.
Kim, Y. G., and Stoloff, N. S. (1971). "Ordering and Strength of an Aligned Ag₃Mg-AgMg Eutectic." Rensselaer Polytechnic Inst., ONR Tech. Rep. 1, Oct.
Kossowsky, R. (1970). *Met. Trans.* **1,** 1909.
Kraft, R. W., Albright, D. L., and Ford, J. A. (1963). *Trans. TMS-AIME* **227,** 540.
Kreider, K., and Marciano, M. (1969). *Trans. TMS-AIME* **245,** 1279.
Kreider, K. G., Dardi, L., and Prewo, K. (1971). AFML-TR-71-204, Dec.
Lager, J. R., and June, R. R. (1969). *J. Compos. Mater.* **3,** 48.
Lawson, W. H. S., and Kerr, H. W. (1971). *Met. Trans.* **2,** 2853.
Lemkey, F., and Salkind, M. (1967). *In* "Crystal Growth," p. 71. Pergamon, Oxford.
Lilholt, H., and Kelly, A. (1969). Quantitative relation between properties and microstructure, *Conf. Proc., Haifa, Israel, 1969.*
Marich, S. (1970). *Met. Trans.* **1,** 2953.
McDanels, D. L., Jech, R. W., and Weeton, J. W. (1965). *Trans. TMS-AIME* **233,** 636.
McDanels, D. L., Signorelli, R. A., and Weeton, J. W. (1967). ASTM Spec. Tech. Publ. 427, p. 124. Amer. Soc. Test. Mater., Philadelphia, Pennsylvania.
McDanels, D. L., Signorelli, R. A., and Weeton, J. W. (1967). NASA-TN-D-4173, Sept.
Metcalfe, A. G., ed. (1974). "Interfaces in Metal-Matrix Composites," Vol. 1, *in* "Composite Materials" (L. J. Broutman and R. H. Krock, eds.). Academic Press, New York.
Morris, A. W. H., and Steigerwald, E. A. (1967). *Trans. TMS-AIME* **239,** 730.
Ohnysty, B., and Stetson, A. R. (1967). AFML-TR-66-156, part II, Dec.
Pattnaik, A. (1972). Ph.D. Thesis, Drexel Univ. Philadelphia, Pennsylvania.
Pattnaik, A., and Lawley, A. (1971). *Met. Trans.* **2,** 1529.
Petrasek, D. W., and Signorelli, R. A. (1970). NASA TN D-5575, Feb.
Petrasek, D. W., and Weeton, J. W. (1964). *Trans. TMS-AIME* **230,** 977.
Petrasek, D. W., Signorelli, R. A., and Weeton, J. W. (1968). NASA TN D-4787, Sept.
Pinnel, M. R., and Lawley, A. (1970). *Met. Trans.* **1,** 1137.

Pinnel, M. R., and Lawley, A. (1971). *Met. Trans.* **2,** 1415.

Quinn, R. T., Kraft, R. W., and Hertzberg, R. W. (1969). *Trans. ASM* **62,** 38.

Rosen, B. W. (1965). *Fiber Compos. Mater. Pap. Semin ASM, Philadelphia, Pennsylvania, October 1964,* p. 37. Amer. Soc. Metals, Metals Park, Ohio.

Rutherford, J. L. (1966). "A Microstrain Analysis of Fiber Reinforced Composites." Res. Rep. 66/RC/3, General Precision Aerospace, July.

Salkind, M. J. (1969). "Interfaces in Composites," ASTM Spec. Tech. Publ. 452, p. 149. Amer. Soc. Test. Mater., Philadelphia, Pennsylvania.

Salkind, M. J., George, F. D., Lemkey, F. D., and Bayles, B. J. (1966). United Aircraft Corp. Final Rep., Contract 65-0384-01.

Salkind, M., Leverant, G., and George, F. (1967). *J. Inst. Metals* **95,** 349.

Salkind, M., George, F., and Tice, W. (1969). *Trans. TMS-AIME* **245,** 2339.

Salkind, M. J., Lemkey, F. D., and George, F. D. (1970). "Whisker Technology" (A. P. Levitt, ed.), p. 343. Wiley (Interscience), New York.

Schverek, H. (1965). NASA-CR-202, April.

Shaw, B. J. (1967). *Acta Met.* **15,** 1169.

Sheffler, K. D., Hertzberg, R. W., and Kraft, R. W. (1969). *Trans. ASM* **62,** 105.

Smartt, H. B., Tu, L. K., and Courtney, T. H. (1971). *Met. Trans.* **2,** 2717.

Soutiere, B., and Kerr, H. W. (1969). *Trans. TMS-AIME* **245,** 2595.

Stuhrke, W. F. (1967). *In* "Advances in Structural Composites." Soc. Aerosp. Mater. Process Eng.

Swanson, G. D., and Hancock, J. R. (1970). NR-031-743, Tech. Rep. No. 2, Sept.

Thompson, E. R., Koss, D. A., and Chestnutt, J. C., (1970). *Met. Trans.* **1,** 2807.

Walter, J. L., and Cline, H. E. (1970). *Met. Trans.* **1,** 1221.

Walter, J. L., Cline, H. E., and Koch, E. F. (1969). *Trans. TMS-AIME* **245,** 2073.

Weatherly, G. C. (1968). *Metal Sci. J.* **2,** 25.

Wilcox, B. A., and Clauer, A. H. (1969). *Trans. TMS-AIME* **245,** 1279.

Yue, A. S., Crossman, F. W., Vidoz, A. E., and Jacobson, M. I. (1968). *Trans. TMS-AIME* **242,** 2441.

7

Effect of Interface on Fracture

*ELLIOT F. OLSTER**

Advanced Composite Applications
Avco Corporation
Lowell, Massachusetts

RUSSEL C. JONES

Department of Civil Engineering
Ohio State University
Columbus, Ohio

I. Introduction

The toughness of composite materials is intimately related not only to the behavior of the individual components but often more importantly to

* *Present address*: Plastics Development Center, Ford Motor Company, Detroit, Michigan.

the materials in concert. There are at least two striking examples of this. The first is fiberglass, a tough composite made up of two brittle constituents; second is wood itself. Ordinary wood, a wood fiber bonded by a matrix of lignin, is quite tough; the same fiber in an epoxy matrix is extremely brittle. In both the fiberglass and the wood composites the interface has a dominant effect upon the toughness. The goal of this chapter is to summarize and to discuss the mechanisms controlling fracture toughness in composites, in particular those mechanisms which depend upon the characteristics of the interface. As composite manufacturing technology advances, this knowledge will allow the tailoring of toughness as well as other mechanical properties.

This chapter begins with a brief discussion of the most widely-used methods for measuring toughness in composite materials. Next the fracture of composite materials is discussed. These composites include fiber-reinforced, particulate-reinforced, and laminated materials with the primary emphasis on unidirectional fiber-reinforced systems. A summary of applicable theoretical analyses along with experimental results are presented for each class of materials. Finally the composite fracture data are reviewed and the effects of the interface are discussed.

Due to the relatively small amount of research documentation available on metal matrix composites, parallels to other composite systems have been drawn upon in several instances to provide completeness of the discussion in this chapter. The effects of interfaces in fiber-reinforced plastics have been extensively studied, so these systems are referred to and parallels drawn when data on metallic systems are not yet available.

II. Fracture Toughness Testing

In order to assess the relative crack resistance of structural materials numerous tests have been developed (Biggs, 1960; Parker, 1957). The two most widely used are the Charpy impact test and the K_c (or G_c) notch toughness tests. In order to more fully appreciate the importance and to allow a more comprehensive understanding of the toughness data the tests briefly are described and parameters affecting the toughness are elucidated.

A. Charpy Impact Test

The Charpy impact test measures the energy required to break a notched flexural specimen. The specimen is impacted with a hammer of known kinetic energy and by determining the energy lost by the hammer a relative measure of the material's toughness can be determined. Although there are standards for hammer energy and specimen geometry these are not often adhered to when testing composites, and as will become clear later, specimen size, especially thickness, can affect the apparent toughness. In addition, some energy is expended in accelerating the broken pieces and possibly absorbed in inelastic impact; generally, however, these are small and not considered, hence all the energy lost by the hammer is associated with crack propagation. It should be emphasized that the toughness measured by the Charpy test is generally reported in units of foot-pounds; in some cases it is given as foot- or inch-pounds per square inch of net section. Even though in this latter case it has the same units as G_c, the critical strain energy release rate, they are not equivalent and data from Charpy tests cannot be used in the fracture equation (which will be defined shortly) for design purposes.

Nevertheless, the Charpy test can be used to determine the relative effects of heat treatment, operating temperature, grain orientation, size effects, and environmental effects, and in addition is a fast and inexpensive test.

B. Notch Toughness Testing

The Charpy test has been shown to be a relative measure of toughness but is not absolute in the sense that it provides no data which could be used to predict at what gross stress level a partially cracked specimen would fail catastrophically. The notch toughness K_c, or its related parameter, the critical strain energy release rate G_c, overcomes this deficiency.

Briefly, the fracture equation is given by

$$K_c = \alpha \sigma_{nc} (\pi a)^{1/2} \tag{1}$$

where K_c is the notch toughness, α the boundary modification factor, σ_{nc} the critical nominal stress, and a the crack length. When performing experiments to determine K_c (see Fig. 1) it is used in the form given above; when used in design, it is inverted so that the allowable stress (σ_{nc}) can be obtained.

Equation (1) is the fundamental relation underlying a body of knowledge known as linear fracture mechanics and, because it provides insight into the toughening mechanisms, it warrants additional attention.

Crack extension occurs when the stresses, or alternatively strain energy density in the region of the crack tip become too great. It has been shown

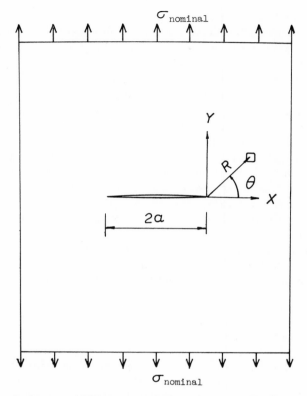

FIG. 1. A central crack in an infinite sheet subjected to uniform tension at its boundaries.

(Irwin, 1957) that only two scalar parameters are required to specify the stress field surrounding a crack tip. For the opening mode of fracture which is shown in Fig. 1, the two-dimensional stress field can be written as

$$\sigma_{ij}(r, \theta) = [\alpha\sigma_n(\pi a)^{1/2}/(2\pi r)^{1/2}]F_{ij}(\theta) \qquad (2)$$

where σ_{ij} is the stress at a point; r, θ the polar coordinates having their origin at the crack tip (see Fig. 1); σ_n the normal stress; and F_{ij} the stress distribution function [fully described by Paris and Sih (1965)]. Defining K as $\alpha\sigma_n(\pi a)^{1/2}$, Eq. (2) can be written as

$$\sigma_{ij}(r, \theta) = KF_{ij}(\theta)/(2\pi r)^{1/2}$$

Here it can be seen that the stress intensity factor K is the crucial parameter required to specify the magnitude of the stress field near the crack tip and herein lies its importance. The primary fracture modes are shown in

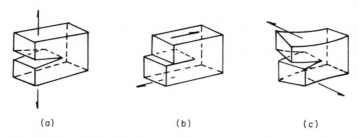

(a)   (b)   (c)

FIG. 2. The fundamental fracture modes: (a) opening mode (mode I), (b) sliding mode (mode II), (c) shearing mode (mode III). [from Corten (1967), courtesy of Addison Wesley].

Fig. 2 and in each case [as discussed by Tetelman and McEvily (1967) and by Paris and Sih (1965)] relations similar to those given in Eq. (2) exist.

When the nominal stress is raised to such a value that K reaches K_c, the critical stress intensity factor or the notch toughness of the material, the crack will extend; hence Eq. (2) can be used to predict fracture.

An alternative approach, the compliance method, has as its primary advantage the versatility of not requiring α, and therefore mathematically more complicated specimen configurations can be handled. It has been shown (Corten, 1967) that the change in strain energy per unit crack extension can be expressed in the form

$$\partial u / \partial a = -\tfrac{1}{2} P^2 \, \partial\lambda / \partial a \,|_a$$

where $\partial u / \partial a$ is the change in strain energy per unit crack extension, P the applied load, and $\partial\lambda / \partial a \,|_a$ the change in specimen compliance per unit crack extension evaluated at the current crack length.

Commonly the term

$$G = -\partial u / \partial a = \tfrac{1}{2} P^2 \, \partial\lambda / \partial a \,|_a \tag{3}$$

is used and is called the strain energy release rate or because of units it is sometimes referred to as the crack extension force. In a given test when P reaches P_c (the subscript c referring to instantaneous values at incipient crack extension) and a reaches a_c, the term G becomes G_c, where G_c, the critical strain energy release rate, is a measure of notch toughness of the material.

Both K_c and G_c are used as measures of the notch toughness. The equivalence of these approaches can be seen following Wu's (1968) presentation. Consider an increment of crack extension β as shown in Fig. 3. The strain energy released associated with the crack extension β is equal to the energy

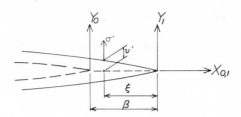

Fig. 3. An incremental advance in the position of the crack front [from Wu (1968)]. Subscript 0 indicates the initial coordinate axes, and subscript 1 indicates the final coordinate axes.

required to close the crack by an amount β; mathematically this can be expressed as

$$G = \lim_{\beta \to 0} \int_0^\beta (1/\beta)\sigma_y' v_y' \, d\xi \tag{4}$$

where σ_y' is the value of σ_{yy} at $r = \beta - \xi$, $\theta = 0°$ and v_y' is the displacement normal to the crack evaluated at the same point as σ_y'; however, since the coordinate axes shift, v_y' is evaluated at $r = \xi$, $\theta = +180°$.
In general

$$\sigma_{yy} = (K/(2\pi r)^{1/2})\{\cos(\theta/2)[1 + \sin(\theta/2)\sin(3\theta/2)]\}$$

and for plane stress

$$v_{yy} = [K(2\pi r)^{1/2}/\pi E(1 - \mu)][2(1 - \mu)\sin(\theta/2) - \sin(\theta/2)\cos(\theta/2)]$$

where E is Young's modulus and μ Poisson's ratio.
Therefore

$$\sigma_y' = \sigma_{yy}\big|_{r=\beta-\xi,\ \theta=0} = K/[2\pi(\beta - \xi)]^{1/2}$$

and

$$v_y' = v_{yy}\big|_{r=\xi,\ \theta=\pi} = 2K(2\pi\xi)^{1/2}/(\pi E)$$

Evaluation of Eq. 4 yields the relation for plane stress

$$G = K^2/E \tag{5}$$

These relationships are valid throughout the range of G and K up to and including the critical values, G_c and K_c.

Both thickness and specimen width have a pronounced effect on the notch toughness of a ductile material. The intrinsic effect of thickness on fracture toughness is described in Fig. 4 which shows the variation in opening mode fracture toughness with thickness. The subscript I in K_{Ic} implies the plane strain fracture toughness value which is generally accepted to be a material constant. With the exception of very thin sheets K_c is greater

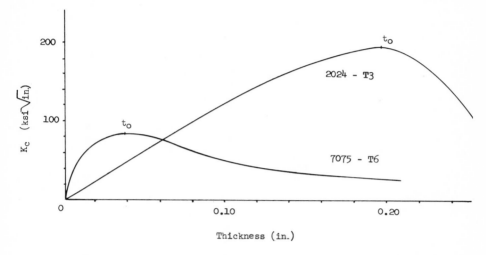

FIG. 4. The variation in notch toughness with specimen thickness for two aluminum alloys.

than K_{Ic}. The values of K_c for thicknesses less than "t_0" are associated with fully slant fracture and the increase of K_c with thickness in this region is generally considered to be caused by an increasing volume of material which is plastically deformed. This portion of highly deformed material is roughly proportional to the square of the thickness of the sheet. The de-

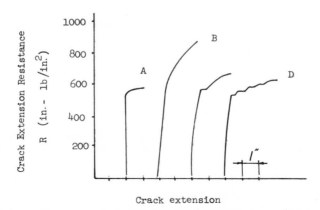

FIG. 5. Several types of crack extension resistance curves [from Srawley and Brown (1965), courtesy of ASTM]. A, sharply defined instability; B, resistance curve for 7075-T6 aluminum; C, pop-in followed by stable crack extension; D, discontinuous crack growth.

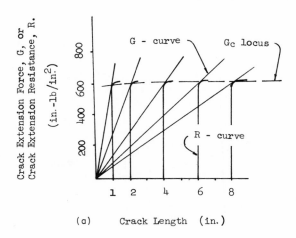

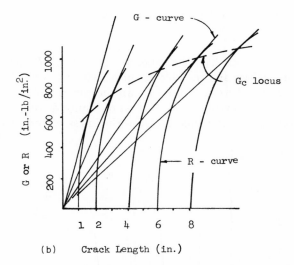

FIG. 6.   Variation in G_c with crack length: (a) for a brittle material, (b) for a ductile material [from Srawley and Brown (1965), courtesy of ASTM].

scending portion of the curve is associated with the occurrence and progressive dominance of square, plane strain fracture. At sufficiently large thickness K_c approaches a limiting value called K_{Ic}, the plane strain fracture toughness. It is usually assumed that the layer of plastically deformed material extends for a constant distance from the fracture surface for plane strain fracture, thus the work per unit thickness G_{Ic} is a constant. However, even though G_{Ic} and K_{Ic} appear to be a material properties and the

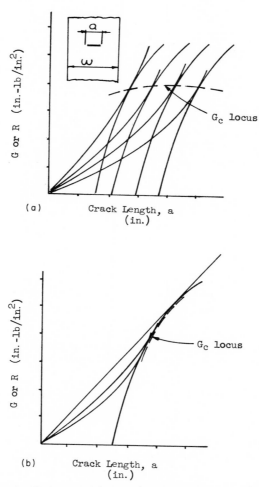

FIG. 7. Variations in G_c for: (a) constant specimen width, crack length varied, (b) constant crack length, specimen width varied [from Srawley and Brown (1965), courtesy of ASTM].

simplified representation in Fig. 4 implies that K_c has a unique value for a given thickness, such is not the case and in general K_c and G_c cannot be thought of as unique material properties. They can be shown, using the concept of a crack extension resistance, to be dependent upon the specimen width and the ratio of crack length to specimen width as indicated in Figs. 5, 6, and 7 (Srawley and Brown, 1965).

The concepts of fracture mechanics have been extended to anisotropic materials (Paris and Sih, 1965; Wu, 1968; Sih and Liebowitz, 1968) and

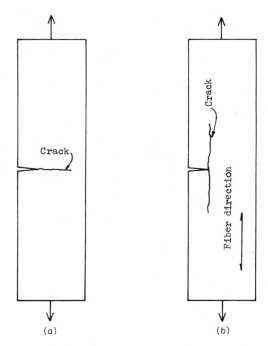

(a)   (b)

Fig. 8. Behavior of (a) a homogeneous, isotropic plate and (b) a unidirectionally-reinforced composite exhibiting longitúdinal splitting.

it has been shown that the stress field can again be represented in the form

$$\sigma_{ij}(r, \theta) = \alpha'(a, w, A)\sigma_n(\pi a)^{1/2}F_{ij}'(\theta, s)/(2\pi r)^{1/2}$$

where α' is the boundary modification, F_{ij}' a stress distribution factor, a the crack length, w the specimen width, A a function of scalars of the compliance matrix of the anisotropic material, and s a function depending upon the orientation of the crack with respect to the plane of elastic symmetry. Hence α' is a boundary modification factor dependent upon geometry as well as material properties and $F_{ij}'(\theta, s)$ is a stress distribution function dependent not only on θ but on the material constants as well as on the orientation of the crack with respect to the planes of elastic symmetry as reflected in the parameter "s." The important thing is that once again the stress intensity factor has the form

$$K = \alpha\sigma_n(\pi a)^{1/2}$$

As long as the crack extends along its original axis the relation between G and K as expressed by Eq. 4 is valid. With the more complex stress and

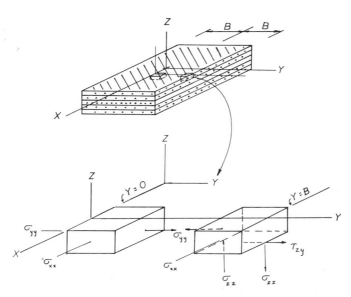

Fig. 9. The development of interlaminar shear τ_{zy} and interlaminar normal stresses σ_{zz} in a composite [from Pagano and Pipes (1971)].

displacement field associated with anisotropic materials Eq. 4 no longer reduces to $G = K^2/E$ but is expressed as $G = CK^2$, where C is a function of the scalars of the compliance matrix (Paris and Sih, 1965).

In general the stress field at the tip of a crack in an anisotropic plate include components of K_I and K_{II}. However, at present most tests are conducted at orientations which eliminate one of these components, specifically on orthotropic materials oriented so that the load is parallel to one principal axis and the crack is parallel to the other principal axis. Under these conditions the high degree of anisotropy present in some composites can result in behavior unknown to common metals. For example, in notched unidirectionally-reinforced tension specimens longitudinal splitting has been observed frequently as shown in Fig. 8. This can be eliminated if the transverse strength (Cook and Gordon, 1964) and shear strength are sufficient, but nevertheless this potential failure mode represents a variation in behavior which must be considered. In addition, the application of uniaxial tension σ_{xx} to a cross plied specimen results in localized interlaminar shear τ_{zy} as well as normal stresses perpendicular to the plane of the specimen σ_{zz} (Pagano and Pipes, 1971) as shown in Fig. 9. The orientation and magnitude of σ_{zz} and τ_{zy} depend upon the stacking sequence, the elastic constants of each lamina, and the longitudinal strain. Significant

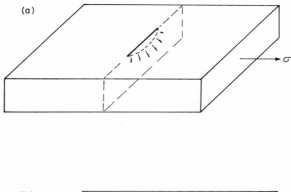

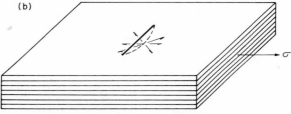

Fig. 10. Propagation of a surface crack in (a) a homogeneous isotropic body showing the coplanar crack growth and (b) a laminate showing the interlaminar failure.

interlaminar tension σ_{zz} and shear τ_{zy} can result in delamination (Pagano and Pipes, 1971; Foye and Baker, 1970) which again is a peculiarity of anisotropic laminates. The last simple example to be illustrated is the behavior of material containing surface cracks. Isotropic materials exhibit a general growth coplanar to the original flaw as shown in Fig. 10a. Laminates, due to the generally low strength bond line, tend to delaminate in subsurface planes as shown in Fig. 10b. Hence these three simple examples illustrate some of the differences between homogeneous isotropic materials and fiber-reinforced composites and serve to point out the importance of obtaining experimental evidence as well as obtaining a complementary analytical description of the behavior of composite materials.

III. Fracture of Composites

Composites are desirable because they can be manufactured such that in certain preferred directions the properties, generally strength and stiffness, are generally greater than those attainable with a homogeneous isotropic material of comparable density. They achieve this effect, however, at the expense of some other properties. For example, most of the so-called advanced structural composites considered for use in the latest military aircraft are supplied in the form of a unidirectional prepreg or preform sheet. These sheets are stacked at various orientations with respect to one another and pressed together at an elevated temperature (300°F to 600°F for resins, 900°F and above for metals) to obtain a fiber-reinforced, laminated composite. These composites, while they have directions and planes of high mechanical strength, also in general possess planes and orientations of weakness, and as discussed earlier the preferential planes of weakness in a composite can alter the behavior from what is found in homogeneous isotropic materials.

A. Fiber-Reinforced Composites

By using idealized composite systems it is possible to study the mechanisms which lead to toughening in fiber-reinforced composites. This in turn allows an assessment of the role of the interface. We consider the contributions of the fiber and matrix separately and finally discuss the effect that the interface has on the relation of these contributions.

1. Crack Perpendicular to the Fibers

Consider first the opening mode of fracture where the crack is perpendicular to the fibers. In unidirectionally-reinforced plastics it may be difficult to obtain this mode since the propensity for longitudinal splitting is great. In the reinforced metals the ratio of transverse tensile and shear strength to longitudinal strength is not so severe and cracks have been forced to propagate perpendicular to the filaments. In multidirectionally-reinforced composites, both with plastic and metal matrices, a crack can often be forced to grow in a direction which is perpendicular to the principal axes of orthotropy and which generally coincides with the direction of one or more layers of fibers. Hence, as crack growth proceeds, fibers are broken.

The toughness of a composite (here it is more convenient to use the critical strain energy release rate which is a scalar and hence additive in the conventional sense) can be written as the sum of the toughnesses of its components

$$(G_c)_{\text{composite}} = (G_c)_{\text{fiber}} + (G_c)_{\text{matrix}} \tag{6}$$

In an ideal composite, fracture proceeds by the advance of the crack tip

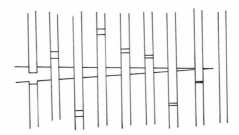

FIG. 11.   Development of fiber pullout energy due to the difference in elevation between the fiber break and the matrix crack.

in the continuous phase (the matrix). Depending upon the specific situation, fiber fracture occurs ahead of, behind, or coincident with the crack in the surrounding matrix. Generally, the fiber–matrix bond is only of moderate strength and the severe stress field at the tip of the crack in the matrix causes a localized debond, thus relieving somewhat the stress state in the fiber. The toughness contribution of the matrix material is similar to its bulk toughness since it is a continuous phase and the presence of the fibers alters the volume of material deformed, but generally not to a truly significant extent. This is subsequently discussed in greater detail. First, however, the behavior of the fibers is discussed.

(a) CONTRIBUTION OF THE FIBERS. The stress field is not nearly so severe as in the surrounding matrix; in fact, there is reason to believe that the stress state approaches a relatively uniform tension across the width of the fiber. Were this not so the shear stresses would also be very high within the fiber. Graphite fibers might then be expected to exhibit longitudinal splitting themselves which is contrary to experimental observations. Assuming, for simplicity, the stresses in the fiber are predominately those corresponding to uniaxial tension, the behavior will be that of a miniature tensile specimen. One way for energy to be absorbed or dissipated is by friction. If the fiber break occurs at some elevation other than at the elevation of the matrix fracture (see Fig. 11) then, as the crack opens up, the fiber will be pulled through a sheath of matrix material. Generally, it is assumed that the frictional force is constant (Tetelman, 1969) and hence the pullout energy is

$$PE = l_f \times \tau_i \pi \, d_f \tag{7}$$

where PE is the pullout energy, l_f the average length of fiber extending from the matrix, d_f the fiber diameter, and τ_i the shear stress at the interface due to frictional forces. Accounting for volume fraction effects, the fiber contribution becomes

$$(G_c)_f = V_f l_f \tau_i \pi \, d_f \tag{8}$$

where $(G_c)_f$ is the toughness due to fibers and V_f the volume fraction of fibers.

On tests with unidirectionally-reinforced carbon fiber-reinforced polyester Harris *et al.* (1971) concluded that four-fifths of the fracture energy was due to fiber pullout. Their experiments involved studying various fiber surface treatments which vary the bond strength (as measured indirectly using the interlaminar shear strength as an indicator). The weakest interface (lowest shear strength) resulted in greater pullout lengths and higher fracture energy. Other researchers (Barker, 1971; Sidey and Bradshaw, 1971) working with carbon fiber-reinforced epoxies found similar results. Fitz-Randolph *et al.* (1971) studying a boron epoxy composite concluded that significant contributions to the total work done during fracture was from the energy absorbed during fiber fracture as well as the work done in pulling broken fibers out of the resin matrix. Metcalf and Klein (1972) argue that, for a given filament strength, as the coefficient of variation of the filament strength increases the fibers will tend to break at widely varying points, hence the toughness should increase (see Fig. 11).

It is natural to inquire whether the fiber contributes in any other way to the toughness. Glass-reinforced plastics exhibit toughness much greater than can be accounted for by the surface energy of each component plus the pullout energy. These materials exhibit significant fiber matrix debonding and even by including the surface energy associated with the debonding

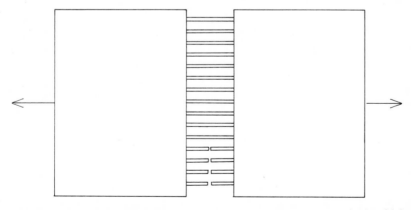

FIG. 12. Schematic showing the idealized behavior of a fibrous composite in which the fiber debonds to approximately a uniform length when the matrix crack approaches. The fibers break as small tensile specimens.

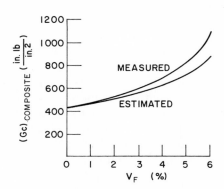

FIG. 13. Toughness of a tungsten–aluminum composite. The debond length increased with greater fiber volume fractions, hence the curves are nonlinear [from Olster and Jones (1971), courtesy of ASTM].

the energy is far short of the experimentally measured values. Outwater and Murphy (1969) proposed that where filament debonding is substantial, a considerable amount of strain energy could be stored in the debonded portions of the fibers. In effect they were miniature tensile specimens. These composites behave as illustrated schematically in Fig. 12. The strain energy in a tensile specimen of length l is $\frac{1}{2}\sigma_f\epsilon_f l$ and hence $(G_c)_f = \frac{1}{2}(\sigma_f\epsilon_f l)V_f$. This relation was substantiated experimentally by various researchers: (Outwater and Murphy, 1963; McGarry and Mandell, 1970; Olster and Jones, 1970 and 1971). The same concepts apply if the fibers are not linearly elastic (Olster and Jones, 1970). The toughness is still proportional to the energy required to break the debonded fibers. Olster and Jones (1970) confirmed this, using a tungsten wire-reinforced aluminum composite which exhibited severe fiber necking prior to failure. As shown in Fig. 13 the toughness increased rapidly with increased fiber volume fraction. This was due to the increased fiber–matrix debonding which was observed to increase as the percentage of fibers increased. The estimated curve was obtained by adding the matrix toughness and the fiber toughness associated with a particular debond length, and as can be seen, the correlation is reasonable. Hence, if debonding is present it is possible to obtain very tough composites.

In the absence of debonding it is also possible to develop tough composites. Boron aluminum is a good example. Two toughening mechanisms have been postulated. First because of the very jagged boron fracture surface an increased amount of surface energy may result (Hancock and Swanson, 1971). Accurate measurements have not been made of the true

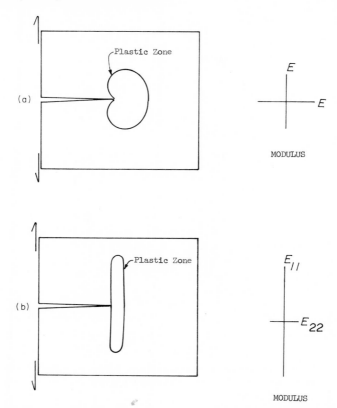

FIG. 14. The plastic zone at the crack tip in (a) a homogeneous isotropic material and (b) an orthotropic material in which the major stiffness is perpendicular to the axis of the crack.

fiber surface area, and so the contribution of this effect has not been quantitatively determined. This, however, is not expected to be the most important source of toughness energy. In metals the stress field ahead of the crack causes a localized yielding phenomena. This so-called plastic zone has a bulbous shape shown schematically in Fig. 14a. In an orthotropic material having its principal axis perpendicular to the crack (as is the case in a unidirectional fiber-reinforced composite), the plastic zone due to the stress field is more constricted, as shown by the residual strains in Fig. 14b. This was verified experimentally (Olster, 1970) by bonding a photoelastic coating to a boron–aluminum composite prior to the introduction of loads. In this composite the fibers are elastic to failure, and the matrix is elasto-plastic. Hence, since the yield strain of the matrix is less than the

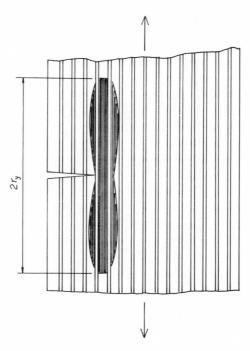

Fɪɢ. 15. A boron–aluminum composite containing a crack perpendicular to the fibers showing the extent of the plastic zone in the aluminum matrix.

ultimate strain of the fiber, matrix yielding will occur. In effect, the fiber will become encased in a plastic sheath of length $2r_y$, as shown schematically in Fig. 15. It is the contention of Olster and Jones (1970) that all the energy in the fiber in this plastic zone contributes to the fracture toughness of the composite. In essence, they argue that the strain energy in the fibers in this region is not completely returned to the system; that it works against the elongated matrix sheath. They conclude that to a first approximation, the toughness contribution is equivalent to the elastic strain energy stored within that portion of the fiber in the plastic zone just prior to fiber fracture. Hence, within the plastic sheath the fiber is under a state of nearly uniform tension, and the strain energy is

$$\mathcal{E}_f = \tfrac{1}{2} \int_{-r_y}^{r_y} \epsilon_f \sigma_f \, dl \tag{9}$$

where $\mathcal{E}_f$ is the strain energy in the portion of the fiber which is within the plastic sheath, ϵ_f the ultimate strain of the fiber, and σ_f the ultimate stress

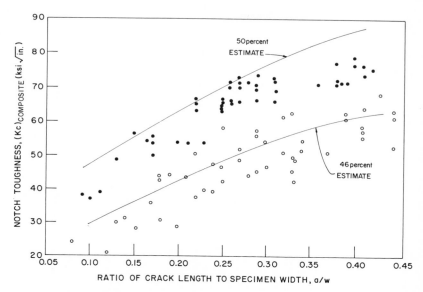

FIG. 16. Comparison between the measured and estimated notch toughness for two types of unidirectionally reinforced boron–aluminum composites [from Olster and Jones (1971), courtesy of ASTM]. Measured values: ●, HA403–50% V_f; ○, HA293–46% V_f.

of the fiber. Therefore

$$(G_c)_f = V_f \tfrac{1}{2} 2r_y (\sigma_f^2/E_f)_{ult} \tag{10}$$

where E_f is the elastic modulus of the fiber. Using Eq. 10 for the fiber contribution and experimentally-determined values for the matrix contribution, Olster and Jones (1971) obtained excellent agreement with measured values on a 46 and 50% fiber content boron–aluminum composite as illustrated in Fig. 16.

Recent studies by Klein and Metcalfe (1972) have experimentally addressed the problem of the effect of the interface on the fracture toughness of metal matrix composites. The composite studied was unidirectionally reinforced 6061 aluminum with a 45% V_f. The interface was altered by soaking for periods of time ranging from 0 to 150 hr at 1000°F. Subsequent to the soak, the specimens were heat treated to either the T-6 condition or put in an air-annealed condition. The authors did not compute the notch toughness, but rather gave the net strength of their notched tensile specimen. From their data, shown in Table I, it appears that the annealed specimens have the same or slightly higher-notched tensile strengths for comparable soak times. This indicates that the toughness is approximately

TABLE I   Notch Tensile Data for Al(6061)–45B Heat-Treated at 1000°F

Specimen number	Time at 1000°F	Specimen thickness (in.)	Specimen width, w (in.)	Notch depth, a (in.)	a/w	Ultimate load (lb)	Net or notched strength, N (ksi)	Unnotched strength, UN (ksi)	Ratio N/UN
						Given the T-6 temper (Quenched into water and aged 7 hr at 350°F after the 1000°F heat treatment)			
108	(As-received)	0.030	0.496	0.126	0.254	1420	132.0	213[a]	0.62
95	0	0.029	0.527	0.137	0.260	1240	110.0	—	—
96	0	0.029	0.527	0.144	0.274	1340	121.0	—	—
83	5 min	0.030	0.524	0.140	0.268	1340	117.0	235[a]	0.50
84	5 min	0.030	0.528	0.139	0.264	1380	118.0	—	0.50
85	15 min	0.030	0.526	0.135	0.257	1080	92.3	243[a]	0.38
87	30 min	0.030	0.490	0.124	0.253	890	84.8	201[a]	0.42
88	30 min	0.030	0.512	0.130	0.254	1100	97.4	—	0.48
89	2 hr	0.030	0.521	0.144	0.276	795	70.4	127[a]	0.55
90	2 hr	0.030	0.528	0.140	0.266	770	67.5	—	0.53
91	10 hr	0.029	0.522	0.133	0.265	450	39.8	100[a]	0.40
92	10 hr	0.030	0.522	0.140	0.268	530	45.7	—	0.46
93	150 hr	0.029	0.524	0.133	0.254	430	37.8	85[a]	0.45
94	150 hr	0.029	0.515	0.138	0.268	440	40.4	—	0.48
						Given the O temper (held 2 hr at 800°F, slow-cooled to 500°F and held 7 hr at 350°F)			
124	0	0.030	0.492	0.129	0.262	1210	111.0	183[b]	0.60
126	30 min	0.030	0.501	0.130	0.260	1160	105.0	190[b]	0.55
127	2 hr	0.0295	0.500	0.1305	0.260	720	66.0	136[b]	0.49
128	2 hr	0.030	0.500	0.131	0.260	810	73.0	136[b]	0.53
130	150 hr	0.030	0.499	0.1295	0.260	472	43.0	72[b]	0.60

[a] Data obtained from Fig. 53, AFML-TR-72-226.
[b] Data obtained from Fig. 56, AFML-TR-72-226.

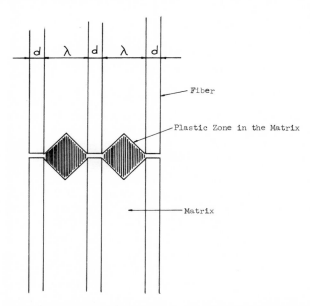

FIG. 17.   A two-dimensional model of the plastic zone in the matrix of a copper–tungsten composite [from Tetelman (1969), courtesy of ASTM].

the same regardless of the temper of the matrix. Apparently a decrease in the toughness of the matrix in the annealed case, as compared to the T-6 heated-treated material, is offset by the increased contribution of the fibers. This is in agreement with the Olster–Jones model in that the plastic zone size r_y is greater for the annealed matrix. Further, as the soaking time increases, the interfacial reaction zone increases. This results in weaker filaments and, as shown by Klein and Metcalfe's data, the notched tensile strength and toughness decrease. This further emphasizes the role of the fiber, particularly its strength as expressed in Eqs. (9) and (10), in determining fracture toughness.

(b) CONTRIBUTIONS OF THE MATRIX. The energy associated with matrix fracture is termed $(G_c)_m$. In many composites, especially the fiber-reinforced plastics, this term is of minor importance. In the metal matrix composites however, it can be significant. Copper–tungsten is perhaps the most widely-studied metal matrix composite. Cooper (1966) and Cooper and Kelly (1967), observing the regions of massive plastic flow (shown in Fig. 17) proposed a two-dimensional model which assumes that the triangular regions have a height and width of λ where λ is the distance between fibers. Letting d be the fiber diameter

$$\lambda = d(1 - V_f)/V_f \tag{11}$$

Applying Rice's (1966) plastic strip model

$$(G_c)_m = (1 - V_f)[\epsilon_m \sigma_m][d(1 - V_f)/V_f]$$

where ϵ_m is the ultimate strain of the matrix and σ_m the ultimate stress of the matrix. Therefore

$$(G_c)_m = \sigma_m \epsilon_m [dV_m^2/(1 - V_m)] \tag{12}$$

where V_m is the volume fraction of matrix.

Both Cooper and Kelly (1967) and Tetelman (1969) cite Eq. (12) as a reasonable estimate of toughness due to the matrix in tungsten fiber-reinforced copper. As pointed out by Gerberich (1971), even though it gave a reasonable quantitative estimate of toughness it is in error since the composite was three-dimensional and had circular, not square fibers. Olster and Jones (1970), in studies of a 0 to 6% V_f tungsten-reinforced aluminum, found that the matrix toughness was not significantly affected by the reinforcement. The same authors in a 50% V_f boron–aluminum composite proposed that the matrix toughness was essentially unaffected by the boron fibers. This can be true only if the deformation of the matrix at the tip of the crack is so highly localized that it is not influenced by the presence of filaments. Hence, each composite must be treated differently depending upon its behavior. The matrix toughness could be as low as that observed in a bulk quantity of the matrix material or it could be somewhat higher depending upon the influence of the filaments as suggested by Cooper and Kelly (1967). If the interface is strong and the fibers have a high co-efficient of variation in strength, fiber breaks would occur at various positions as discussed by Metcalfe and Klein (1972), this in turn could result in increased matrix deformation, which would be reflected in an increase in toughness.

2. Crack Parallel to the Fibers

Generally in unidirectionally-reinforced materials it is common to find failure occurring parallel to the filaments as shown in Fig. 18. The opening mode toughness is dependent upon either the transverse filament strength, the interface strength or the matrix properties. For example in notched tensile tests on Borsic–aluminum, Kreider *et al.* (1970) found evidence of filament splitting. More common however is an interface or a matrix failure. Gerberich (1971) studied both of these situations.

The bond in diffusion-bonded stainless steel–aluminum may be primarily

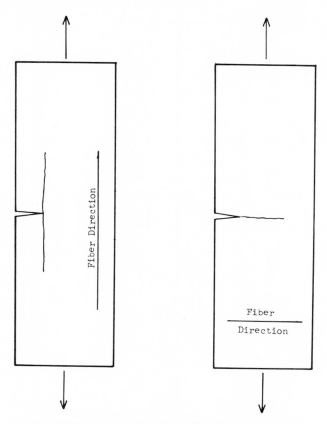

FIG. 18. Modes of failure in unidirectional composites.

mechanical and hence transverse loading results in debonding. Gerberich (1971) felt that the Rice (1966) plastic strip model is a realistic idealization and proposed that the plastic strip height h be taken as the distance between rows of filaments indicated by $\lambda' - d$ in Fig. 19. Since

$$\lambda' - d = [(d/2)(\pi/V_f)^{1/2} - d]$$

It follows that

$$G_{cm} = \sigma_m \epsilon_m h = \sigma_m \epsilon_m d[\tfrac{1}{2}(\pi/V_f)^{1/2} - 1] \tag{13}$$

This relation was in excellent agreement with his measurements (Fig. 20) over the range of 6 to 40% fiber volume fraction.

When significant interfacial bonding exists, as in the case of boron–aluminum, Gerberich (1971) proposed that the plastic strip height extend

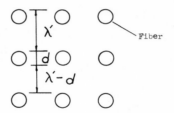

Fɪɢ. 19. Geometrical parameters affecting transverse fracture toughness.

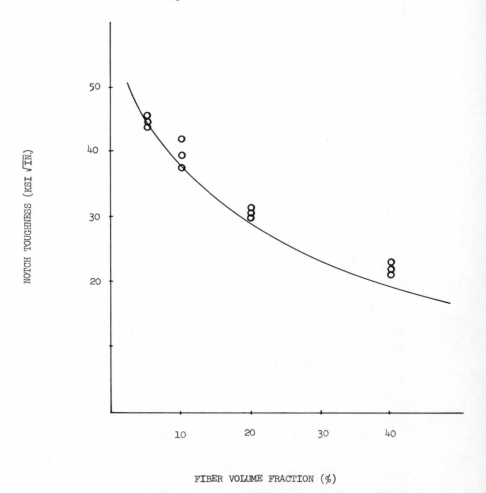

FIBER VOLUME FRACTION (%)

Fɪɢ. 20. Transverse fracture toughness of aluminum–stainless steel composites [from Gerberich (1971), courtesy of University of California, Lawrence Berkeley Laboratory].

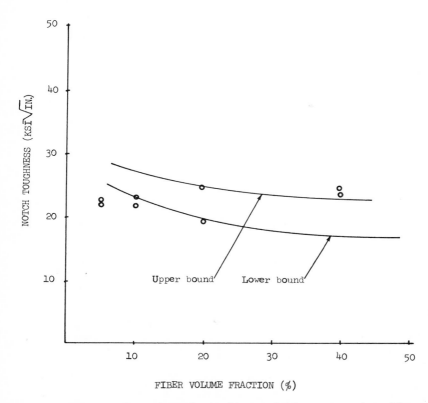

FIG. 21. Transverse fracture toughness of boron–aluminum composites with upper and lower bounds depending upon the estimate of the composite modulus [from Gerberich (1971), courtesy of University of California, Lawrence Berkeley Laboratory].

to the extremities of the fiber, i.e., $\lambda + d$. However, he further modifies this to account for the region occupied by the filaments themselves which are assumed to be rigid. Thus, based again on geometrical considerations, he concludes that the effective plastic strip height h' is

$$h'_{\text{effective}} = d[\tfrac{1}{2}(\pi/V_f)^{1/2} + 1 - (V_f\pi)^{1/2}] \tag{14}$$

Thus the toughness becomes

$$(G_c)_m = \sigma_m \epsilon_m h' \tag{15}$$

and using $K_t = E_{ct}G_c$ where E_{ct} is the transverse composite modulus

$$K_t = E_{ct}\sigma_m\epsilon_m d[(\pi/4V_f)^{1/2} + 1 - (V_f\pi)^{1/2}]^{1/2}$$

Gerberich (1971) (Fig. 21) did not find such good agreement, but rather

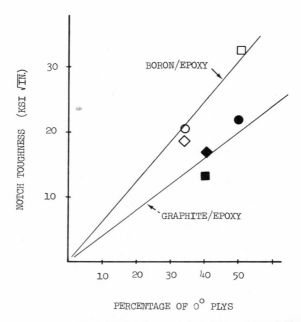

FIG. 22. Relation between notch toughness and the percentage of fibers perpendicular to the crack [from Olster and Woodbury (1972)]. Closed symbols, graphite epoxy; open symbols, boron epoxy.

found that K_t was relatively invariant with V_f. The predictions are, however, of the proper order of magnitude and as can be seen, the toughness is primarily matrix controlled; the filaments affect to varying degrees the amount of the matrix material involved in the massive plastic flow. The resin matrix composites have significantly lower toughness than do the metal matrix materials. The toughness of the resin alone ranges from $\frac{1}{2}$ to 15 in.-lb/in.² (Mostovoy and Ripling, 1965, 1966; Larsen, 1971). At normal volume fractions (60%) the filaments are close and the matrix ligaments are small. With glass and graphite reinforcement (Sanford and Stonesifer, 1970, 1971; Konish *et al.*, 1972) the interface is relatively weak and the measured toughness is of the order of 1 in.-lb/in.² Sanford and Stonesifer (1971) performed experiments using filaments which ranged from 0.4 to 5.0 mil diam. Gerberich's (1971) plastic strip model indicated an increase in G_c with increased fiber diameter. This was not found by Sanford and Stonesifer and perhaps the Gerberich model must be applied rather restrictively and may be limited to matrices with a substantial amount of ductility. Where the bonding is significantly better, as in boron–epoxy (Fitz-Randolf *et al.*, 1971), the transverse toughness has been found to be

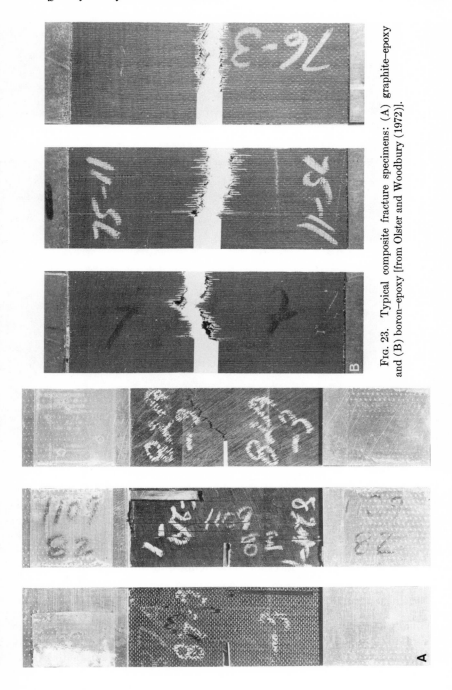

Fig. 23. Typical composite fracture specimens: (A) graphite–epoxy and (B) boron–epoxy [from Olster and Woodbury (1972)].

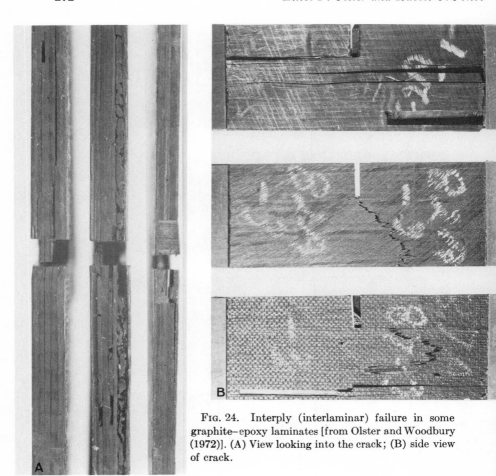

FIG. 24. Interply (interlaminar) failure in some graphite–epoxy laminates [from Olster and Woodbury (1972)]. (A) View looking into the crack; (B) side view of crack.

28 in.-lb/in.^{2}. Again it is found that the maximum transverse toughness is of the same order of magnitude as for the bulk resin. Work by Sanford and Stonesifer (1970) indicates that in some systems water immersion can reduce the transverse toughness to effectively zero which gives some appreciation for the effect of the interface on toughness in the transverse direction.

3. Multidirectionally Reinforced Fibrous Composites

Since the majority of usages require composites with reinforcement in several directions the unidirectional composites are used primarily to study fundamental behavior. Very little work has been done to characterize the toughness of multidirectionally-reinforced composites. Work cited earlier indicates that the toughness when the crack is advancing perpendicular to the fibers is much greater than when the crack travels along a fiber. These studies also show that the toughness is linearly dependent upon fiber volume fractions. In a recent study (Olster and Woodbury, 1972) using a 0/±45, a 0/±45/90, and a 0/±60 lay up with both boron and graphite reinforcement in an epoxy matrix, the opening mode fracture toughness varied linearly with fiber volume fraction as shown in Fig. 22. The modulus and ultimate tensile strength are also approximately linearly dependent on fiber volume and hence either of these parameters could also be used to indicate the relationship. An increase in toughness and an increase in strength are common in fiber-reinforced materials; however an inverse relationship is normally found with structural metals. The interface strength in the graphite composites was less than in the boron and as a result two phenomena were observed. First the fractures were not as uniform in the graphite as indicated in Fig. 23 and secondly interply failure shown in Fig. 24 was observed in the graphite. This is one of the additional modes of failure open to laminated materials and was most pronounced when the layups accentuated high interlaminar shear.

B. Laminated Composites

1. Through the Thickness Cracks

The simplest laminates are composed of homogeneous isotropic plates bonded together. Because of the preferential planes of weakness these materials can be designed to be extremely tough. Consider the idealized laminate shown in Fig. 25. The stress field ahead of the crack is given by

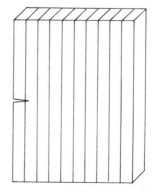

FIG. 25.   Crack arrester laminate.

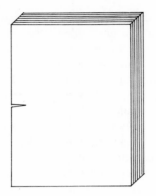

FIG. 26. Crack divider laminate.

Eq. (2). A short distance ahead of the crack tip transverse tensile stresses (σ_{xx}) develop. These in combination with the shear τ_{xy} which exists for all angles θ except $\theta = 0°$ can cause failure along the bond line. McCartney *et al.* (1967) studied crack arrest behavior of high strength (290 ksi) steel laminated with low, medium, and high strength bonds. The low strength bonds (5–10 ksi) consisted of epoxies, lead–tin solder and tin solder; the medium strength bonds (55–85 ksi) were formed using silver solder; and high strength bonds (200 ksi) were formed by diffusion bonding the layers together. In all cases the notched Charpy test specimens fractured only up to the first bond line. The remainder of the specimen underwent massive deformation and debonding along this interface, but did not fracture. Similar results were obtained by Embury *et al.* (1967). Whenever the bond is weaker than the parent material the so-called crack arrester is effective. Leichter (1966) has found, however, that in the presence of an embrittling phase, as results from certain brazing alloys, the toughness can be substantially reduced.

Both Arnold (1956) and Ianelli and Rizzitano (1962) have shown that the toughness of a laminate tested in the direction shown in Fig. 26, which has been called the crack divider mode by Embury *et al.* (1967), is equal to the sum of the individual sheet toughnesses. This allows a designer to maximize the toughness by choosing sheets which are themselves the toughest possible for a given material as was previously indicated in Fig. 4, where the thickness-toughness relations are presented. Even if these sheets are bonded, as was proven by Embury *et al.* (1967) in their studies of mild steel joined with lead–tin solder, silver solder, or copper, the sheets debond upon fracture and the toughness is as predicted by Arnold (1956). Leichter (1966) has found, when a reaction zone is formed which contains brittle intermetallic compounds, that the toughness is somewhat

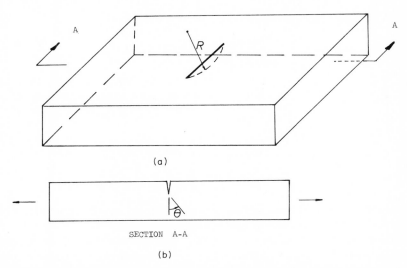

Fɪɢ. 27. Schematic showing (a) surface crack in a plate and (b) its analogy to a single-edge notch specimen.

reduced. The work by Olster and Woodbury (1972) on fiber-reinforced graphite–epoxy laminates serves as a further example since interply failure (Fig. 24) is an interface phenomena peculiar to laminated materials composed of anisotropic layers.

2. Surface Cracks (Gouges)

In addition to thickness cracks, laminates can contain surface cracks such as gouges or debond zones. Both in-plane and transverse loads can produce a delamination initiating from the gouge. This was described earlier (Fig. 10) and is common behavior with laminated materials. The behavior can be explained using the stress field equations, but a simple analogy is also sufficient. Take a cross section through the gouge as indicated in Fig. 27. The resulting body is similar to a single edge notch specimen. The shear stress along plane m–m has the form shown in Fig. 28. In addition, a transverse tensile stress σ_{zz} exists (Fig. 28). The shear stresses are introduced by extensional (or flexural) loads which in turn define K_I. Hence, even if delamination is the dominant behavior, K_I could possibly be used to predict onset of this failure. There will, of course, be some dependence on the layup, but in general the greatest propensity for delamination is expected at the deepest portion of the gouge where, using the single edge notch specimen analogy, the stress intensity is greatest. Additional

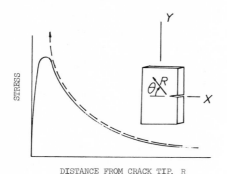

FIG. 28.   Stress distribution in a single-edge notch specimen at $\theta = 90°$ (—, shear stress; – – –, normal stress).

propagation of the delamination will be governed by the toughness of the composite in the forward sliding mode (Mode II).

There is another interesting effect. Consider the size of the crack at several elevations for example, at each ply. The surface ply has the largest crack length; each subsequent ply has a smaller crack until there is a point at which the gouge ends. Between any two plys where the crack length is changing there exists interlaminar shear forces. This can be seen by examining the relative displacements of the two plys. The top layer would tend to undergo larger displacements since K is larger. Thus to maintain

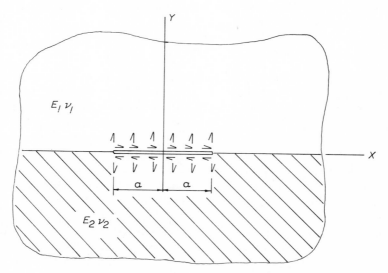

FIG. 29.   A model of a debonded region between two semiinfinite plates subjected to normal and shear tractions [from Sih *et al.* (1970), courtesy of Lehigh University].

continuity, interlaminar shear is developed. This can be aggravated or alleviated in a composite depending upon the layup of the plys and the ratios of the elastic constants. Nevertheless it offers the possibility of a delamination at some position other than at the deepest part of the gouge.

3. Delamination

A delamination can propagate under both in-plane and transverse loading. Consider first the effect of in-plane loads. As shown in Fig. 9a a laminate composed of plys having different Poisson ratios will develop both interlaminar shear stresses, τ_{zy}, and in-plane normal stresses, σ_{yy}. Along the plane $y = 0$ the interlaminar shear stresses τ_{zy} vanish, whereas at $y = b$ they reach their maximum value. These shear stresses are significant only in the vicinity of the boundary (generally considered to be of the order of the thickness of the specimen (Pagano and Pipes 1971). The distribution of stresses along the y axis for the uppermost lamina resulting from a strain in the x direction is given in Fig. 9b. At $y = 0$ only σ_{yy} is present; at $y = b$ the normal force cannot be supported and becomes a shear stress resultant τ_{zy}. The lamina supports no shear in the z direction

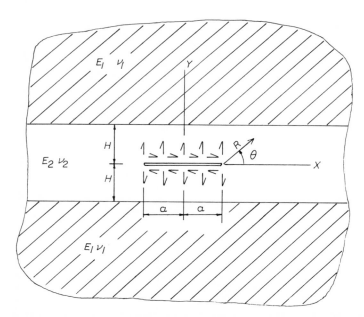

FIG. 30. A debond zone in a layered composite subjected to normal and shear tractions [from Sih *et al.* (1970), courtesy of Lehigh University].

and hence the couple caused by τ_{zy} and σ_{yy} is resisted by normal stresses σ_{zz}. The sign of σ_{zz} depends upon the relative Poisson's ratios. When σ_{zz} is tensile it, in combination with the shear stresses τ_{zy}, tends to cause delamination. This is the so-called stacking sequence argument given by Pagano and Pipes (1971) and in part explains the behavior observed experimentally by Foye and Baker (1970).

Transverse flexural loadings can also be critical since they develop significant interlaminar shear stresses. The shear stress level is dependent upon both the depth or position of the delamination as well as the layup of the laminate. To some degree this problem has been studied. Analytical

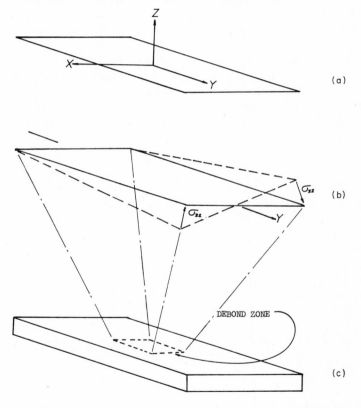

Fig. 31. (A) A portion of an undeformed lamina. (b) Warping of the lamina due to extension in the x direction (shown by the dotted lines); the normal stresses σ_{zz} are generated to resist the warping in a balanced laminate. (c) A composite containing a debonded portion of its outer layer; along this boundary, σ_{zz} stresses are developed if warping is prevented by the rest of the laminate.

solutions describing the stress distribution in the vicinity of a through delamination between two materials have been determined for semi-infinite media as shown in Fig. 29, and for layered media described in Fig. 30. The stress intensity has been shown to be affected by the material constants and for layered media by the thickness of the material containing the defect. In effect these parameters affect the boundary modification factor in Eq. (1). This form of fracture has direct application to adhesively-bonded joints and has been studied experimentally by Patrick *et al.* (1964) and Jermian and Ventrice (1969).

It should be noted, if the debond occurs between sections which themselves show a coupling between twisting and extension, then normal stresses σ_{zz} will develop (see Fig. 31) at the boundary of the debond. At portions of the boundary where these are tensile there will exist an additional driving force for further delamination. These phenomena are of particular importance in fatigue loadings and in fact initiate interlaminar failure. Hence it is not only of importance to further study this behavior but it is necessary to develop the capability to predict under what loadings conditions propagation will occur.

C. Particulate Reinforcement

On a fine enough scale virtually any material can be considered to be a particle-reinforced composite. An example of this extreme would include impurity atoms present in an otherwise perfect crystal. The particulate composites under consideration here cover only those containing reinforcing phases of 1 μ or greater. The most commonly used particulates, excluding those with macroscopic phases such as either portland or bituminous concrete, are the dispersion-strengthened metals and the filled or modified plastics.

Consider first a hard, brittle particle in a relatively ductile matrix. The particle size, volume fraction, and interface strength will directly affect the composite behavior. The particle acts as a stress riser. Its size and proximity to its neighbors determine the degree of interaction between the stress fields of adjacent particles. At fracture, if the bond at the matrix–particle interface is weak, the crack in the continuous phase (the matrix) will frequently encounter what appears to it as voids. That is, regions which are incapable of carrying tensile stresses and which are relatively blunt in comparison to the crack front. As a result it is possible to obtain an increase in toughness. Some support of this can be obtained from data on reinforced plastics. In these materials the interface strength can be controlled somewhat by the surface treatment given to the reinforcement.

In studies by Broutman and Sahu (1971) and by Wambach *et al.* (1968) on glass sphere-reinforced epoxy, polyester, and polyphenylene oxide composites (PPO), the toughness was found to increase as the interface strength decreased. The crack blunting phenomenon is not the only toughening mechanism proposed. Broutman attributes the increased toughness to increased subsurface cracking and deformation. The variation of toughness with volume fraction of glass beads depends strongly upon the system studied. The PPO showed a monotonic decrease in toughness with additional glass beads whereas the polyester and epoxy composites showed a maximum toughness with approximately 20% glass by volume.

The crack blunting mechanism is substantiated by the behavior of certain metals also. Small amounts (2 to 5%) of fine (1 to 5 μm diam), hard, spherical particles which are weakly bonded to a material which exhibits predominately low energy cleavage fracture can result in a significantly tougher composite. The weak interface permits cavities to form which are blunt and cannot support a tensile stress; hence the crack decelerates due to the reduction in the localized tensile stress and in addition is blunted by the cavity. In this manner the work of fracture is substantially increased (Johnston *et al.*, 1961).

With most rigid fillers the toughness decreases with V_f when the interface is strong. As the V_f increases so does the constraint and plastic flow of the matrix. In the tungsten carbide-reinforced–cobalt system, which has been widely studied, the matrix constraint at 80% V_f is great enough to prevent appreciable matrix flow; hence fracture takes place almost exclusively by linking of cracked carbide particles which are contiguous with one another. In this kind of a situation the fracture strength depends, to a large degree, on the same statistical functions which describe failure in a fibrous composite, i.e., when the number of broken particles is great enough so that the load carrying capacity of the remaining particles is exceeded, fracture will occur. At lower volume fractions the matrix properties have a more significant role (Tetelman and McEvily, 1967).

When the fillers have a lower modulus than the matrix a good bond may in fact increase the toughness. McGarry and Willner (1968) and Sultan and McGarry (1967) present excellent discussions of the toughening mechanism operable in rubber-modified plastics. The rubber particles act as stress concentrators within the plastic matrix. As load is applied to the composite the stress concentration effect of the rubber spheres tends to cause the matrix to deform and yield. Spherical voids would have the same initial effect. With further loading the rubber, which is well-bonded to the matrix, begins to deform but while doing so also imposes restraint on the matrix. The local deformation pattern becomes complex and the rubber particles experience a state of triaxial tension. This causes increased deformation of all of the matrix material affected by the particle. In this manner the composite, due to the presence of the rubber particles, is toughened by forcing more matrix material to undergo deformation. A tenfold increase in toughness has been obtained in this manner (Sultan and McGarry, 1967). If the bond is poor, the particles fail to carry load and toughness is only slightly increased due to the additional matrix deformation at the stress concentration sites.

Thus, the interface strength is seen to again greatly influence the toughness. The choice of a strong or weak bond depends upon the specific composite.

IV. Summary

In fiber-reinforced composites it has been shown that when the crack propagates across the filaments the toughness increases in proportion to the debonding. Hence a weak interface is desirable for increased toughness. However, when the crack in the matrix is parallel to the fibers a strong interface is required to prevent a low energy failure along the interface. Some variations have been noted; when the crack is perpendicular to the fibers it is possible to develop high toughness in several ways. One is by fiber pullout where the frictional forces and the pullout length control the toughness. In composites where debonding is not present and where pullout is not observed it is still possible to obtain high toughness. An example is boron–aluminum where the toughness is attributed primarily to strain energy in the fiber within the plastic zone of the composite just prior to fiber fracture. The matrix toughness generally has little effect on the toughness of unidirectional composites unless the crack is parallel to the fibers and the interface is strong. Then the toughness is comparable to that of bulk matrix material. If the interface is weak (assuming the volume fraction to be reasonably high) it controls the toughness. In multidirectionally-reinforced composites the toughness is primarily controlled by those filaments most nearly perpendicular to the crack which must be broken for the crack to advance. The mechanisms operable in unidirectional composites apply here also but must be adjusted to the proper volume fraction.

Laminates have been shown to take advantage of the preferential planes of weakness and, for through cracks, can be used in a crack arrester mode or a crack divider mode. In the crack divider mode the toughness can be optimized by making use of the thickness–toughness relationships. In both of these modes the interface could be nearly as strong as the parent material without affecting the observed behavior. However, a significantly weaker interface would result in a reduction in other properties.

In particulate reinforced composites the interface has a very significant effect on the toughness. Where the bond is weak and the particles are much more rigid than the matrix, the toughness increase is primarily due to crack blunting effects and it appears that the same effects could be obtained by a similar dispersion of voids. Where the bond is strong and the particles are less rigid than the matrix, significant increases in toughness can be obtained. In this latter case, the particles toughen by causing substantial increases in the amount of material undergoing massive plastic deformation.

References

Arnold, S. V. (1956). "A Proposed Method for Impact Testing of Sheet Metal." Watertown Arsenal Lab., WAL TR 112/90, July.

Barker, A. J. (1971). Charpy-notched impact strength of carbon-fiber/epoxy-resin composites, *Int. Conf. Carbon Fibers, Their Composites and Appl.*, *London*, Paper No. 20.

Biggs, W. D. (1960). "Brittle Fracture of Steel." MacDonnell and Evans, London.

Broutman, L. J., and Sahu, S. (1971). The effect of interfacial bonding on the toughness of glass filled polymers, *Annu. Reinforced Plastics Conf.*, *26th SPI*, Sec. 14 C.

Cook, J., and Gordon, J. E. (1964). *Proc. Roy. Soc. London* **A282,** 508–520.

Cooper, G. A. (1966). "Fracture of Fiber Reinforced Materials." Ph.D. Thesis, Trinity Hall, Cambridge, England.

Cooper, G. A., and Kelly, A. (1967). *J. Mech. Phys. Solids* **15,** 279–297.

Corten, H. T. (1967). *In* "Modern Composite Materials" (L. J. Broutman and R. H. Krock, eds.), p. 45. Addison-Wesley, Reading, Massachusetts.

Embury, J. D., Petch, N. J., Wraith, A. E., and Wright, E. S. (1967). *Trans. Metall. Soc. AIME* **239,** 114–118.

Fitz-Randolph, J., Phillips, D. D., Beaumont, P. W. R., and Tetelman, A. S. (1971). Fracture energy and acoustic emission studies of a boron–epoxy composite, *St. Louis Symp. Advan. Fiber Composite, 5th, St. Louis, Missouri.*

Foye, R. L., and Baker, D. J. (1970). Design of orthotropic laminates, *AIAA Conf.*, *11th, Denver, April 1970.*

Gerberich, W. W. (1971). "Crack Growth Mechanisms in Fibrous Composites." Ph.D. Thesis, UCRL-20524, Lawrence Radiat. Lab., Univ. of California, Berkeley, California.

Hancock, J. R., and Swanson, G. P. (1971). Toughness of filamentary boron–aluminum composites, *in* "Composite Materials: Testing and Design." ASTM STP 497. Amer. Soc. Test. Mater., Philadelphia, Pennsylvania.

Harris, B., Beaumont, P. W. R., and Moncunill de Ferran, E. (1971). *J. Mater. Sci.* **6,** 238–251.

Iannelli, A. A., and Rizzitano, F. J. (1962). "Charpy Impact Tests of 4340 Steel and 6 Al–6V–25n Titanium Alloy Using Standard and Thin Charpy Specimens." Watertown Arsenal Lab., WAL TR-112.5/3, January.

Irwin, G. R. (1957). *J. Appl. Mech.*, 361–364.

Jermian, W. A., and Ventrice, M. B. (1969). *J. Adhesion* 1, 180–207.

Johnston, T. L., Stokes, R. J., and Li, C. H. (1961). *Trans. AIME* 221, 792.

Klein, M. J., and Metcalfe, A. G. (1972). Air Force Mater. Lab., sponsored work in progress.

Konish, H. J., Jr., Swedlow, J. L., and Cruse, T. A. (1972). *J. Composite Mater.* 6, 114–124.

Kreider, K. G., Dardi, L., and Prewo, K. (1970). "Metal Matrix Composite Technology." Tech. Rep. AFML-TR-70-193, July, pp. 12–31.

Larsen, J. V. (1971). Fracture energy of CTBN/epoxy—carbon fiber composites, *Annu. Tech. Conf., Reinforced Plastics, 26th*, Sect. 10D. Composites Div., Soc. Plastics Ind.

Leichter, H. L. (1966). *J. Spacecraft Rockets* 3, No. 7, 1113–1120.

McCartney, R. F., Richard, R. C., and Trozzo, P. S. (1967). *Trans. ASM.* 60, 384–394.

McGarry, F. J., and Mandell, J. F. (1970). "Fracture Toughness of Fiber Reinforced Composites." Massachusetts Inst. of Technol. Tech. Rep. R70-79, December.

McGarry, F. J., and Willner, A. M., (1968). "Toughening of an Epoxy Resin by an Elastomeric Second Phase." Massachusetts Inst. of Technol. Tech. Rep. R68-8, March.

Metcalfe, A. G., and Klein, M. J. (1972). Solar Division of International Harvester, personal communication.

Mostovoy, S., and Ripling, E. J. (1965). "Factors Controlling the Strength of Composite Bodies–Interphase Fracturing of Composite Bodies." Bur. of Naval Weapons Contract NOW 64-0410-c.

Mostovoy, S., and Ripling, E. J. (1966). *J. Appl. Polym. Sci.* 10, 1351–1371.

Olster, E. F. (1970). Unpublished data.

Olster, E. F., and Jones, R. C. (1970). "Toughening Mechanisms in Fiber Reinforced Metal Matrix Composites." Massachusetts Inst. of Technol. Tech. Rep. R 70-75, November.

Olster, E. F., and Jones, R. C. (1971). Toughening mechanisms in continuous filament unidirectionally reinforced composites, *in* "Composite Materials: Testing and Design," ASTM STP 497. Amer. Soc. Test. Mater., Philadelphia, Pennsylvania.

Olster, E. F., and Woodbury, H. (1972). "Evaluation of Ballistic Damage Resistance and Failure Mechanisms of Composite Materials." Techn. Rep., AFML-TR72-79.

Outwater, J. O., and Murphy, M. C. (1963). On the fracture energy of unidirectional laminated, *Tech. Conf., Reinforced Plastics, Composites Div., 24th*, Soc. Plast. Ind.

Pagano, N. J., and Pipes, R. B. (1971). *J. Composite Mater.*, 5, 50–57.

Paris, P. C., and Sih, G. C. (1965). Stress analysis of cracks, *in* "Fracture Toughness Testing and Its Applications," ASTM STP 381, p. 30. Amer. Soc. Test. Mater., Philadelphia, Pennsylvania.

Parker, E. R. (1957). "Brittle Behavior of Engineering Structures." Wiley, New York.

Patrick, R. L., Ripling, E. J., and Mostovoy, S. (1964). Fracture Mechanics Applied to Heterogeneous Systems, *Annu. Reinforced Plastics Conf., 19th* Sect. 3B, Soc. Plast. Ind.

Rice, J. R. (1966). *Proc. Int. Conf. Fracture, 1st* (T. Yokobori, M. Kawasaki, and M. Swedlow, eds.), p. 309. Jap. Soc. for Strength and Fracture of Mater.

Sanford, R. J., and Stonesifer, F. R. (1970). "Fracture Toughness of Filament Wound Composites." Naval Res. Lab., NRL Report 7112, July.

Sanford, R. J., Stonesifer, F. R. (1971). *J. Composite Mater.* **5.**

Sidey, G. R., and Bradshaw, F. J. (1971). Charpy-notched impact strength of carbon-fiber/epoxy-resin composites, *Int. Conf. Carbon Fibers, Their Composites Appl., London,* Paper No. 25.

Sih, G. C., and Liebowitz, H. (1968). "Mathematical Theories of Brittle Fracture" (H. Liebowitz, ed.), Vol. 2, pp. 108–114. Academic Press, New York.

Sih, G. C., Hilton, P. D., and Wei, R. P. (1970). "Exploratory Development of Fracture Mechanics of Composite Systems." AFML-TR-70-112, June.

Srawley, J. E., and Brown, W. F., Jr. (1965). Fracture toughness testing methods, *in* "Fracture Toughness Testing and Its Applications," ASTM STP 381, pp. 52–56. Amer. Soc. Test. Mater., Philadelphia, Pennsylvania.

Sultan, J. N., and McGarry, F. J. (1967). "Toughening Mechanisms in Polyester Resins and Composites." Massachusetts Inst. of Technol. Tech. Rep. R67-66, December.

Tetelman, A. S. (1969). Fracture processes in fiber composite materials, *in* "Composite Materials: Testing and Design," ASTM STP 460, pp. 495–498. Amer. Soc. Test. Mater., Philadelphia, Pennsylvania.

Tetelman, A. S., and McEvily, A. S. (1967). "Fracture of Structural Materials," pp. 44–45, 650–659. Wiley, New York.

Wambach, A., Trachte, K., and DiBenedetto, A. (1968). *J. Composite Mater.* **2,** No. 3, 266–283.

Wu, E. M. (1968). Fracture mechanics of anisotropic plates, *in* "Composite Materials Workshop, (S. W. Tsai, J. C. Halpin, and N. J. Pagano, eds.), pp. 27–28. Technomic Publ., Stamford, Connecticut.

8

Interfaces in Oxide Reinforced Metals†

RICHARD E. TRESSLER

Department of Material Sciences
The Pennsylvania State University
University Park, Pennsylvania

In the preceding chapters orderly development of the specific subjects from analytical models through experimental verification of specific prop-

† This chapter was drafted while the author was affiliated with the Advanced Metallurgical Studies Branch, Metals and Ceramics Division, Air Force Materials Laboratory, Wright-Patterson AFB, Ohio.

erties was possible largely because adequate supply of uniform test specimens permitted statistically valid property determinations. In the case of oxide-reinforced metals, none of the candidate materials systems have been developed to the point where sufficient quantities of reproducible test specimens are available for exhaustive parametric investigations. This relative infancy stage stems largely from the lack of a suitable filament–fiber supply. However, the fundamental fabrication studies, limited bond strength investigations, and mechanical property data form a significant body of basic information and general guidelines on these promising composite materials such that a systematic presentation of the state-of-the-art is deemed valuable.

The chapter coverage begins with a discussion of bulk thermodynamic properties of metals and oxides, emphasizing oxides of technological importance as fibers or coatings. The importance of solid solution thermodynamics for a realistic evaluation of composite stability is reviewed. The extensive literature on liquid metal–oxide interactions stemming from fabrication studies for cermets and liquid-infiltrated whiskers is reviewed for the principles and data on wetting of oxides by liquid metals. The relationships between wetting and bonding in composite systems are treated. Solid state bonding and solid state reactions with attention to kinetics of oxide–metal reactions and detailed interface characterization are examined from the standpoint of fabrication and use regimes. Finally, the available data on interface sensitive mechanical properties in the few oxide-reinforced metal systems which have been studied to date are examined.

I. Bulk Thermodynamic Considerations for Oxide–Metal Compatibility

A. Phase Equilibrium

The obvious starting point in evaluating a composite system is to assess the chemical compatibility of the filament–matrix combination. Systems in which there is no thermodynamic driving force for reaction are the most desirable. For example, in the class of composite materials discussed in Chapter 9, the phases are in thermodynamic equilibrium at the solidification temperature. The first step in the evaluation of a particular composite system is to accumulate the phase equilibrium data for the materials in question over the range of temperatures including fabrication and use temperatures. The phase diagram, a graphical integration of the thermodynamic functions which are necessary to define chemical stability, may provide immediate answers. They are easy to use and excellent compilations are available [see, for example, Levin *et al.* (1964, 1969), Rudy (1969), and Hansen (1958)]. However, when the requirements of technologically useful alloys as matrix metals, and very high specific strength and modulus filamentary materials are imposed on candidate composite systems, one finds a paucity of useful phase diagrams. Particularly in oxide systems, where the composition of the vapor phase is an important variable, carefully-determined multicomponent phase diagrams are scarce. However, often thermodynamic data are available which may provide answers or insight to the stability question.

B. Free Energy Considerations

Apart from mechanical property considerations, the initial step in filament selection has been an evaluation of the calculated free energy changes for possible reactions between filament and matrix, as presented by Lynch and Burte (1968). Figure 1 depicts the free energy changes as a function of temperature for several alumina–metal reactions. Metals that readily reduce alumina to aluminum and the metal oxide have negative values of ΔF for the reaction as written. However, important reactions such as ternary compound formation, intermetallic phase formation, simple solution of the filament in the matrix (and vice versa), and the change in ΔF with solid solution in the matrix are all ignored in this preliminary analysis. Often data are lacking for calculations of these other cases.

As an example of the use of thermodynamic data to treat metal–oxide reactions more realistically, consider the case of Ti–Al_2O_3. The reaction plotted in Fig. 1 with the reactants and products in their standard states

$$Ti + \tfrac{1}{3} Al_2O_3 \rightarrow TiO + \tfrac{2}{3} Al \tag{1}$$

has a positive ΔF indicating that Ti is stable in contact with Al_2O_3 over the temperature range plotted. Similar conclusions can be drawn when one considers exchange reactions in the system Ti–ZrO_2 and rare earth sesquioxides in equilibrium with titanium with reactants and products in their standard states. However, when one considers the reaction

$$Ti + X Al_2O_3 \rightarrow Ti(\underline{O})_{3x} + 2X Al \tag{2}$$

where oxygen goes into solution in titanium, one finds that the calculation more closely fits the experimental fact that Ti reduces Al_2O_3. Considering now the reduction of Al_2O_3 and solution of oxygen in Ti one writes the

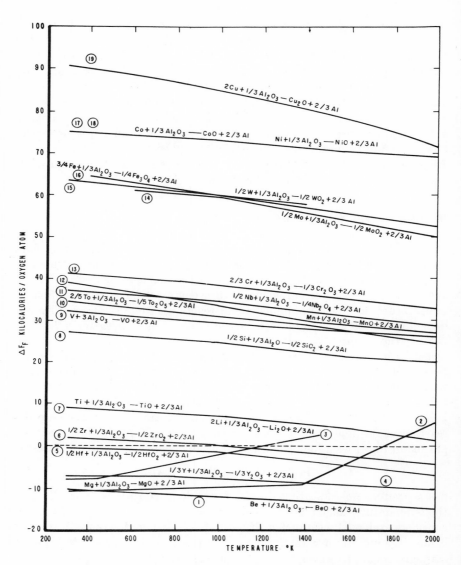

Fig. 1. Thermodynamic values for metal–alumina reactions [from Lynch and Burte (1968)].

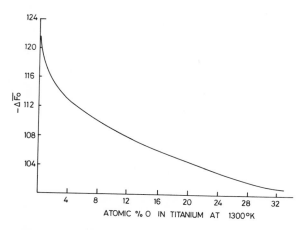

FIG. 2.   Variation of partial molal free energy of oxygen in titanium at 1300°K [from Komarek and Silver (1963)].

following reactions:

$$\tfrac{2}{3}\text{Al} + \tfrac{1}{2}\text{O}_2 \rightarrow \tfrac{1}{3}\text{Al}_2\text{O}_3 \tag{3}$$

$$\tfrac{1}{2}\text{O}_2 \rightarrow \underline{\text{O}}(\text{Ti}) \tag{4}$$

The standard free energy of reaction for reaction Eq. (3) then is the standard free energy of formation of Al_2O_3

$$\Delta F_f{}^\circ = -RT \ln K = RT \ln (P_{\text{O}_2})^{1/2} \tag{5}$$

The relative partial molar free energy of oxygen in the alloy is the free energy for the second reaction

$$\bar{F}_0 - F_0{}^\circ = \Delta \bar{F}_0 = RT \ln a_0 = RT \ln (P_{\text{O}_2})^{1/2} \tag{6}$$

At equilibrium

$$\Delta \bar{F}_0 = RT \ln (P_{\text{O}_2})^{1/2} = RT \ln a_0 = \Delta F_f{}^\circ \tag{7}$$

By comparing the values of $\Delta F_f{}^\circ$ and $\Delta \bar{F}_0$, Tressler *et al.* (1970) have estimated the equilibrium concentrations of O in Ti as a function of temperature for Al_2O_3 in equilibrium with titanium, assuming the reaction occurs as in reaction Eq. (2) above.

In Fig. 2 the change in $\Delta \bar{F}_0$ with oxygen concentration in titanium at 1300°K is plotted. At that temperature, $\Delta F_f{}^\circ$ of Al_2O_3 is approximately

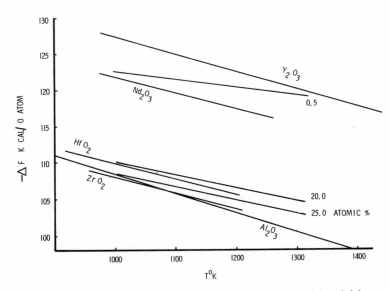

FIG. 3. Comparison of oxide free energies of formation with partial molal free energies of oxygen in titanium [from Tressler *et al.* (1970)].

-100 K cal/g-atom of oxygen indicating that Al_2O_3 is unstable with respect to oxygen dissolved in titanium at 1300°K.

Similar arguments can be used to discuss in a preliminary fashion the stability of other oxides in contact with Ti. In Fig. 3 the ΔF_f°'s of a few oxides are superimposed on the $\Delta \bar{F}_0$'s for various oxygen contents in titanium. In all of these cases but Y_2O_3 (at temperatures below 1350°K) one would conclude that the Ti will reduce the oxide to form a solid solution with oxygen if other reactions are not preferred and kinetically more feasible. Experimental studies by Tressler and Moore (1971) and Lyon (1971) have verified the thermodynamic instability of Al_2O_3 and ZrO_2, respectively, in contact with Ti over a broad spectrum of temperatures. The details of the Ti–Al_2O_3 case are covered in Chapter 3 and later in this chapter. After heat treatment for 60 hr at 950°C in vacuum the ZrO_2–Y_2O_3(ss) is reduced by commercially pure Ti to Zr($\underline{O}$) + Y_2O_3. The oxygen diffuses very rapidly into the titanium causing a band of oxygen-stabilized α-Ti. The growth of the α-Ti band into the β-Ti (at the heat treating temperature) can be directly related to the oxygen diffusion rate in α-Ti.

From Brentnall and Metcalfe's (1968) discussion of the $\Delta \bar{F}_0$ in Cb and the ΔF_f° of Al_2O_3 at 1500°K, one expects less than 100 parts per million of oxygen dissolved in Cb in Cb/Al_2O_3 mixtures equilibrated at this temperature. The stability of these materials has been demonstrated by the lack of reaction after heat treatment for 100 hr at 1500°C in vacuum (Tressler *et al.*, 1970). As a composite system, this particular combination provides the opportunity to develop the desired degree of chemical interaction through reactive metal additions to the Cb matrix.

From this discussion, Y_2O_3 appears to be a possible candidate for an inert diffusion barrier for Ti matrix composites, since very low levels of oxygen will be present in Ti equilibrated with Y_2O_3 below 1350°K.

In the case of the $Ni(O)-Al_2O_3$ system, calculations by a number of investigators have yielded the oxygen partial pressure and temperature regimes for $NiAl_2O_4$ spinel formation at the interface. At 1000°C with a $P_{O_2} > 7 \times 10^{-13}$ atm, $NiAl_2O_4$ will form at the interface. Below this region Ni and Al_2O_3 are in equilibrium. Moore (1969) and Mehan and Harris (1971) demonstrated $NiAl_2O_4$ spinel formation at the $Ni-Al_2O_3$ interface at $P_{O_2} > P_{O_2}$ critical. The kinetics of this reaction are discussed later in the chapter. The effect of metal additives to the Ni which have a greater afinity for oxygen than does Ni has been experimentally demonstrated. Chromium at the 20 wt % level lowers the P_{O_2} to maintain a single phase solid solution (free of Cr_2O_3) to 10^{-23} atm at 1000°C, not sufficient for spinel formation. Titanium has a similar effect, but also reduces the Al_2O_3 as discussed above.

Obviously, to treat the above cases quantitatively, thermodynamic data for possible ternary compounds, ternary solid solutions, and other situations as described in Chapter 3 would be necessary. These examples give only a first-order idea about the chemical stability of these systems. Clearly, the reaction kinetics must be evaluated in reactive systems since useful composites can be chemically unstable but kinetically stable as discussed in Chapter 3.

II. Liquid Metal–Oxide Interactions

The early oxide-reinforced metal composites literature is largely involved with liquid infiltration technology and basic studies of liquid metal wetting of and bonding to oxide materials. This predominance of liquid metal–solid oxide studies was necessitated by the oxide raw material forms, very fine Al_2O_3 whisker mats, and spun glass fibers. The most obvious way to permeate the exceedingly small interfiber passages was by liquid infiltration. As a result, a considerable body of basic information of lasting technical importance was generated and is reviewed here. The intent of this discussion is to outline the general principles that apply to wetting of

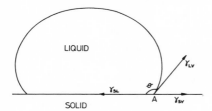

Fig. 4. Schematic diagram of interfacial tension forces acting on a sessile drop.

solid oxides by metals, liquid infiltrating and bonding, and to review the work in specific metal–oxide systems.

A. Principles of Liquid Wetting of Solid Surfaces

Physical wetting of a solid oxide by a liquid metal is a major factor in composite fabrication by liquid infiltration and liquid phase consolidation as well as in producing an effective bond. The degree of wetting is determined by the forces acting at the interface of a drop of liquid on a smooth solid. As depicted in Fig. 4, the sessile drop technique is a convenient way to observe the degree of wetting and relate the forces using Young's equation

$$\cos \theta = (\gamma_{sv} - \gamma_{sl})/\gamma_{lv} \tag{8}$$

where γ_{sl} is the interfacial tension, γ_{sv} the surface tension of the solid, γ_{lv} the surface tension of the liquid metal, and θ the angle of contact or wetting angle. If γ_{sl} is larger than γ_{sv}, the contact angle is larger than 90°, the condition of nonwetting. If the reverse is true, θ is acute and wetting occurs. If $\gamma_{sv} = \gamma_{sl} + \gamma_{lv}$ the liquid will spread over the solid. Put in terms of a spreading coefficient S_{ls}, spreading will occur if S_{ls} is negative

$$S_{ls} = \gamma_{sv} - (\gamma_{lv} + \gamma_{sl}) \tag{9}$$

In pure metal–solid oxide systems, in general θ is greater than 90°. Livey and Murray (1956) have rationalized this behavior on the basis that the surface of the oxide is negative in character. The negative electron cloud of a nonreactive metal is repelled when it comes in contact with the oxide surface resulting in high γ_{sl} and large contact angles. This hypothesis was supported by their results with MO oxides where the electronegativity of the cations varies greatly from MgO to CdO. With this condition prevailing, fabrication is very difficult. Sutton *et al.* (1965) have analyzed the problem for a system of aligned fibers in a tube which is oriented horizontally to minimize gravitational forces. The capillary pressure P_c, which is proportional to the driving force of the infiltrant, can be expressed as

$$P_c = 4\,\gamma_{lv}\,(\cos\theta)\,V_f/d_f \qquad (10)$$

where d_f is the fiber diameter and V_f is the fiber volume fraction. With P_c positive ($\theta < 90°$), the liquid metal penetrates the fiber array. If P_c is negative ($\theta > 90°$), then external pressure is required ($P_e > P_c$) to effect penetration. The situation is illustrated in Fig. 5 for the liquid Ni–Al$_2$O$_3$ whisker case. However, under these conditions, undesirable consequences such as dewetting in small channels during solidification shrinkage necessitate wetting conditions during the entire composite fabrication process. It is this problem as well as that of bonding which has precipitated the era of concerted study to develop wetting in specific metal–oxide systems which are covered later in this section.

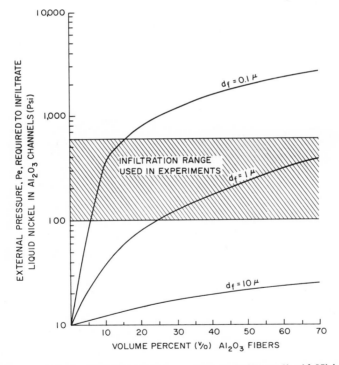

FIG. 5. Calculated external pressures necessary to infiltrate liquid Ni into channels between Al$_2$O$_3$ fibers [from Sutton *et al.* (1965)]. Conditions: temperature 1500°C, surface tension 1800 dyn/cm, contact–angle 112°.

One of the approaches to reduce θ in nonwetting systems has been to take advantage of selective adsorption of electropositive metal solute atoms at the liquid metal–oxide interface thus reducing the interfacial tension as given by the Gibbs adsorption equation

$$\Gamma = -d\gamma/[RT\, d\,(\ln C)] \tag{11}$$

where Γ is the excess surface concentration of solute, γ is the interfacial tension, R is the gas constant, T is the absolute temperature, and C is the bulk concentration of solute. As demonstrated later for specific systems, the interfacial tension decreases over a narrow concentration range, which corresponds to the formation of a monolayer of adsorbed atoms on the solid at the interface.

In applying Young's equation to calculate the interfacial tension (energy), sessile drop experiments give directly the contact angle θ. The value of γ_{lv} is determined from the drop shape (Allen and Kingery, 1959). However, γ_{sv} must be derived by other means. In the case of Al_2O_3, the value has been measured directly as 905 erg/cm^2 at 1850°C (Kingery, 1954). Eberhart (1967) has used the Zisman relationship

$$\cos\theta = 1 + A(\gamma_{lv} - \gamma_c) \tag{12}$$

where A is an empirically determined constant and γ_c is the critical liquid surface tension for spreading, to calculate the surface tension of solid alumina. The treatment assumes that γ_c is a property of the solid surface only. His value calculated from γ_{lv} versus $\cos\theta$ for liquid transition metals on Al_2O_3 is 1050 erg/cm^2 which agrees reasonably with Kingery's value of 940 erg/cm^2 (corrected to 1500°C using $d\gamma_{sv}/dT = -0.1$ erg/cm^2 deg). However, Rhee's (1971) recent treatment of the Cu–O/Al_2O_3 system and the analysis by Brennan and Pask (1968) of the nature of sapphire surfaces as a function of temperature and gaseous environment indicate that these values are not constants. The γ_{sv} of Al_2O_3 is dependent on temperature and gaseous environment, and must be determined for the actual experimental conditions.

The theoretical work of adhesion, the work required to pull apart a unit area of interface creating a metal and oxide surface of unit area, is given by the relation

$$W_{ad} = (\gamma_{lv} + \gamma_{sv}) - \gamma_{ls} = \gamma_{lv}(1 + \cos\theta) \tag{13}$$

Correcting each γ to the test temperature from the temperature at which the γ's were determined, one can obtain a theoretical number which estimates the attraction between the different materials, as demonstrated by

Allen and Kingery (1959). Attempts to relate W_{ad} to the interfacial strength of these materials at room temperature have not been completely successful as described in the discussion of specific systems.

B. Experimental Results for Specific Metal–Oxide Systems

1. Silver–Alumina

Sutton and Chorne (1964) investigated the sapphire whisker–reinforced Ag system early in their composite program to demonstrate the possibility of whisker reinforcement in a chemically stable system. Their sessile drop experiments in Ar and air demonstrated marginal wetting in air and non-wetting in an Ar atmosphere. The dissolution of oxygen in Ag was thought to cause a lower γ_{lv}, thus causing wetting in the air experiments. However, composite properties indicated poor bonding under these conditions. A duplex coating of titanium–nickel promoted wetting and drastically improved the bonding as evidenced by the improvement in tensile strength of composites fabricated with coated and uncoated fibers (Fig. 6). The fracture mode changed from interfacial failure and whisker pullout in the uncoated condition to shear failure through the metal with coated whiskers. These experiments demonstrated the feasibility of manufacturing whisker-

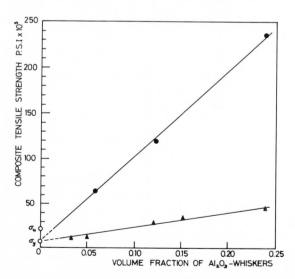

FIG. 6. Tensile strength of silver reinforced with Ti–Ni-coated and uncoated sapphire whiskers; ●, whiskers coated; △, whiskers uncoated [from Noone et al. (1969a)].

reinforced metal matrix composites and showed the potential for reinforcement in these materials.

2. *Aluminum–alumina*

The wetting behavior of Al_2O_3 by Al has been the subject of several
investigations. Carnahan *et al.* (1958) found that in the course of sessile
drop experiments the drop spread and contracted repeatedly. Wolf *et al.*
(1966) observed partial dissolution of sapphire in sessile drop experiments
with Al at temperatures over 925°C. In a more recent study, these phe-

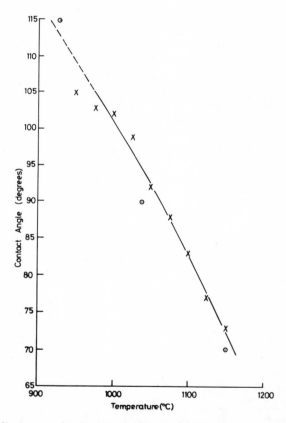

FIG. 7.   Contact angle versus temperature for liquid aluminum on single crystal
sapphire with data points ⊙, [from Wolf *et al.* (1966)], and ✕, [from Champion *et al.*
(1969)].

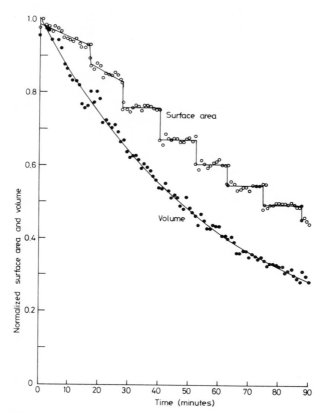

FIG. 8. Variation of free surface area and volume of a drop of liquid aluminum on single-crystal sapphire at 1350°C versus time, normalized to $t = 0$, [from Champion *et al.* (1969)].

nomena were elaborated on in sessile drop experiments with Al, Al–Ni, Al–Cu, Cu, and Au on sapphire, ruby, and recrystallized polycrystalline alumina (Champion *et al.*, 1969). At temperatures below 1150°C the contact angle decreased continuously with increasing temperature as shown in Fig. 7. The contact angle decreased with time to the equilibrium values plotted in Fig. 7. Wetting ($\theta = 90°$) occurs above 1050°C.

At temperatures above 1150°C the spreading followed by sudden contraction, continuously repeated, phenomenon occurred. During this process the drop decreased in area and volume as shown in Fig. 8, apparently due to evaporation. The same general observations were true for the aluminum alloys but not for Cu and Au. A series of reaction rings which corresponded directly to the number of contractions were invariably observed on the substrates.

These observations are consistent with the observations of Brennan and Pask (1968) who also examined wetting of sapphire by aluminum in the temperature range 660°C to 1250°C. They divided the observed behavior into three regions: large obtuse angles $T \leq 870$°C, small obtuse angles 870°C $< T < 950$°C and acute angles $T \geq 950$°C.

The detailed observations were rationalized in terms of the structure of the sapphire surfaces. At low temperatures where large obtuse angles are observed, the sapphire surface is hydroxylated. Van der Waals forces provide poor bonding, i.e., large γ_{sl}. This interface breaks down to form a stable, chemically-bonded interface with small obtuse contact angles. The sapphire surface structure approaches that of bulk sapphire. In this region γ_{sl} approaches γ_{sv}; and, when $\gamma_{sl} = \gamma_{sv}$, $\theta = 90$°. Here γ_{sv} must be less than γ_{lv}, thus much less than 905 erg/cm^2, which corresponds to the third region of behavior.

In the region of acute angles which corresponds to the spreading–contracting region described by Champion *et al.* (1969), the γ-alumina surface is reduced to an oxygen-deficient spinel-type surface containing Al^{2+} ions. This type of surface develops in low partial pressures of oxygen and at high temperature where the kinetics of the reaction are rapid ($T >$ 900–1000°C). The γ_{sv} in this regime corresponds to Kingery's (1954) γ_{sv} of 905 erg/cm^2. The surface is reactive with molten Al. In agreement with Champion *et al.* (1969), Brennan and Pask (1968) suggest that the volatile species Al$_2$O is formed. The mechanism is not clear. Conceivably, the AlO surface of the oxygen-deficient, spinel-type surface combines with Al to form Al$_2$O and causes spreading. The Al^{2+} supply is then replenished giving rise to the contraction in ever decreasing diameters as the Al volatilizes as Al$_2$O. In these investigations the pure Al–Al$_2$O$_3$ bonding was good except in the instances of large obtuse angles. Wetting was not necessary to effect bonding. The interfaces for the alloys in the Champion *et al.* (1969) study were poorer with bubble formation and cracking in the sapphire.

Wetting, thus liquid infiltration, is made possible in the pure Al–Al$_2$O$_3$ system by heating to elevated temperatures. However chemical reaction results, indicating that an optimum temperature must be determined for incorporating corundum whiskers into liquid aluminum without excessive chemical attack.

In a practically-oriented study to establish potential for sapphire wool–reinforced Al, the investigators adopted the approach of using thin coatings to promote wetting (Mehan *et al.*, 1968) which had been successful

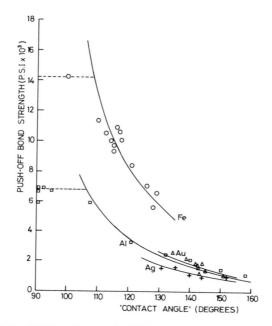

FIG. 9. The pushoff bond strengths of samples of four metal–alumina systems plotted as a function of their contact angles [from Nicholas (1968)].

in the $Ag–Al_2O_3$ system. In this way lower temperatures, nearer the melting point of Al, could be used for liquid infiltration. The duplex coating of Ti–Ni was found to provide wetting and good bonding between Al and Al_2O_3. Nichrome was equally successful as a coating and was used to fabricate most of the specimens for which mechanical property data are discussed later in this chapter.

Recently, Nicholas (1968) has used an analysis (see $Ni–Al_2O_3$ discussion) which assumes an interfacial tensile strength U, which can be calculated from bond strength versus contact angle curves. His data for Al–polycrystalline alumina are presented in Fig. 9. The observed data fit his model showing a transition from tensile failure to shear failures at a contact angle of 109°. The analysis indicates that good interfacial tensile strength is not dependent on wetting.

3. Nickel–alumina

The early work on wetting and bonding in this system was stimulated by efforts to develop a high temperature "cermet" with nickel–alloy in-

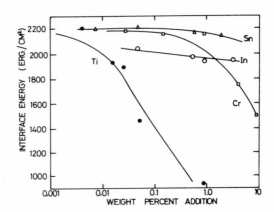

FIG. 10. Effect of Sn, Ti, Zn, and Cr on the Ni(l)–Al$_2$O$_3$(s) interfacial energy at 1475°C. [Reprinted from C. R. Kurkjian and W. D. Kingery, *J. Phys. Chem.* **60,** 961 (1956). Copyright 1956 by the American Chemical Society. Reprinted by permission of the copyright owner.]

filtrated alumina. The pioneering work of Kingery and coworkers is largely summarized in a review paper by Humenik and Whalen (1960). These studies showed that substantial decreases in γ_{sl} are possible by alloying with interfacially active metals such as Ti. The values of γ_{lv} are altered very slightly. The observation that certain alloy compositions could change the Ni–Al$_2$O$_3$ system from nonwetting to wetting was of major importance for liquid infiltration manufacturing processes.

The effect of solute atoms in the metal on the interfacial energy for some nickel alloy–Al$_2$O$_3$ couples is shown in Fig. 10. The more highly electropositive constituents tend to concentrate at the interface. The concentration at which essentially complete coverage of the oxide surface at the interface occurs decreases with increasing affinity of the atom for oxygen (as implied from free energies of formation of the respective oxides). The solute atoms which form less stable oxides than the solvent affect the interfacial energy very little and are apparently not adsorbed at the interface. The Gibb's adsorption equation gives the excess interfacial concentration from the change in γ_{sl} with activity of the solute. Figure 11 demonstrates, with the linear portion of the γ_{sl} versus $\ln(\%$ Ti in Ni), the selective adsorption of Ti at the interface. In the region 0.1 to 1.0% Ti, a monolayer of Ti forms on the Al$_2$O$_3$. At higher concentrations γ_{sl} becomes constant at about 400 erg/cm^2, probably corresponding to multilayer adsorption. In this region the contact angle of the liquid metal drop is less than 90° ($\simeq$70°) indicating that liquid infiltration is possible.

In a similar study, Armstrong *et al.* (1962), examined the mechanisms of bond formation between Ni–Ti and Ni–Cr alloys, and Al_2O_3. Using the sessile drop technique, selective absorption of Ti and Cr at the liquid metal Al_2O_3 interface was observed. However, the linear portions of the γ_{sl} versus % solute curves for both solute atoms occurred at higher concentrations than in the work of Kurkjian and Kingery (1956). This result points out the extreme sensitivity of the selective adsorption phenomenon to oxygen content in the liquid metal. Clearly, the difference could be due to the tie-up of Ti and Cr in the solution by dissolved oxygen, effectively lowering the activity of those constituents in the liquid metal. The two bonding mechanisms identified were selective adsorption of atoms at the interface in the absence of third phase formation in the case of Cr (Cr^{3+} dissolved in the Al_2O_3) and a reaction between the excess atoms and alumina to form a third phase in the case of Ti ($\alpha - Ti_2O_3$ was observed in the alumina in the Ni–Ti/Al_2O_3 samples).

Subsequent studies have concentrated on the details of the wetting and adherence of Ni-alloys on Al_2O_3 and on the optimization of processing parameters, including composition, to achieve useful composite properties. With the advent of sapphire whiskers, Sutton (1964) and Sutton *et al.* (1965) began an extensive study to develop whisker-reinforced metals. Because of the technological importance of nickel, considerable effort was focused on Ni-base alloys reinforced with sapphire. They confirmed much of the earlier work concerning the effect of interfacially active metals on the wetting behavior of binary alloys in sapphire (Sutton and Feingold, 1966). Table I summarizes some results of sessile drop and bonding studies

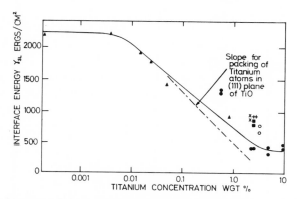

Fig. 11. Interface energies between Ni–Ti alloys and Al_2O_3 with data points ▲, Ni–Ti; ●, Ni–Ti; +, Inconel X550; ×, Inco 700; ○, superior Waspaloy; ■, M252 [from Allen and Kingery (1959)].

TABLE I

RESULTS OF SESSILE-DROP TESTS AND SHEAR TESTS PERFORMED ON Ni-Al$_2$O$_3$ SPECIMENS[a]

Specimen	Contact angle, θ (degrees)	Nickel surface tension, γ_{lv} (erg/cm^2)	Ni-Al$_2$O$_3$ interfacial energy, γ_{sl} (erg/cm^2)	Work of adhesion, W_{ad} (erg/cm^2)	Ni-Al$_2$O$_3$ shear apparent strength (psi)	Type of fracture
High-purity nickel	100.7	1770	1290	1390	17,800	Essentially interfacial separation
High-purity nickel +1.35 at % chromium	107.7	1670	1470	1170	20,000[b]	Essentially interfacial separation
High-purity nickel +0.99 at % titanium	94.5	1770	1090	1640	4800	Fracture in Al$_2$O$_3$ substrate
High-purity nickel +1.00 at % zirconium	136.2	1770	2220	500	0[c]	Fracture in Al$_2$O$_3$ substrate

[a] All data for 1500 ± 20°C and 5 × 10^{-6} Torr except for shear tests, which were conducted at room temperature [from Sutton and Feingold (1966)].

[b] Exceeded limit of test device.

[c] No shear strength; plaque cracked under drop on cooling.

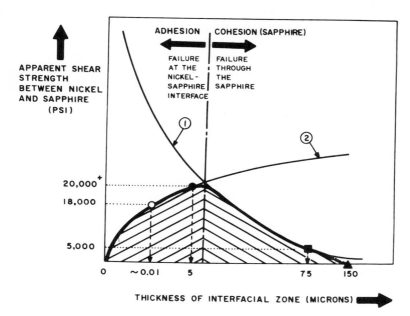

FIG. 12. Apparent shear strength between nickel and sapphire as a result of two competing effects with data points O, HPNi; ●, HPNi + Cr; ■, HPNi + Ti; ▲, HPNi + Zr [from Sutton and Feingold (1966)].

on sapphire. One of the most significant observations of these early composites studies was that the highly reactive metals, Ti and Zr, produced a distinctly undesirable interface causing fracture in the substrate at very low "apparent shear strengths." The authors depict their interpretation of the bonding scheme in the systems considered here in Fig. 12. The competing effects of weakened sapphire and enhancement of interfacial bonds produce a maximum in bond strength which does not correlate with the theoretical work of adhesion W_{ad}, calculated using the parameters determined at the test temperature.

The degree of reaction experienced during contact with the liquid metal of the sessile drop containing reactive metals was enough to destroy sapphire filaments up to 5 μm diam in a matter of a few minutes at 1500°C (Noone *et al.*, 1969a). Concurrent studies also indicated that the sapphire strength could be seriously degraded without appreciable reaction since their high strength is dependent on surface perfection (see Section IV,A.). An extended study of coatings which permitted wetting without sapphire degradation was undertaken. The coating which promoted wetting and provided adequate protection was determined to be elemental tungsten.

However, the rapid dissolution rate of tungsten in the liquid matrix necessitated a coating thickness of 10 μm to survive the fastest vacuum infiltration cycle. Clearly, the volume fraction of fine whiskers with a 10 μm coating would be much too low to effectively reinforce a metal matrix. The effort to use sapphire whiskers in a Ni-base matrix by liquid infiltration techniques was abandoned.

Subsequent liquid infiltration experiments with nickel matrices focused on large diameter sapphire rods, 20 mil diam, coated with various coatings (Noone *et al.*, 1969b). Even with thick coatings of tungsten, these experiments were unsuccessful due to damage to the sapphire rods (see Section IV,A.). Further composite development programs using electroplating and solid state bonding are discussed in Section III.

Other studies in the areas of wetting and adherence have contributed to the understanding of the interfacial characteristics in Ni–alloy/Al_2O_3 systems. Ritter and Burton (1967) investigated the effect of test atmosphere and alloy additions of Ti and Cr on the surface tension and contact angle of nickel on sapphire substrates at 1500°C. The atmosphere had no discernible effect on the γ_{lv} or contact angle in the case of pure Ni–sapphire. The alloy results were consistent with previous investigations. The decrease in contact angle for an alloy in pure argon versus pure hydrogen atmosphere is probably due to greater oxygen content in the argon. The shear test results indicated greater bond strengths for Ni melted in oxygen-containing atmospheres than for Ni melted in oxygen-free atmospheres. The possibility of spinel formation at the interface causing this effect was suggested.

In the Ni–Ti alloy tests a roughened interface resulted with no interfacial compound observed. The purer atmospheres produced the maximum

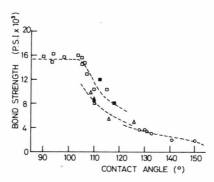

FIG. 13. Correlation between the wetting and bonding behavior of three metal–alumina systems □, Ni–Al_2O_3; ■, Ni–sapphire; △, HPC Cu–Al_2O_3; ○, Cu–Al_2O_3 [from Nicholas *et al.* (1968)].

bond strengths. When oxygen was present a black reaction product was present and resulted in very low strengths. The bonding mechanism could not be clearly deduced. With the Ni–Cr alloys, an optimum (but unknown) oxygen content for maximum bond strength was suggested on the basis of the various test atmospheres. Solution of Cr^{3+} in the Al_2O_3 was suggested as a bonding mechanism. No correlation between W_{ad} and bond strength could be made.

Nicholas *et al.* (1968) have examined the "apparent shear strength" of metal–alumina inferfaces using polycrystalline alumina plaques. In the Ni–alumina system they developed a range of contact angles for bond strength tests by doing sessile drop experiments at a variety of temperatures and times. The data are presented in Fig. 13. They analyzed the forces and bending moments during the strength test to interpret their data. The four assumptions involved were:

(1) The drop surface was spherical.

(2) The ultimate tensile strength U of the interface was uniform over the entire contact area.

(3) The interface was planar throughout the test.

(4) The interface was only elastically-strained prior to fracture.

From the analysis, they predicted that interfaces of drops having $\theta >$ 107° would fail in tension rather than shear if the shear strength of the interface is 0.8 U. With $\theta <$ 107°, shear failure will occur. The dashed lines in Fig. 13 are calculated assuming $U = 17,000$ lb/in.² for the pure Ni–Al_2O_3 specimens. The observed variation of bond strength with contact angle was accounted for with a single U value. The interfacial tensile strength need not be a function of contact angle. The analysis was verified using a model system of ball bearings bonded to brass plates. The analysis was used to calculate a U value for Sutton's data (1964) and Ritter and Burton's data (1967) yielding values of 18,000 lb/in.² and 13,500 lb/in.², respectively, for comparable purity Ni. The value of this analysis, that is, U independent of wetting, seems to be in explaining the great variability and apparent contradictions with respect to bond strength versus contact angle plots for various compositions. From the practical standpoint, wetting is essential for manufacturing if not for strong interfaces. This analysis has been applied to other metals bonded to Al_2O_3 as discussed below.

Using experimental techniques very similar to those of Sutton (1964) and normalizing "apparent shear strengths" to "bond strengths" after Nicholas *et al.* (1968), Rossing (1971) verified Sutton's model presented in Fig. 12. With both single crystal and polycrystalline alumina substrates the interfacial strengths were shown to go through maxima with increasing

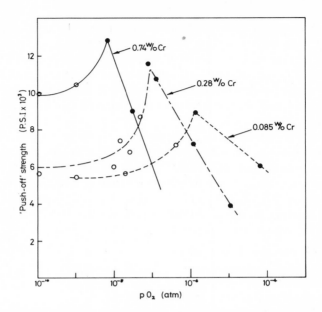

Fɪɢ. 14. Push-off bond strengths of Ni–Cr alloys on sapphire as a function of oxygen partial pressure environment of liquid drop: ○, interfacial failure; ●, substrate failure [from Rossing (1971)].

Cr content in Ni with the failure mode changing from interfacial separation to substrate fracture at the maxima.

The most interesting aspect of Rossing's work is the study of bond strengths as a function of oxygen partial pressure environment of the drop for the various Cr contents in Ni. A wide range of oxygen partial pressures was possible through the use of a Y_2O_3-stabilized ZrO_2 gauge. In Fig. 14 the observed strengths are plotted as a function of oxygen partial pressure. In contrast to earlier investigations the results were very reproducible suggesting that much of the scatter in previous studies was due to variability in oxygen concentration in a given alloy.

Oxygen analyses of the solidified drops showed a linear dependence of bond strength on oxygen concentration in the interfacial failure regime. A qualitative correlation between the calculated solubility limits of oxygen in Ni–Cr and Ni–Ti as a function of Cr and Ti contents and the positions of the maxima in Fig. 14 is presented in Fig. 15. Apparently, the formation of the respective oxides causes weakening of the sapphire–metal interface. The appearance of Cr^{3+} in the interface of the specimens which failed in the substrate is suggested by green coloration in the sapphire. This study documents the suggestion of various earlier investigators that third phase

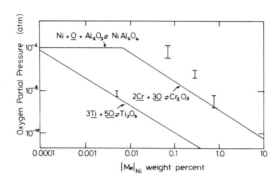

FIG. 15. Metal-oxygen-oxide equilibrium. Bars represent maxima in Fig. 14 and for a Ni–Ti alloy [from Rossing (1971)].

formation at the interface weakens the interfacial region in sessile drop specimens.

4. Copper–alumina

The nature of the bond between Cu–O alloys and sapphire was the subject of investigation by Chaklader *et al.* (1968). Using the sessile drop technique they determined the variation of contact angle with oxygen percentage. At all compositions with greater than 0.18% oxygen, $CuAlO_2$ was detected at the interface. Wetting occurred at very low concentrations of oxygen (0.18%). The possibility of incomplete solution of CuO in the Cu was not ruled out.

In their investigation of the adhesion of metals to polycrystalline alumina, Nicholas *et al.* (1968) fabricated sessile drop specimens of varying contact angle using spectroscopically pure Cu and commercial high conductivity Cu (all in the nonwetting and tensile failure region of their model). Their correlation of bond strength versus contact angle to a tensile strength of the interface demonstrated a strength of 10,800 lb/in.² for the purer Cu and 12,300 lb/in.² for the less pure Cu. These data indicate that a high tensile strength bond does not depend on wetting, is independent of contact angle, and is influenced by the metal purity.

5. Other Liquid Metal–Crystalline Oxide Systems

Nicholas (1968) extended his considerations of the strength of metal polycrystalline alumina interfaces to a wide variety of metals searching for correlations between physical properties and interface strengths, and between the interfacial strength and the W_{ad} values for the interfaces.

TABLE II

Solid–Solid Strengths and
Liquid–Solid Energetics of
Metal–Alumina Interfaces[a]

Metal	U (ksi)	W_{ad} (erg/cm^2)
U	0	429
Co	$0 \to 7.65$	619
Pd	$0 \to 12.9$	477
Ag	8.25	100
Cu	10.5	166
hc Cu	12.4	475
Au	12.5	114
Al	13.0	63
Ni	16.8	962
Fe	27.2	603

[a] From Nicholas (1968).

The data for pushoff bond strength versus contact angle are presented in Fig. 9 for Fe, Au, and Ag, with the lines representing the fit of his model to the data with the interfacial tensile strengths U, presented in Table II. The metals U and Co were essential nonadherent, apparently due to a solid state martensitic transformation which caused the interface to fail from the volume changes associated with the transformation. The iron–alumina interfaces remain intact apparently because the $\gamma \to \alpha$ trans-

TABLE III

The Work of Adhesion W_{ad} of Solid–Solid
Metal–Alumina Interfaces[a]

Metal	Environment, U (ksi)		Temperature (°C)	W_{ad} (erg/cm^2)
Ag	H$_2$	7.65	700	435
Cu	H$_2$	10.5	850	475
Au	Air	12.5	1000	530
Ni	H$_2$	16.8	1000	645
	Ar		1400	518
Fe	H$_2$	27.2	1000	810

[a] Nicholas (1968).

formation occurs by a process of nucleation and diffusion-controlled growth during slow cooling. From Table II there appears to be no correlation between the interfacial tensile strengths and W_{ad} calculated from liquid–solid data using Eq. (14).

In Table III the correlations between some high temperature W_{ad} for solid–solid interfaces and the interfacial strengths are apparent. This agreement in ranking must be viewed with some reserve since other factors may play more prominent roles in the room temperature fracture processes. However, the interfaces appeared to be of comparable perfection.

6. *Aluminum–silica glass*

Composites have been successfully fabricated from silica glass which retained high strength after a rapid liquid Al coating process (Arridge *et al.*, 1964). The deleterious Al–SiO$_2$ reactions were minimized by the rapid freezing of the aluminum coating. Consequences of the reactions in secondary fabrication and use are covered in the section on interface-sensitive mechanical properties.

III. Solid State Bonding and Interface Development

From the foregoing discussion of liquid infiltration and bonding, it is obvious that several very difficult problems are associated with that process. Wetting of the filament by the metal is essential for manufacturing of composites with liquid metals. Marginal wetting can be achieved in systems of technological importance by use of interfacially active metals or temperatures much higher than the metal melting point (in the case of Al). These approaches cause unacceptable mechanical property degradation of the filament, hence the composite. Coatings such as tungsten promote wetting but do not provide sufficient coating life in contact with the liquid metal to be useful with small filaments. Large diameter filaments (>0.010 in.) are mechanically damaged, i.e., twinned and slipped, during cool-down from infiltration temperatures in strong matrices which are technologically interesting.

The alternative approach of solid state fabrication using metal in foil, powder, or electroplated forms holds the promise of eliminating several of the difficulties. Certainly in reactive systems, as many commercial alloy–oxide systems are, the added advantage of lower processing temperatures provides a reduced degree of reaction.

In this discussion the available knowledge concerning solid state reaction

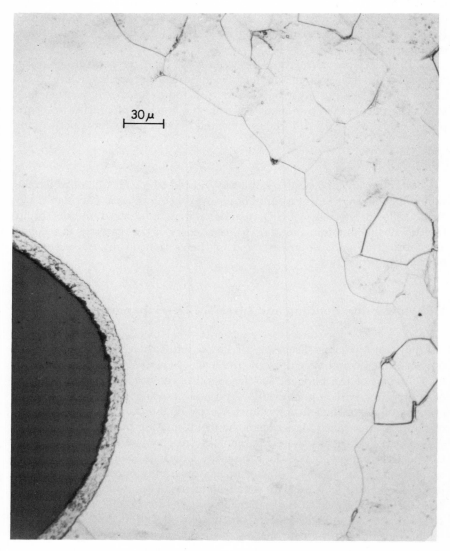

FIG. 16.   Ti–Al₂O₃ reaction zone after 60 hr at 1600°F [from Tressler and Moore (1971)].

kinetics and mechanisms at potential fabrication and use temperatures, and interfacial strengths in systems of interest for useful composite systems is reviewed.

A. Specific Solid Metal–Oxide Systems

1. Titanium–alumina

Tressler *et al.* (1971, 1973) have studied the solid state reactions between pure titanium, Ti–6Al–4V, Ti–8Al–1Mo–1V, and Ti–6Al–2Sn–4Mo–2Zr alloys, and single-crystal sapphire as part of a study to determine the feasibility of fabricating useful Al_2O_3-reinforced titanium matrix composites. The samples were prepared by vacuum diffusion bonding 0.010 in. diam *C*-axis sapphire single-crystal filaments between foils of the titanium alloys at 1500–1550°F for 15–30 min. Subsequent heat treatments were performed in a vacuum of $\sim 1 \times 10^{-5}$ torr. Figure 16 illustrates the microstructure resulting from 60 hr heat treatment at 1600°F of a pure Ti–Al_2O_3 couple. Adjacent to the filament is a gross reaction zone consisting of two distinct regions. The broad, high reflectivity region between the Ti matrix and the gross reaction zone is hardened, recrystallized α-Ti. From replica electron micrographs isolated particles of a second phase were detected in the inner zone adjacent to the filament. The microstructural features of the composites studied were quite similar, apparently consisting of the same phases.

In the temperature range 1200–1600°F, titanium reduces Al_2O_3 to form a gross reaction zone comprised of two distinct zones, an inner zone adjacent to the Al_2O_3 of a TiO phase containing isolated particles of (Ti, Al)$_2O_3$, as determined by phase compatibility studies in the Ti–Al–O system, and an outer zone of a Ti_3Al phase adjacent to the recrystallized, Al- and O-hardened α-Ti band. The isothermal growth of the gross reaction zone follows a parabolic rate law indicating a diffusion-controlled process. The temperature dependence of the rate constants fit an Arrhenuis relationship with activation energies of 50–52 kcal/mole as shown in Fig. 17. The high Al alloys, except Ti–6Al–2Sn–4Mo–2Zr, reacted more rapidly than pure titanium, indicating that possibly Al diffusion through the reaction zone is the rate limiting step, a case for which the activation energies would be reasonable. Both Al and O liberated in the reduction reaction embrittled the titanium matrix. Tentative deductions concerning the influence of the reaction on mechanical properties are discussed later in this section.

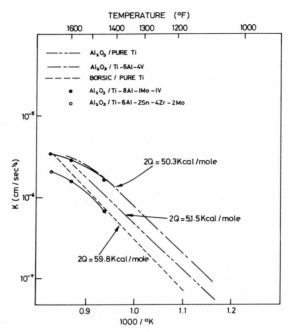

FIG. 17. Parabolic growth rate constants of gross reaction zone versus reciprocal temperature for various titanium matrix composites [from Tressler *et al.* (1973)].

2. *Nickel–alumina*

Moore's (1969) study of nickel–alumina composites identified combinations of materials and oxygen environments in which the formation of NiAl$_2$O$_4$ spinel is favored. Metcalfe has, in Chapter 3, discussed the kinetics of reactions in the oxygen-saturated Ni–Al$_2$O$_3$ system proceeding from the work of Mehan and Harris (1971). It is not clear from these studies whether in fact the formation of NiAl$_2$O$_4$ promotes bonding in this system, as suggested in some of the sessile drop experiments. Further discussion of reaction effects on mechanical properties of these composites is presented in the next section.

In the absence of sufficient oxygen activity to form spinel, Moore (1969) observed a distinctly altered Al$_2$O$_3$ surface from which it is clear that material transfer is occurring at the Ni–Al$_2$O$_3$ interface. Feingold (1967) has addressed the problem of material transfer from the Al$_2$O$_3$ phase through the Ni matrix by a coupled diffusion process to sites of lower chemical potential, i.e., convex surfaces to flat surfaces or small particles to larger particles. Stapley and Beevers (1969) observed spheroidization and in some cases strong crystallographic morphologies in hot-pressed sapphire whisker–nickel composites annealed at 1400°C. In the first case, the driving force for mass transfer is lowered interfacial energy of the sphere, and in the latter case, growth of lower interfacial energy planes at the expense of higher energy planes. None of these explanations fits the observations in the sapphire-reinforced nickel case of Moore (1969). Perhaps as Mehan and Harris (1971) suggest, the effect is the dissolution of Al and oxygen into the matrix. However, the pitting does not correspond with nickel grain boundaries. Apparently the pitting is associated with compositional, structural, or residual stress fluctuations on the surface of the sapphire.

Feingold (1969) measured a solid–solid interfacial tension value for the Ni–Al$_2$O$_3$ interface of 2500 erg/cm^2. This value was lowered by the addition of aluminum.

The reaction kinetics of other metal–oxide solid state reactions have been studied but from a composites standpoint are only of peripheral interest and not included.

B. Strength of the Interface in Solid State Bonded Metal–Oxide Systems

Calow et al. (1971) and Calow and Porter (1971) have performed the only direct measure of bond strength between technologically interesting metals for composite matrices and sapphire crystals. In the case of nickel, powders with 1000 to 2000 ppm oxygen were hot press-bonded to small sapphire disks at various pressures, temperatures, times, and atmospheres. The vacuum-bonded specimens exhibited the best bonding. The high purity powder used in screening studies did not bond in any of these conditions. Vacuum bonding was used for all other parametric test specimens. Full consolidation of the nickel powder occurred at temperatures greater than 1100°C. The data for the parametric studies are shown in Figs. 18 and 19. In apparent disagreement with some earlier studies on pure Ni–Al$_2$O$_3$ bonding using sessile drop techniques, specimens in which NiAl$_2$O$_4$ could be detected by reflection electron diffraction demonstrated the no-bond condition. The well-bonded specimens showed no reaction product at the interface. A surface condition exhibiting X-ray diffraction patterns reminiscent of NiAl$_2$O$_4$ was also observed to give a no-bond interface. The observed bonding time versus shear strength curves were rationalized on the basis of reaction rate effects, sapphire degradation, removal of porosity,

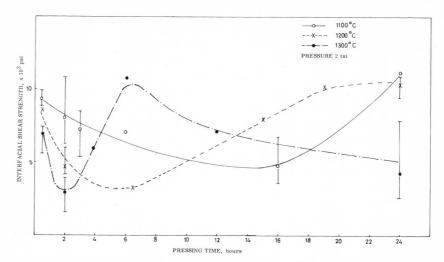

FIG. 18. Shear strength of the Ni–Al₂O₃ interface as a function of pressing time [from Calow and Porter (1971)].

and residual stress relief. Long term heat treatments at 1100°C affected the bond strength very little but changed the fracture mode from interfacial to through-the-sapphire due to the sapphire degradation. The measured shear strengths are adequate for composite reinforcement using the shear lag treatment for current sapphire whiskers.

In a similar study, the solid state bonding of Cr and 80% Ni–20% Cr to sapphire was investigated. Curves very similar to those for Ni–sapphire were observed, with the maximum bond strength for Cr–sapphire 13×10^3 psi and for Ni–Cr 20×10^3 psi. The maximum was achieved for Cr at 1100°C in a slightly oxidizing atmosphere and at 1300°C for Ni–Cr in a similar atmosphere. The bonding mechanism is suggested to be that of solid solution of Cr^{3+} in the Al_2O_3 surface, since segregation of Cr appears in the metal at the interface in short time heat treatments. The affected zone in the sapphire would be too narrow to detect Cr^{3+} in solution according to the authors. This mechanism has been espoused by other investigators studying Cr–Al₂O₃ cermets (Humenik and Whalen, 1960) and Ni–Cr/ sapphire sessile drop specimens (Ritter and Burton, 1967). Clearly, from Rossing's (1971) study for the case of sessile drop specimens, chromium oxide formation at the interface signals the weakening of the interface. Further careful analysis of the solid state bonded system is necessary to resolve this apparent discrepancy.

In these studies, the interfacial shear strength developed between these

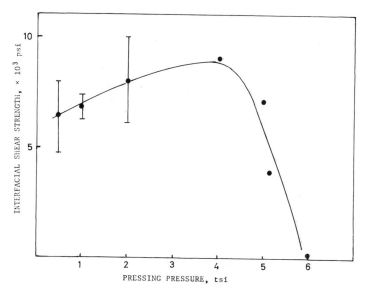

FIG. 19. Shear strength of the Ni–Al$_2$O$_3$ interface as a function of pressing pressure at temperature 1100°C for 2 hr [from Calow *et al.* (1971)].

materials under these conditions are adequate to observe reinforcement of the respective metals if one uses the shear lag analysis of Kelly and Davis (1965). However, the questions of fiber strength retention under these fabrication conditions, and the expected result for mechanical properties that are more dependent on interfacial bonding, i.e., transverse strength, have not been addressed for these systems. As discussed in the following section on mechanical properties, answers are beginning to take form, at least for the Ni–Cr case, from the work of Mehan and Harris (1971).

IV. Interface Sensitive Mechanical Properties

A. *Effect of Chemical Interaction on Fiber Strength*

1. *Aluminum–silica glass*

Silica glass fibers coated with Al by the rapid freeze coating process mentioned in the liquid infiltration discussion were tested by Baker *et al.* (1966), in tension and fatigue, after simulated composite manufacturing

TABLE IV

EFFECT OF MANUFACTURING CONDITIONS ON THE STRENGTH
OF SILICA FIBERS IN ALUMINUM[a]

Treatment	Percent broken fibers after treatment	Mean fatigue life of unbroken fibers (cycles)	Mean breaking stress of fibers unbroken (ton/in²)
Untreated	—	3750	282
450°C for 1 hr in powder	0	3300	198
6000 lb/in² for 1 hr in powder	10	4600	334
450°C for 1 hr at 6000 lb/in² in powder	30	2650	186

[a] Fatigued at 60 ton/in.² [from Baker *et al.* (1966)].

cycles. The data are presented in Table IV. General points emerge from this data:

(1) Fiber breakage of the weaker fibers is a significant problem and accounts for the apparent anomalous increase in breaking stress after the pressure treatment.

(2) The chemical interaction due to temperature application alone causes a significant decrease in breaking stress.

(3) The application of temperature and pressure results in a marked increase in fiber breakage and decrease in breaking stress.

This composite system is discussed further in the treatment of composite strength and fatigue behavior.

2. Nickel–alumina

One of the main justifications for developing metal matrix composites is the potential for greatly improved tensile strength over the metal at least in the filament direction. From the model of Sutton and Feingold (1966) which explained the observed bond strengths and failure modes in sessile drop experiments (Fig. 12) there were substantial reasons to be concerned about fiber strength retention during the fabrication cycle and subsequent use environment. To measure quantitatively the degree of degradation in nickel alloy–Al_2O_3 composites, Noone *et al.* (1969b) resorted to 0.020 in. diam centerless ground "60°" (C-axis 60° from rod axis) rods

TABLE V

BEND STRENGTH OF HIGH QUALITY FLAME-POLISHED SAPPHIRE RODS AFTER
RECEIVING VARIOUS COATINGS AND HEAT TREATMENTS[a]

History of specimens	Mean bend strength (10^3 psi)	Number of results
1. As polished (earlier work)	618	4
2. As polished and heated to 1420°C for "0" min in H_2	635	5
3. Ni–Cr–Fe coated, as sputtered	804	1
4. Ni–Cr–Fe coated, then coating removed in aqua regia	820	1
5. Ni–Cr–Fe coated and heated to 725°C in molten aluminum, then metal removed in acids	846	3
6. Ni–Cr–Fe coated and heated to 1000°C for 18 hr in H_2	245	4
7. As 5 above but metal removed in aqua regia before testing	375	2
8. Ni–Cr–Fe coated and heated to 1450°C for 1 hr in H_2	204	2
9. Ni–Cr–Fe coated and heated to 1420°C for "0" min in H_2	138	4
10. Ni–Cr–Fe coated and heated to 1000°C for 66 hr in air	208	4
11. Ni–Cr–Fe coated and heated to 1000°C for 16 hr in air	215	5
12. Ni–Cr–Fe coated and thick electro-plated nickel coat, then heated to 1000°C for 18 hr in H_2, then metal removed in acids	180	1
13. As polished (recent work)	835	12
14. Ni–Cr–Fe coated, as sputtered	884	4
15. Ni–Cr[b] coated and heated to 1000°C for 23 hr in H_2	227	2
16. Ni–Cr[b] coated and heated to 1000°C for 16 hr in H_2	236	4
17. Ni–Cr[b] coated and heated to 1450°C for 1 hr in H_2	229	3
18. Ni–Cr[b] coated and heated to 1000°C for 16 hr in air	145	2
19. Ni coated and heated to 1000°C for 16 hr in H_2	200	2
20. Ni coated and heated to 1450°C for 1 hr in H_2	238	2

TABLE V *Continued*

History of specimens	Mean bend strength (10^3 psi)	Number of results
21. Ni coated and heated to 1000°C for 16 hr in air	197	4
22. Ti coated and heated to 1300°C in carbon-rich atmosphere to form TiC for 3.5 hr in H_2	120	8
23. TiC coated, as sputtered	724	2
24. TiC coated by sputtering, then heated to 1420°C for "0" min in H_2	286	4
25. HfC coated and heated to 1420°C for "0" min	577	5
26. W coated and heated to 1420°C for "0" min	467	13
27. W coated and heated to 1000°C for 16 hr in H_2	398	7

[a] From Noone *et al.* (1969a).
[b] 80Ni:20Cr.

which could be tested much more easily in bend tests than the small irregular whiskers. In Table V the strength data for filaments with many different coating, heating, etching, etc. treatments are presented. From these data the authors conclude that sapphire filaments cannot be fabricated or used at temperatures greater than 1000°C in these nickel matrices with coatings such as W or monocarbides. Subsequent efforts by this team of investigators followed two divergent approaches. Noone (1970) adopted the approach of optimizing coatings to minimize filament degradation in the manufacture of nickel alloy composites. Mehan and Harris (1971) adopted the approach of acquiring a detailed understanding of the strength degrading mechanism in Ni–sapphire composites to determine the maximum filament strength one can realize in an oxidation resistant matrix. At this point, limited quantities of 0° filaments (0.010 and 0.020 in. diam rods) became available for experimentation. The coating system of Y_2O_3, W, and Ni on the 60° sapphire filaments resulted in substantial increases in room temperature strength as compared with W coatings when exposed at high temperatures to a Ni–Cr matrix. Indeed, filaments coated with the above scheme and incorporated into the Ni–Cr (80–20) matrix results in

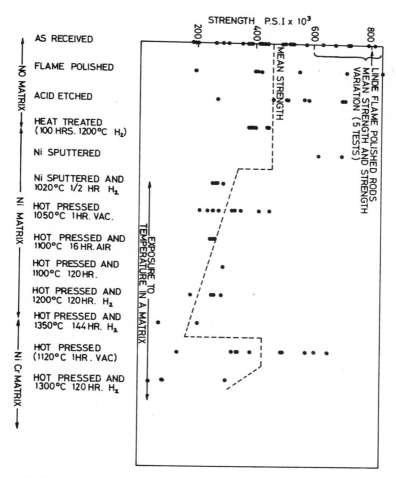

Fig. 20. Bend strength of sapphire filaments after various thermochemical treatments [from Mehan (1970a)]. Dashed line is to show trend only.

extracted filament 4-point bond strengths of 292,000 psi. Discussion of composite properties is deferred until later in this section.

Mehan (1970a) bend-tested 0° filaments after a variety of heat treatments with Ni and Ni–Cr matrices (Fig. 20). Apparently, the Ni–Cr degradation of sapphire is less severe. Subsequent tensile tests of 0° filaments extracted from Ni–Cr composites showed average strengths well over 200 ksi and high values near 270 ksi. In that investigation, the *in-situ*

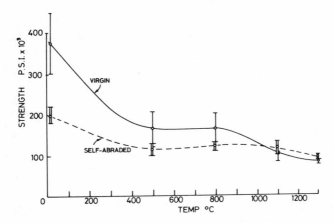

FIG. 21. Tensile strength as a function of testing temperature for virgin and abraded 0° sapphire filaments [from Crane and Tressler, (1971)].

strength of the filaments was estimated by testing the composite in tension and instrumenting the specimen for acoustic emission monitoring, thereby detecting the strain at which the filament fractured. Assuming equal strain in the matrix and filament, the in-situ strength of the weakest fiber was 324,000 psi which is considerably higher than the comparable value for extracted filaments.

From the multitude of photographs and observations of surface structure of filaments degraded by metal attack, the reduction in strength can be attributed to the notch effect of surface flaws caused by selective pitting of the sapphire surface. The effect of this surface damage on the high temperature strength of the filaments in the projected use temperature regimes for the various composites is one of the interesting questions raised by the great volume of degraded filament test data.

Crane and Tressler (1971) tested 0° filaments received with a protective acrylic coating. The surface damage characteristic of filaments attacked by metals was synthesized by self-abrading the filaments (with the coatings removed in vacuum at 800°F) in a paper ball mill. The filaments were tested in air at a cross-head speed of 0.05 in./min. The test data are presented in Fig. 21. The abraded filaments with room temperature strengths approaching 200 ksi showed the same strengths at 1100 and 1300°C as virgin filaments. The significance of this result to high temperature sapphire-reinforced composite mechanical properties is that composites demonstrating filament strengths of 200 ksi will have high temperature

($>1100°C$) tensile strengths similar to those containing filaments with pristine strength levels. For composite materials whose critical parameters for application (such as gas turbine engine hot components) are high temperature strength or stress-rupture strength, concentration on composite fabrication process modifications to improve filament strength retention to levels higher than those observed in nichrome matrix composites (see discussion above) is not warranted with current sapphire filaments.

B. Interface Effects on Tensile Strengths of Composites

1. Longitudinal Strength (*Test Axis Parallel to Fiber Axis*)

(a) SAPPHIRE WHISKER COMPOSITES. The work by Noone *et al.* (1969a) on the model system $Ag–Al_2O_3$ dramatically demonstrated the effect of bond strength on longitudinal tensile properties of discontinuous fiber composites (Fig. 6). The effect of a Ni–Ti coating on the sapphire whiskers incorporated into the Ag matrix was to increase drastically the shear strength of the interface. This effect in terms of the shear lag theory was to lower the critical aspect ratio L_c/d_f to a value less than the aspect ratios of the sapphire whiskers used in these composites. The failure mode was altered from pull out of fibers in the uncoated whisker composites to matrix shear failure in the coated whisker composites.

The extensive development work in sapphire whisker-reinforced nickel alloys was not successful in producing economical, reproducible, and useful composite materials (Noone, 1970). Many of the problems were the result of whisker uniformity (size and quality). However, the great difficulty in fabricating reproducible test specimens by the liquid infiltration and electroplating followed by hot pressing technique (EP/PB) was the major hindrance to further development. The liquid infiltration studies revealed that coatings were necessary to promote wetting. However, a stable coating for use at the infiltration temperature ($\simeq 1450°C$) was not found. Wetting conditions were hard to achieve and in most cases tensile tests were not performed because of gross porosity.

The electroforming technique resulted in composite specific strengths as high as 340,000 in. at 250°C with 12 vol % whiskers. Reproducibility was poor and fiber breakage during the pressing operation was a problem due mainly to the wide variations in whisker size. Furthermore, at high temperature (1800°F) the composite strengths dropped to $\simeq 3000$ psi because of very poor coating–whisker bond strength resulting in gross pullout. The critical fiber transfer length was much greater (due to the low bond strength) than the actual fiber length resulting in very poor tensile properties. The poor bond strength resulted from coating instability, i.e., the W coating diffused into the Ni matrix during bonding.

During the course of the nickel alloy composite work a side study was conducted to evaluate the usefulness of a coating–matrix combination in which the constituents are mutually insoluble. Tungsten-coated sapphire whiskers were liquid infiltrated with Cu. However, familiar problems of coating removal and alignment resulted in composites with as high as 116,000 psi tensile strength with 30 vol % whiskers but extreme variability and inexplicably low high temperature strengths (Chorne *et al.*, 1968).

In a study of sapphire whisker-reinforced aluminum composites (Mehan, 1970b), nichrome and 1020 carbon steel sputtered coatings were used to promote wetting during liquid infiltration. From the very high elevated temperature tensile strengths relative to pure Al (30,000 psi versus 1000 psi at 500°C) and the relative lack of fiber pullout, excellent bonding was achieved with this combination of materials. The creep–rupture curves (to 1000 hr at 400 and 500°C) are quite flat indicating that fiber strength and matrix filament bonding are not degraded in the time and temperature limits of the tests.

(b) CONTINUOUS ALUMINA FILAMENT COMPOSITES. The available literature for continuous sapphire filament-reinforced metals does not allow clear separation of filament degradation and interfacial bonding effects on longitudinal tensile strength. Some of the general features of longitudinal tensile properties of the very limited systems studied to date are presented.

Tressler and Moore (1971) concluded from their longitudinal strength and modulus values for Ti–6Al–4V/Al$_2$O$_3$ (22 vol %) that the modulus approximated the "rule of mixtures" value and the tensile strength contribution from the filaments was approximately 300 ksi indicating a minor degradation of filament tensile strength. The lack of filament pullout indicated adequate interfacial bond strength. They also came to the tentative (since the specimens were not ideally bonded) conclusion, based on tensile tests after heat treatment, that Al$_2$O$_3$ filaments in titanium are less sensitive to the degree of reaction at the interface than are boron or Borsic filaments in titanium.

In subsequent studies in attempts to fabricate 40 vol % plates of the same materials, the tensile results were disappointingly low. Extraction of filaments revealed flaws regularly spaced along the filament length. The flaws were coincident with rhombohedral planes. The flawed planes which are very weak spots in the filaments have their origin in rhombohedral twins. The further assessment of interfacial effects on Ti–6Al–4V/Al$_2$O$_3$ composites awaits the fabrication of specimens without twinning flaws in the sapphire.

In the nickel alloy, continuous sapphire system (Noone *et al.*, 1969b) electroplated nickel–sapphire composites showed longitudinal strengths higher than the whisker-reinforced Ni but failure surfaces were replete with filament pullout indicating a less than ideal bond. Extending the work to Y_2O_3-coated sapphire-reinforced Ni and Ni–Cr fabricated by hot pressing foil–filament arrays, Noone (1970) found evidence of filament strength degradation using 0° filaments. Concurrently, Mehan and Harris (1971) were hot pressing uncoated sapphire-reinforced Ni and Ni–Cr. In low volume fraction composites they measured indirectly *in situ* filament strengths of over 300,000 psi while the extracted filaments averaged slightly over 200 ksi.

(c) Silica Glass Fiber-Reinforced Alumina. The key to successful fabrication of silica fiber reinforced aluminum was the development of a rapid freeze coating process for aluminum coating the fiber prior to hot press consolidation as discussed in Section II,B,6. Chemical reactions during consolidation can drastically reduce the strength of silica fiber as discussed in Section IV,A,1. Morley (1964) implies that these two sources of filament strength reduction control the ultimate tensile strength of these composite materials. The nature of the interface in this material most seriously limits the fatigue behavior as discussed in Section IV,B,3.

2. Transverse Tensile Strength

Since the longitudinal strength of the class of materials discussed in this chapter historically has been used as a measure of composite quality when compared with the "rule of mixtures" calculated strengths, a body of data is available for most materials chosen for study. Only recently have off-axis properties of oxide reinforced metals in the research stage been measured. Consequently, transverse (tensile test direction perpendicular to the filament alignment direction) strength data have only been reported for continuous sapphire filament-reinforced titanium and nickel alloys.

In their preliminary assessment of sapphire-reinforced titanium composites Tressler and Moore (1971) measured values of transverse tensile strength for 22 vol % sapphire composites approaching 60 ksi which compared favorably with other titanium composites. More interesting was the fact that the interface appeared very well bonded as seen in Fig. 22, where the fracture parted the filaments, not the interface. This potential for a very strong interface has spurred interest in this system.

In their study of sapphire reinforced Ni and Ni–Cr, Mehan and Harris (1971) investigated the transverse strength of reinforced Ni–18.5% Cr as a function of volume fraction and processing conditions. In Fig. 23 their

FIG. 22. Scanning electron micrograph of fracture surface of a transverse tensile specimen of Ti–6A1–4V/Al₂O₃ [from Tressler and Moore (1970)].

data is compared with two theoretical treatments of transverse strength as a function of volume fraction for ideal bonding and no bonding conditions as discussed in Chapter 5.

For all preparation conditions, the transverse strengths were consistent with the no bond strength condition. Metallographic observations of the fractured specimens were consistent with this result, i.e., the specimens failed by interface separation and through the specimen cross section of minimum metal volume fraction. Even with the application of the interfacially active metal Ti, the bond strength was not improved. However, this result is quite tentative and the approach of additions of minor amounts of interfacially active metals should be studied more carefully. The filament damage problem during hot pressing as discussed in Section IV,A complicates the findings of Mehan and Harris (1971).

3. Fatigue Behavior

The fatigue behavior of oxide-reinforced metals has been studied to a very limited extent, and only in the case of the silica fiber–Al system,

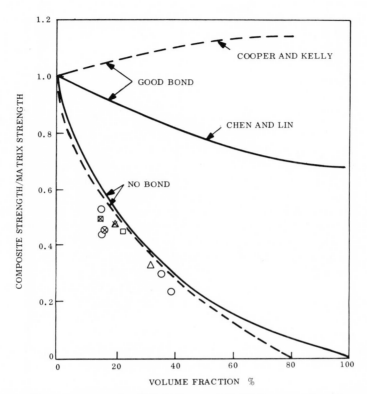

FIG. 23. Transverse strength of Ni–18.5 percent Cr/Al₂O₃ composites: ○, as-pressed; □, sputtered NiCr; △, sputtered Ti; ×, heat treated [from Mehan and Harris (1971)]

sufficiently to establish interfacial effects on the observed phenomena. Mehan (1970b) reported a limited set of fully-reversed constant load fatigue data for 20 vol % Al₂O₃–Al composites. The results indicated lower fatigue strengths than corresponding values for aluminum alloys. No systematic evaluation of the fatigue failure process was performed.

Baker *et al.* (1966) have quite exhaustively studied the serious fatigue effect in silica fiber-reinforced aluminum composites. These experiments demonstrated that uncoated fibers did not experience a fatigue effect while Al-coated fibers showed the same fatigue effect as the composite materials. Detailed examination of fatigued specimens ruled out the possibility of a matrix crack-initiated failure of the glass fibers. Their evidence all points to a mechanism of hard particles in the interface causing abrasion damage to fibers while moving during cyclic loading of the specimen. The origin of the hard particles is probably the reaction of Al with the SiO₂ fibers to form Al_2O_3 and Si. The movement of these particles apparently results in progressive damage to the fibers resulting in the very sharply sloping S–N curves for the composite materials.

V. Conclusion

Although oxide-reinforced metals have been researched since the earliest days of ceramic fiber-reinforced metals, as a class of materials they are still in a technological infancy stage. The clearest potential for high modulus, high strength refractory oxide-reinforced metals is in the very high temperature regime in applications such as hot components of gas turbine engines. The very high temperature stability demands of these applications complicate the development cycle drastically, compared with the lower temperature metal matrix composites such as Al–B. Much of the emphasis to-date on interfacial studies in these materials has been in the area of compatibility under fabrication conditions where the reactivity problem is most acute. The equally important question of interfacial integrity during service (presumably at lower temperatures than during fabrication) has only recently been carefully studied in the continuous sapphire filament–metal systems.

The approach of liquid-infiltrating sapphire whisker arrays has been very productive from the standpoint of understanding of liquid metal–oxide interfacial phenomena and demonstrating reinforcement potential in low melting temperature matrices. But useful composites have not resulted largely because of problems of filament supply, uniformity, and difficulty in manufacture of composites with the available whisker geometries. The use of continuous Al_2O_3 filaments circumvents many of these problems, and at this stage, offers the best potential for fabricating useful high temperature metal matrix composites. Although the fundamental scientific background, as reviewed in this chapter, necessary for understanding interface development and stability in oxide-reinforced metals during liquid phase or solid state fabrication and subsequent service is in fairly good order, considerable detailed knowledge must be developed to understand the behavior of specific alloy–filament combinations. The chemical and mechanical interactions between specific alloy–filament combinations during fabrication and service must be examined systematically using the scientific background developed to date.

As reproducible composite specimens become available, emphasis must be placed on the effect of service conditions on interfacial integrity and interface-sensitive mechanical propreties. As has been the case with Al–B composites, feedback from these studies is very important in establishing precise fabrication conditions for developing a specific interfacial condition for a particular application. In this vein, the designer must establish realistic property requirements for these anisotropic, limited ductility materials in applications such as gas turbine vanes and blades so that properties which are interface sensitive, such as transverse strength, can be efficiently developed for a particular application. At the current state-of-the-art as reflected in this chapter, the principles and basic knowledge are available for altering interfacial structure in oxide-reinforced metals; but, because of the lack of reproducible composite specimens, the effects of changes in interfacial structure and composition on composite mechanical properties are virtually unknown.

References

Allen, B. C., and Kingery, W. D. (1959). *Trans. AIME* **215**, 30.

Armstrong, W. M., Chaklader, A. C. D., and Clarke, J. F. (1962). *J. Amer. Ceram. Soc.* **45**, 115.

Arridge, R. G. C., Baker, A. A., and Cratchley, D. (1969). *J. Sci. Instrum.* **41**, 259.

Baker, A. A., Mason, J. E., and Cratchley, D. (1966). *J. Mater. Sci.* **1**, 229.

Brennan, J. J., and Pask, J. A. (1968). *J. Amer. Ceram. Soc.* **51**, 569.

Brentnall, W. D., and Metcalfe, A. G. (1968). Air Force Mater. Lab. Tech. Rep. 68–82.

Calow, C. A., and Porter, I. T. (1971). *J. Mater. Sci.* **6**, 156.

Calow, C. A., Bayer, P. D., and Porter, I. T. (1971). *J. Mater. Sci.* **6**, 150.

Carnahan, R. D., Johnston, T. L., and Li, C. H. (1958). *J. Amer. Ceram. Soc.* **41**, 343.

Chaklader, A. C. D., Armstrong, W. M., and Misra, S. K. (1968). *J. Amer. Ceram. Soc.* **51**, 630.

Champion, J. A., Keene, B. J., and Sillwood, J. M. (1969). *J. Mater. Sci.* **4**, 39.

Chen, P. E., and Lin, J. M. (1969). *Mater. Res. Std.* **2**, 29.

Chorne, J., Bruch, C. A., Jakas, R., and Sutton, W. H. (1968). Final Rep. Contract N00019-67-C-0243, Naval Air Syst. Command.

Cooper, G. A., and Kelly, A. (1969). *In* "Interfaces in Composites," ASTM STP 452, p. 90. Amer. Soc. Test. Mater., Philadelphia, Pennsylvania.

Crane, R. L., and Tressler, R. E. (1971). *J. Comp. Mater.* **5**, 537.

Eberhart, J. G. (1967). *J. Phys. Chem.* **71**, 4125.

Feingold, A. H. (1967). Ph.D. Thesis, Cornell Univ., Ithaca, New York.

Hansen, M. (1958). "Constitution of Binary Alloys." McGraw-Hill, New York.

Humenik, M., and Whalen, T. J. (1960). *In* "Cermets" (J. R. Tinklepaugh and W. B. Crandall, eds.), pp. 6–79. Van Nostrand Reinhold, Princeton, New Jersey.

Kelly, A., and Davies, G. J. (1965). *Metall. Rev.* **10**, 37.

Kingery, W. D. (1954). *J. Amer. Ceram. Soc.* **37**, 42.

Komarek, K. L., and Silver, M. (1963). *Proc. Symp. Thermodynamic Nucl. Mater., Vienna, 1962*, pp. 749–773.

Kurkjian, C. R., and Kingery, W. D. (1956). *J. Phys. Chem.* **60**, 961.

Levin, E. M., Robbins, C. R., and McMurdie, H. F. (1964). "Phase Diagrams for Ceramists." Amer. Ceram. Soc., Columbus, Ohio.

Levin, E. M., Robbins, C. R., and McMurdie, H. F. (1969). "Phase Diagrams for Ceramists, Supplement." Amer. Ceram. Soc., Columbus, Ohio.

Livey, D. T., and Murray, P. (1956). *Plansee Proc., 1955*, pp. 375–404. Pergamon, Oxford.

Lynch, C. T., and Burte, H. M. (1968). *In* "Metal Matrix Composites," ASTM STP 438, pp. 3–25. Amer. Soc. Test. Mater., Philadelphia, Pennsylvania.

Lyon, S. R. (1971). Private communication.

Mehan, R. L., (1970a). Air Force Mater. Lab. Tech. Rep. 70-160.

Mehan, R. L. (1970b). *J. Comp. Mater.* **4,** 90.

Mehan, R. L., and Harris, T. A. (1971). Air Force Mater. Lab. Tech. Rep. 71-150.

Mehan, R. L., Jakas, R., and Bruch, C. A. (1968). Air Force Mater. Lab. Tech. Rep. 68-100.

Moore, T. L. (1969). *Proc. TMS Symp. Metal Matrix Composites, 1969*, DMIC Memo. 243, p. 47.

Morley, J. G. (1964). *Proc. Roy. Soc.* **A282,** 43.

Nicholas, M. (1968). *J. Mat. Sci.* **3,** 571.

Nicholas, M., Forgan, R. R. D., and Poole, D. M. (1968). *J. Mater. Sci.* **3,** 9.

Noone, M. J. (1970). Final Rep. Contract N0019-69-C-0310, Naval Air Syst. Command.

Noone, M. J., Feingold, E., and Sutton, W. H. (1969a). *In* "Interfaces in Composites," ASTM STP 452, pp. 59–89. Amer. Soc. Test. Mater., Philadelphia, Pennsylvania.

Noone, M. J., Mehan, R. L., and Sutton, W. H. (1969b). Final Rep. Contract N00019-68-C-0304, Naval Air Syst. Command.

Ritter, J. E., and Burton, M. S. (1967). *Trans. AIME* **239,** 21.

Rhee, S. K. (1971). *J. Amer. Ceram. Soc.* **54,** 376.

Rossing, B. (1971). Private communication.

Rudy, E. (1969). Air Force Mater. Lab. Tech. Rep. 65-2, Part V.

Stapley, A. J., and Beevers, C. J. (1969). *J. Mater. Sci.* **4,** 65.

Sutton, W. H. (1964). Rep. R-64SD44, G.E. Space Sci. Lab., Philadelphia, Pennsylvania.

Sutton, W. H., and Chorne, J. (1964). *In* "Fiber Composite Materials," p. 173. Amer. Soc. Metals, Metals Park, Ohio.

Sutton, W. H., and Feingold, E. (1966). *In* "Materials Science Research" (W. W. Kriegel and Hayne Palmour, III, eds.), Vol. 3, pp. 577–611. Plenum Press, New York.

Sutton, W. H., Chorne, J., Gatti, A., and Sauer, E. (1965). Final Rep. Contract N0w 64-0540c, Bur. of Naval Weapons.

Tressler, R. E., and Moore, T. L. (1970). *Proc. Ann. SEM Symp.*, 3rd *1970* IITRI, Chicago, Illinois.

Tressler, R. E., and Moore, T. L. (1971). *Metals Eng. Quart.* **2,** 16.

Tressler, R. E., Lyon, S. R., and Gegel, H. L. (1970). Presented at Metal Matrix Composites Working Group Meeting, USAFA, Colorado Springs, Colorado.

Tressler, R. E., Moore, T. L., and Crane, R. L. (1973). *J. Mater. Sci.* **8,** 151.

Wolf, S. M., Levitt, A. P., and Brown, J. (1966). *Chem. Eng. Progr.* **62,** 74.

9

Interfaces in Directionally Solidified Eutectics

RICHARD W. HERTZBERG

*Department of Metallurgy and Materials Science
and Mechanical Behavior Laboratory,
Materials Research Center,
Lehigh University,
Bethlehem, Pennsylvania*

I. Introduction

While much research and development work was being conducted toward the development of high strength filamentry particles and their use in composite structures, a unique approach was being investigated. By unidirectional solidification of a eutectic alloy in the Al–CuAl₂ system, an aligned two-phase structure was produced (Kraft and Albright, 1961). While the achievement of an aligned microstructure, in itself, certainly did not mean that a reinforced composite material was produced, it remained for Lemkey and Kraft (1962) to show that Cr whiskers, grown during the unidirectional solidification of the Cu–Cr eutectic alloy, could exhibit tensile strenghts in excess of 1.2×10^6 psi and associated elastic strains of about 3%. Later, Hertzberg and Kraft (1963) demonstrated that a strong bond was developed between the Cr whisker and the Cu matrix. Summarizing these three observations and generalizing their existence to many other eutectic systems (on the basis of subsequent work to be discussed below) one may conclude that through the unidirectional solidification of eutectic alloys a high strength phase (rods or lamellae) can be simultaneously produced, aligned parallel to one direction, the ingot axis, and properly bonded to the matrix phase.

The three-in-one processing operation for the preparation of a eutectic composite is highly efficient in that problem areas peculiar to each processing step are eliminated. For example, there is no need to handle individual fibers as they would have to be in the alignment process of a conventional composite; also, bonding difficulties such as inadequate wetting or oxide film formation at the interphase interface are eliminated. In fact, one extraordinary feature of controlled eutectic microstructures is the interphase interface, its morphology, crystallography, stability, and overall response to internal and external stress fields. This important characteristic of the eutectic composite will be the focus of discussion in this chapter.

With consideration given to the classification of interfaces, as discussed earlier in this text, eutectic interfaces would logically belong to Class I: Filament and matrix mutually nonreactive and insoluble (or soluble to a small degree).

II. Characterization of Eutectic Interfaces

A. Nature of the Solidification Process

The key factor behind unidirectional solidification of a eutectic alloy, or any melt for that matter, is that of uniaxial heat flow. All that is needed is a furnace and travelling mechanism to effect relative movement of the ingot and furnace. Let us assume that the ingot is to be withdrawn from the furnace as shown in Fig. 1. Since most of the ingot will remain within

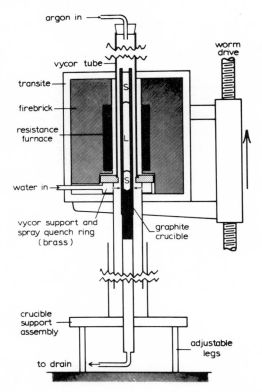

argon in

worm drive

vycor tube

transite

firebrick

resistance furnace

water in

vycor support and spray quench ring (brass)

graphite crucible

crucible support assembly

adjustable legs

to drain

FIG. 1. Cross section of typical unidirectional solidification apparatus [from Eckelmeyer and Hertzberg (1972)].

the furnace for a long period, little lateral heat flow is possible. Rather, the heat flow is directed along the ingot axis toward that segment that has exited the furnace. This axial heat flow toward the colder part of the ingot soon produces an equilibrium spatial position of the solid–liquid interface; this interface is found to be normal to the axis of heat flow, the ingot axis. As the ingot is slowly withdrawn from the furnace, the solid–liquid interface moves up until all the material has solidified. For an alloy of eutectic composition, the solidification reaction is†

$$L \rightarrow A + B$$

† It is appropriate to note at this juncture that the two solids, A and B, need not be one-component phases, but could involve a multicomponent system. Therefore, A and B, the eutectic mixture, is intended to reflect other phase combinations such as: AB + B, AB + AC, A + BC, etc. As will be discussed later, much attention is being given to these multicomponent eutectic systems.

Since the two solid phases (A and B) grow simultaneously perpendicular to the solid–liquid interface, they are, in fact, parallel to one another. The process of unidirectional solidification is, therefore, responsible for the development of the aligned microstructure.

The solidification details of the eutectic reaction have been critically examined by Scheil (1959), Tiller (1958), Jackson and Hunt (1966), and many others as extensively reviewed by Hogan *et al.* (1971). Scheil and Tiller showed that for stable growth of a lamellar eutectic to occur, some measure of undercooling below the equilibrium eutectic temperature was necessary. For one thing, the interfacial energy between the two solid phases must come from the energy released during solidification of the melt; consequently, the degree of undercooling was set by the subsequent interfacial energy necessary between the solid phases, which reflects the free energy difference between the solid and liquid phases (Zener, 1946). Also, some undercooling is necessary to achieve a balance between local diffusion rates of atoms at the interface with the overall velocity of the interface. Combining these two factors and minimizing the degree of undercooling with respect to the interlamellar spacing λ, Tiller showed that a relationship of the form

$$\lambda^2 R = \text{const} \tag{1}$$

existed between the degree of undercooling, the solidification rate R, and the resulting interlamellar spacing. As reviewed by Hogan *et al.*, many subsequent experimental studies have proven Eq. (1) to be valid in lamellar structures where λ is the interlamellar spacing, and in rod-like structures where λ represents the interfiber spacing.

B. Morphology of Eutectic Microstructures

Not all eutectic alloys are desirable from the standpoint of the microstructures resulting from the directional solidification process. Sheil (1946) showed that eutectic microstructures could be separated into two major categories: "normal" and "abnormal." "Normal" microstructures are produced by the simultaneous formation and growth of the two solid phases at the solid–liquid interface. A critical condition for this type of solidification is that adjacent units of the two phase structure grow with the same velocity into the receding melt. Consequently, a planar solid–liquid interface is developed. Typical "normal" microstructures are either in the form of alternate lamellae (Fig. 2) or parallel rod-like particles embedded within a continuous matrix (Fig. 3). By contrast, "abnormal" eutectic microstructures are the result of solidification wherein each solid phase does not

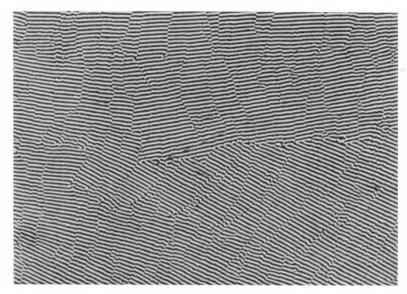

FIG. 2. Typical "normal" microstructure of Al–CuAl₂ in the form of alternate lamellae [from Kraft (1966)].

grow into the melt at the same rate. Since the leading phase with the higher growth velocity may move freely into the melt, it will usually do so by a complex dendrite and branch formation process. The lagging phase can then only form in the interdendritic regions resulting in the development of a nonuniform microstructure (Fig. 4). While some eutectic systems prefer either "normal" or "abnormal" growth, other eutectic alloys will be bimorphic in that either "normal" or "abnormal" microstructures could be developed depending upon the growth conditions. Almost without exception, "normal" microstructures rather than "abnormal" types are being considered for reinforced composite potential; consequently, further discussion will relate only to eutectic alloys possessing "normal" microstructures.

In their review of eutectic grains,† Hogan *et al.* (1971) reported the microstructure of "normal" eutectic alloys to be predominately of the rod or lamellae type. After many studies of different eutectic systems, it remained for Cooksey *et al.* (1964) and Jackson and Hunt (1966) to define the condition wherein the transition from a rod to lamellar microstructure

---

† A eutectic grain describes an oriented position of a eutectic specimen which is comprised of many crystals of each of the two eutectic phases.

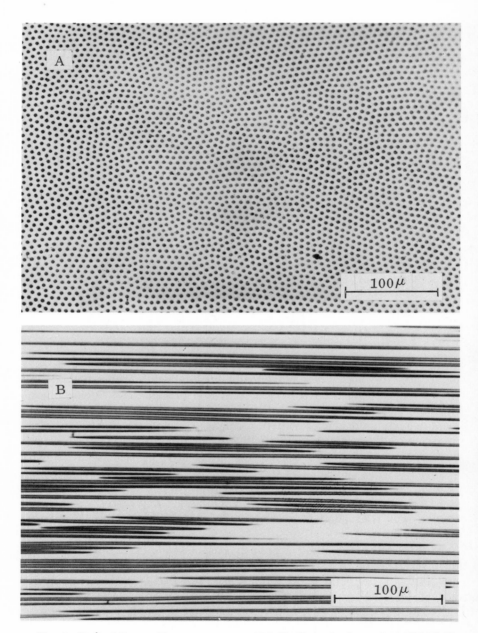

FIG. 3. Typical "normal" microstructure of MnSb–Sb in the form of parallel MnSb whiskers embedded within a continuous matrix of Sb: (A) transverse section and (B) longitudinal section [from Jackson (1967)].

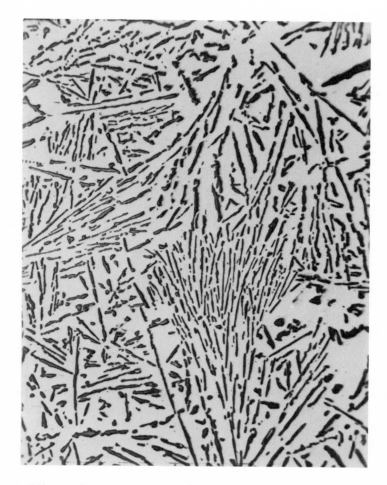

FIG. 4. "Abnormal" microstructure in Al–Si eutectic. Growth direction bottom to top [from Day and Hellawell (1968)].

would occur. By considering the interfacial energy per unit volume for both the rod and lamellar morphology it was shown that the transition would occur at a volume fraction of $(1/\pi)$ (vol $\%$ = 32%) (Fig. 5). Therefore, with all other factors equal, a rod-like morphology is preferred when the minor phase is present in amounts less than $1/\pi$ of the total volume (the minor phase would assume the fiber morphology within a continuous matrix of the major constituent.) When the minor phase constitutes more than $1/\pi$ of the total volume, a lamellar structure consisting of alternate platelets of the two solid phases would be preferred.

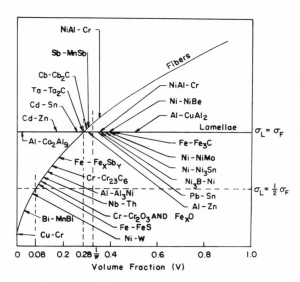

FIG. 5. Rod to lamellar morphology change in terms of interfacial area/unit volume [from Hogan *et al.* (1971)].

C. Crystallography of the Interface

The plane front growth characteristics associated with the development of "normal" microstructures reflect the presence of a low solid$_1$–solid$_2$ interfacial energy as compared to the solid$_1$–liquid and solid$_2$–liquid interfacial energies, respectively. The achievement of such a minimization of surface energy would suggest a preferential interfacial relationship between the two solid phases. Indeed, unique crystallographic relationships have been found in many eutectic systems wherein specific planes in each phase are parallel to one another and to the lamellar habit plane while certain directions in these planes are mutually parallel [see review by Hogan *et al.* (1971)]. That such crystallographic relationships are reproducible in different eutectic grains and separate ingots strongly indicates the associated interface to be preferred and, therefore, of low energy. It remains, then, to show that the reported unique crystallographic relationships are reasonable in terms of their potential for minimization of interfacial energy.

Using metallographic and X-ray diffraction techniques, Kraft (1962) observed a unique crystallographic relationship in the Al–CuAl$_2$ eutectic alloy of the form

$$\text{Interface} \quad || \ (\bar{1}11)_{Al} \ || \ (\bar{2}11)_{CuAl_2}$$
$$[101]_{Al} \ || \ [120]_{CuAl_2}$$

By analyzing the structure of the two phases, Kraft was able to show that the two parallel planes comprising the presumably low energy interface were widely-spaced and, most importantly, of nearly equivalent atomic density. (Kraft's analysis was somewhat complicated by his definition of the ($\bar{2}$11) planes in the CuAl$_2$ phase. They were, in fact, defined as "puckered" planes containing several very closely-spaced layers of Al and Cu atoms. The density and spacing of these clusters were taken to be the density of the clustered planes taken together and the distance between the clusters, respectively.) For such planes of comparable atomic density, a small misfit and low interfacial energy are to be expected. In a critical survey of crystal interface models, Fletcher (1971) demonstrated that the interfacial energy rises sharply with increasing misfit. Using similar reasoning, Kraft (1963) was able to rationalize the observed crystallographic relationship in the Mg–Mg$_2$Sn eutectic system. Here again, those planes found parallel to one another were of similar atomic density. More recently, the presence of matching planes with comparable atom density has been verified in the Ni$_3$Nb–Ni$_3$Al (Thompson and Lemkey, 1969), Ni–Ni$_3$Nb (Quinn *et al.*, 1969), Ni–Cr (Hopkins and Kossowsky, 1971), and NiAl–Cr (Walter *et al.*, 1969), eutectic systems, as well as in many other alloys (Hogan *et al.*, 1971).

Reconsidering the role of volume fractions on the rod to lamellar transition, we are now in a position to explain some anomalous results to be noted in Fig. 5. For example, it is seen that the Al–Co$_2$Al$_9$ system develops a lamellar structure upon directional solidification even though the volume fraction of the Co$_2$Al$_9$ phase is less than 3%. Since the two phases possess a unique crystallographic relationship one may presume that the associated interfacial energy is substantially reduced, thereby stabilizing the lamellar morphology even at a low volume fraction.

D. Interface Dislocations

The preferred orientations found in the eutectic structure were based on the minimization of lattice misfit through mating of planes of similar atomic density. As previously cited, Fletcher (1971) showed the interfacial energy to increase rapidly with increasing misfit. To eliminate the lattice mismatch between two phases with different lattice parameters would require the presence of a uniform strain field in both phases so as to eliminate the lattice size difference (Brooks, 1952). Large elastic strain fields would exist with strain energy levels proportional to the modulus of elasticity in each phase. The total energy of such a system could be reduced by the introduction of dislocations at the solid–solid interface which

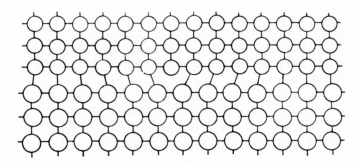

FIG. 6. Edge dislocation model for phase interface.

would accommodate the misfit and localize the strain energy to the proximity of the boundary. An appropriate dislocation network would be introduced to the interface so as to minimize the strain energy of the system.
For the simple model shown in Fig. 6, an edge dislocation would take up
the misfit between two crystals of different lattice parameters. It is clear
that more dislocations are needed for conditions of greater lattice misfit.
It may be shown that the spacing between such interface dislocations, S, is
given by

$$S = \tfrac{1}{2}(d_1 + d_2)\delta^{-1} \tag{2}$$

(Brooks, 1952), where d_1 and d_2 are the lattice spacings of the two phases,
respectively, and δ is the misfit as given by

$$\delta = 2(d_1 - d_2)/(d_1 + d_2) \tag{3}$$

Using Eqs. (2) and (3), Hopkins and Kossowsky (1971) computed the
dislocation network spacing in the Ni–Cr system to be within 10% of the
experimentally-observed value (84 Å versus 93 Å). The excellent agreement between theoretically-predicted and experimentally-observed values
led the authors to attempt a computation of interfacial energy based on
interface dislocation considerations due to Brooks (1952). The results
were found to be in good agreement with other surface energy computations
based on solidification data.

Walter *et al.* (1969) also computed network dislocation spacings in
NiAl–Cr. They found the experimentally-determined dislocation spacing
of 775 Å to be a reasonable value in view of the extremely small misfit
between the NiAl and Cr lattice dimensions. An example of a dislocation
network at the semicoherent interface is shown in Fig. 7. Note from Eq. (2)
that dislocation spacing will rise sharply as misfit decreases. In the limit,
when the misfit is zero, the interface dislocation spacing becomes infinite;

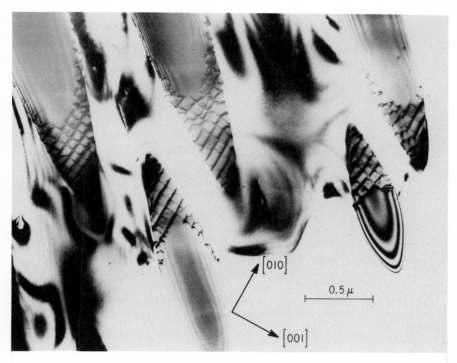

Fig. 7. Dislocation network at semicoherent interface in NiAl–Cr [from Walter *et al.* (1969)].

that is, the dislocation network disappears and the interface becomes fully coherent.

In a related study, Davies and Hellawell (1969) measured dislocation spacings between lamellae in the Al–CuAl₂ eutectic alloy and found them to be in reasonably good agreement with predicted values based upon lattice misfit. It was also observed that rapid cooling from a high temperature produced an increase in the hardness level. This was attributed to internal strains resulting from increased lattice misfit upon quenching from high temperatures, the larger lattice misfit being associated with differences in the respective coefficients of thermal expansion of the Al and CuAl₂ phases. Similar residual stress effects were observed in the (Co, Cr)–(Cr, Co)₇ C₃ system by Koss and Copley (1971). Considerable tension–compression yield strength anisotropy was related to differences in thermal coefficients of expansion. This subject is discussed more fully in an earlier chapter by Ebert and Wright.

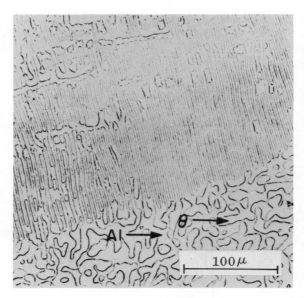

FIG. 8. Micrograph of Al–CuAl₂ heated to 400°C for 600 hr revealing spheroidization in the grain at the bottom and general stability in the grain at the top [from Kraft *et al.* (1963)].

E. Interface Thermal Stability

The total interfacial energy, E, in a two-phase solid is related to the total surface area A of the interphase boundaries and the specific surface energy γ at the interface as given in Eq. (4)

$$E = \sum_i A_i \gamma_i \tag{4}$$

For long time high temperature annealing treatments it is common to note a lowering of the total surface energy of the system by an elimination of interphase grain boundary area. While there might be a strong tendency for similar response by the fine particles found in controlled eutectic micro-structures, there is a counter-balancing factor, the presumably low energy interfaces associated with the preferred crystallographic relationships be-tween the two phases, which would tend to stabilize the interfaces in their original form. The possibility for reduction of surface area through spheroid-ization in one given eutectic grain and stability of microstructure due to an initially-preferred interface in another grain was clearly demonstrated in the Al–CuAl₂ system by Kraft *et al.* (1963) (Fig. 8). They argued that those grains which exhibited thermal stability possessed a preferred crystallography and low value of γ_i; consequently, they would be stable. Bayles *et al.* (1967), arrived at a similar conclusion based upon their investigation of the Al–Al$_3$Ni eutectic alloy. A preferred crystallographic relationship between the two phases was observed which persisted even after long time high temperature heat treatments. This they associated with a stable low energy interface. While these latter results revealed the eutectic microstructure to be stable in the sense that the rod-like morphology was preserved, the Al$_3$Ni fibers were found to coarsen. Therefore, it is necessary to explore those mechanisms associated with both shape stability and coarsening.

Cline (1971) described three models which would account for either spheroidization or coarsening of a rod microstructure. Allowing for small periodic variations in rod diameter along the rod axis, diffusion calculations revealed that the perturbed fiber could break down into a linear array of spheres (also described by Nichols and Mullins, 1965). Since the process of spheroidization would introduce additional interphase plane combinations of presumably higher energy levels than the originally-preferred one, the spheroidization process would not be favored. The previous statement has been proven for a number of eutectic systems that develop, upon directional solidification, a preferred crystallographic relationship between the two phases. On the other hand, when preferred crystal plane matching is not achieved, spheroidization will occur. Marich and Jaffrey (1971) observed that Cu$_2$S rods in the Cu–Cu$_2$S-controlled eutectic microstructure would spheroidize sometime after passage of the solid–liquid interface (Fig. 9). Consequently, the globular structure seen at room temperature was actually developed by a rod spheroidization process at temperatures close to the melting point. It is important to note that no unique crystallographic relationship was observed in this system, consistent with the necessary conditions for spheroidization.

The last two models proposed by Cline (1971) were intended to describe the mechanism of fiber coarsening in stable microstructures. In one model, a composite was considered to contain fibers of an average radius $\bar{R}$, but with a Gaussian distribution of sizes as shown in Fig. 10a. With time at high temperatures, a diffusion gradient would be set up so that small diameter rods would dissolve at the expense of larger rods (Greenwood, 1956); this would produce a spreading of the rod size distribution at constant $\bar{R}$ until the smallest rods would begin to completely dissolve (Fig. 10b). After this incubation time, $\bar{R}$ would rise and coarsening would occur (Fig. 10c). It is readily seen that the rate of coarsening should increase with decreasing average fiber size $\bar{R}$, and increasing rod size distribution.

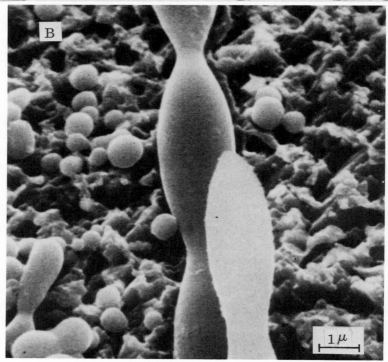

342

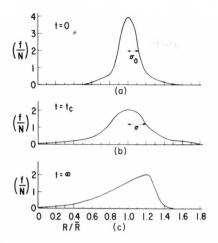

FIG. 10. Two-dimensional coarsening model for fibers in a continuous matrix revealing a normalized distribution function f/N versus the rod size ratio $R/\bar{R}$. (a) Assumed gaussian distribution of fiber sizes at time $t = 0$. (b) Diffusion-enhanced growth of larger fibers at expense of smaller fibers at constant $\bar{R}$. (c) Continued coarsening with elimination of small fibers so that R increases [from Cline (1971)].

While the suggested role of $\bar{R}$ on coarsening was verified by Bayles *et al.* (1967), other experimental observations could not be rationalized in light of the two-dimensional coarsening model. They found coarsening to occur immediately, without any evidence for an incubation period. Cline attributed this initial coarsening response to diffusional processes associated with fault† migration and annihilation. As the rod density decreases due to fault elimination, the remaining rods would then obviously thicken. The concept of fault migration as a mechanism to coarsen morphological stable eutectic microstructures was first proposed by Graham and Kraft (1966) for the lamellar Al–CuAl₂ system. Since the effective radius of the sides of any lamellae is infinitely large, two-dimensional coarsening is not possible. Consequently, lamellae coarsening was made possible only by fault migration and annihilation. (Spheroidization and subsequent coarsening did not occur in this system due to the presence of a low specific surface

† A fault in a rod eutectic may consist of termination of two rods into one or branching of one rod into two (see Fig. 11).

FIG. 9. Illustration of necking down and pinching off of Cu_2S in $Cu–Cu_2S$ eutectic: (A) solidification direction from right to left with quenched solid–liquid interface defined by vertical arrows and (B) scanning electron micrograph of Cu_2S fiber necking down, pinching off and spheroidization [from Marich and Jaffrey (1971)].

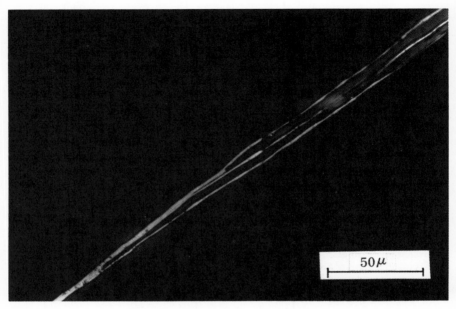

FIG. 11. Branching in Al₃Ni whiskers

energy interface, inferred from the preferred crystallographic relationship before and after heat treatment.)

Coupling the observations that heavily-faulted ingots coarsened more rapidly with the finding that the Al₃Ni fiber size distribution never spread to the point where fibers of vanishingly small size would appear, Nakagawa and Weatherly (1972) concluded that the fault migration and annihilation mechanism was dominant in the Al–Al₃Ni eutectic system. Finally, the competitive nature of the two-dimensional and fault annihilation coarsening mechanisms may be inferred from the results by Smartt *et al.* (1971), Fig. 12. The Al₃Ni rod density is seen to decrease initially (i.e., rod coarsening), stabilize for a short time, and then resume the trend toward fiber coarsening. One may interpret this behavior as a series of the following events: initial coarsening due to fault annihilation until fault migration was exhausted, stabilization of $\bar{R}$ while the rod size distribution spread, and resumption of coarsening after an incubation time due to two-dimensional geometrically dependent diffusional processes. While this line of reasoning appears plausible, there are some inconsistencies to be reconciled. For example, the coarseness of the structure at the time of fault exhaustion might likely be different for different degrees of faulting.

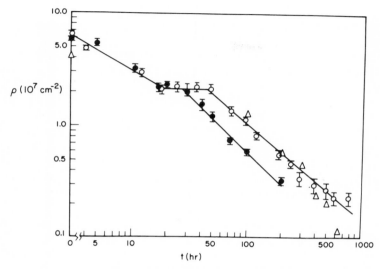

FIG. 12. Variation of Al$_3$Ni whisker density resulting from 600°C prolonged heat treatment [from Smartt *et al.* (1971)]. Data points: ○, normal material; ●, faulted material; △, Bayles *et al.* (1967).

F. Summary

Unidirectional solidification of eutectic alloys can yield materials with aligned high strength filamentary or plate-like particles properly bonded to a matrix phase. The growth of these controlled eutectic microstructures is often related to a minimization of solid–solid interfacial energy as evidenced by unique crystallographic relationships between the solid phases in many eutectic systems. Computation of interfacial dislocation spacing based upon a simple model provides good agreement with experimentally-determined values.

It is readily apparent that the interface in eutectic composites has been more extensively characterized than interfaces in other composite systems. From the standpoint of a minimum desired internal energy for the material, a highly reproducible interface is achieved from sample to sample for any given eutectic system.

Furthermore, eutectic interfaces in systems possessing unique crystallographic relationships are found to be exceptionally stable. Various models suggest that resistance to spheroidization or particle coarsening is enhanced by the presence of an initially low interphase interfacial energy, increased perfection of the structure (i.e., minimization of faulted regions) and increased starting size of the microstructural constituents.

III. Interface Control of Mechanical Behavior

The first eutectic composites showing reinforcement to be prepared by unidirectional solidification were the Al–Al₃Ni and Al–CuAl₂ systems (Hertzberg *et al.*, 1965). Proper load transfer was shown in these materials, i.e., composite failures took place by whisker or lamellae fracture and subsequent void coalescence rather than by interfacial decohesion. Only under loading conditions associated with large shear stresses parallel to the interface did decohesion occur. Since this initial study, many other eutectic systems have been shown to possess reinforcing capabilities (Thompson and Lemkey, 1972). Here again, the observed dominant fracture mechanisms reflect the presence of strong matrix-reinforcing phase interfacial bonding.

Section III will now focus on the role played by the interface in controlling the mechanical response of eutectic composites. Particular attention will be given to the effect of temperature and the nature of applied stress on interface behavior.

A. Effect of Interface on Yielding Behavior

The deformation behavior of aligned eutectic composites will depend upon the separate deformation characteristics of the two phases and the reaction of the interface region to the movement of dislocations. For purposes of clarity, a discussion of deformation models in these materials will be divided into two major categories: eutectic alloys capable of plastic flow in both matrix and reinforcing phases and eutectic alloys capable of plastic flow only in the matrix phase.

1. Plastic–Plastic Eutectic Systems

For a eutectic system with two deformable phases, composite mechanical properties (e.g., yield strength) will depend upon the magnitude of several potential strengthening mechanisms based upon dislocation movements in each phase and at the interface.

One such strengthening mechanism is related to the ability of dislocations to move across a grain boundary or the nucleation of additional dislocations in the second phase as a result of a dislocation pile up at the boundary in the first phase. On the basis of this model, the strength of the composite should increase with decreasing lamellar or rod size since the pile up distance would be decreased. Several investigators have shown that a Petch-type relationship does exist between flow stress and the inverse

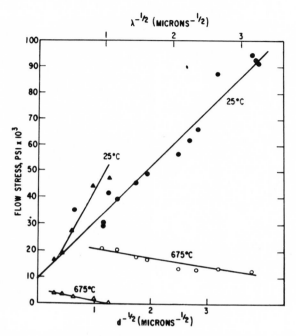

FIG. 13. Flow stress (0.5% offset) versus structure size ($d^{-1/2}$) in equiaxed (extruded) and lamellar (directionally-solidified) Ag–Cu eutectic [from Cline and Lee (1971)]. Data points: $\bigcirc$, $\bullet$, directionally solidified; $\triangle$, $\blacktriangle$, extruded.

square root of lamellar size or interrod spacing (Shaw, 1967; Cline and Stein, 1969; Cline and Lee, 1970; Thompson *et al.*, 1971).

Also, by modifying the rule of mixtures formula to account for an additional strength contribution due to a Petch-type size effect, Kossowsky *et al.* (1969) showed good agreement between measured and computed composite strength in the Ni–Cr eutectic alloy.

The size effect should be sensitive to the nature of the interface. It would be expected that incoherent boundaries would prove to be more effective barriers to dislocation movement than semicoherent boundaries resulting from a preferred crystallographic relationship. To wit, Cline and Lee (1970) showed a stronger flow stress dependence of (size)$^{-1/2}$ for equiaxed Ag–Cu eutectic with noncoherent boundaries than for the lamellar structure with semicoherent boundaries (Fig. 13). Furthermore, Stoloff and Kim (1971) have shown the flow stress–structural size dependence to vary with the nature of deformation in the second phase. By selective heat treatment it was possible to develop eutectic microstructures containing AgMg with either an ordered or unordered Ag_3Mg second phase. Stoloff

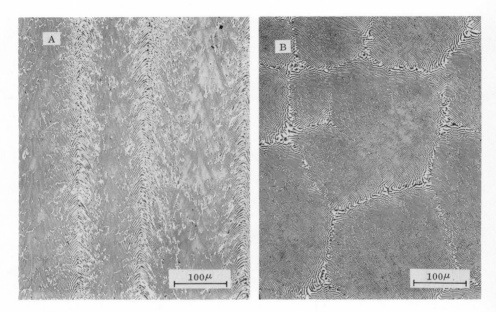

FIG. 14. Directionally-solidified $Mg-Mg_{17}Al_{12}$ eutectic with colony structure. (A) Longitudinal section (growth direction from bottom to top). (B) Transverse section [from Kraft (1966)].

and Kim related the stronger flow stress dependence on lamellar $(size)^{-1/2}$ in the structure containing ordered Ag_3Mg to the necessity for antiphase boundary formation as a result of dislocation movement in the ordered phase. Recalling that the lamellar spacing of a eutectic phase is proportional to the inverse square root of the solidification rate, the flow stress should, therefore, be proportional to the inverse fourth root of the growth rate. While it is tempting to consider high growth rate conditions in the preparation of controlled eutectic ingots so as to minimize λ and thereby maximize the flow stress, there are limitations to this gambit. At high solidification rates the solid–liquid interface often breaks down to a cellular front causing the rods or plates to fan out (Fig. 14); this breakdown of ideal alignment into a colonied structured causes serious loss of composite strength. Also, the preferred interface relationship may not develop under very high growth rate conditions, thereby causing degradation of mechanical properties at elevated temperatures.

A second potential strengthening mechanism in eutectic alloys that deform in both phases is associated with dislocation image forces acting at the interface; these forces act to resist the movement of a dislocation from a phase with a lower modulus into a phase with a higher modulus and are related to the change in dislocation strain energy due to the change in the modulus (Weertman and Weertman, 1964). Using the relationship

$$\tau(\Delta G) = \Delta G/(8\pi) \tag{5}$$

as proposed by Fleischer (1969), where $\tau(\Delta G)$ is the image force strengthening contribution and ΔG the modulus difference between the two phases, Cline and Lee (1970) showed image force strengthening to be an important contribution in the Ag–Cu system.

An additional strengthening contribution is possible due to the interaction of slip dislocations with the network of interface dislocations at semicoherent boundaries. As discussed above, these interface dislocations are generated to relieve the long range elastic stress fields associated with the lattice misfit of the two phases. Determining the distance between interface dislocations

$$S = b^2/\Delta b \tag{6}$$

from Eqs. (2) and (3), and approximating the stress necessary to cross the network by

$$\tau(\Delta b) = Gb/(2\pi S) \tag{7}$$

(Friedel, 1964), where $\tau(\Delta b)$ is the slip dislocation–interface dislocation strength contribution, Friedel found this strengthening contribution to be approximated by

$$\tau(\Delta b) = G\,\Delta b/(2\pi b) \tag{8}$$

Using Eq. (8), Cline and Lee (1970) and Walter *et al.* (1969) determined the interface dislocation strengthening contribution to be 100,000 psi in the Ag–Cu eutectic alloy but only 10,000 psi in the NiAl–Cr alloy. This large difference was related to differences in atomic mismatch in the two eutectic systems examined.

Setting Petch-type grain refinement relationships aside as empirically-determined quantities, it is not possible to predict the strength of controlled eutectic alloys from the above theoretical considerations. The computed values of image force strengthening and interface dislocation strengthening in the Ag–Cu system can not be related directly to experimental findings, while the image force value for the NiAl–Cr system are found to be very low. As such, these calculations serve only to provide a feeling for the relative importance of any particular strengthening mechanism in a given eutectic alloy system.

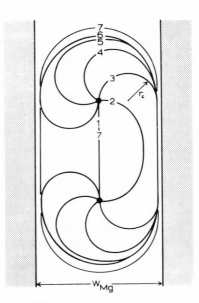

FIG. 15. Schematic showing stages of movement of dislocation line segment [from Eckelmeyer and Hertzberg (1972)].

2. Elastic–Plastic Eutectic Systems

When one phase (usually the reinforcing phase) does not deform, a Petch-type relationship based upon dislocation pile up-induced slip in the second phase is essentially inappropriate. To account for plastic flow in the Mg deformable phase of the Mg–Mg$_2$Ni eutectic alloy, Eckelmeyer and Hertzberg (1972) considered deformation to be controlled by dislocation barriers contained within individual Mg lamellae. From Fig. 15, it was shown that the critical stage in yielding was related to the movement of a dislocation line segment to a radius equal to one fourth the thickness of a Mg lamellae on the slip plane. Since the lamellae width is dependent upon growth rate, the flow stress was defined as

$$\sigma_{\text{flow}} = \sigma_0 + \sigma(R) \tag{9}$$

where σ_0 is the growth rate independent parameter and $\sigma(R)$ the growth rate dependent parameter. Relating $\sigma(R)$ to the critical stress necessary to bow out the dislocation loop shown in Fig. 15, the flow stress was re-defined as

$$\sigma_{\text{flow}} = \sigma_0 + 2Gb/W_{\text{Mg}} \tag{10}$$

where W_{Mg} is the Mg lamellae width on the slip plane. By recalling that the lamellae spacing is proportional to $R^{-1/2}$ and determining all constants associated with the resolved shear stress on the slip plane necessary for slip, Eckelmeyer and Hertzberg (1972) were able to show that $\sigma(R) = 1.8R^{1/2}$, which is in very good agreement with their experimental observations. By using the same analysis procedure for the Al–CuAl₂ alloy, another elastic–plastic eutectic system, computed yield strength values at different growth rates were in reasonably good agreement with experimental results by Grossman *et al.* (1969).

B. *Effect of Interface on Fatigue Response*

The response of eutectic composites to cyclic loading patterns has received scant attention compared to their static and creep responses. This in unfortunate since many potential applications for these materials may, in fact, be fatigue-limited. Predicting the fatigue behavior of eutectic composites poses three intriguing questions:

(1) Since fatigue damage involves the accumulation of plastic deformation damage, how does one predict the behavior of a two-phase material when the reinforcing phase is often completely elastic?

(2) What will happen to a propagating crack when it reaches an interface?

(3) Will the propagating crack traverse the boundary into the second phase or be deflected parallel to the loading direction and whisker axis?

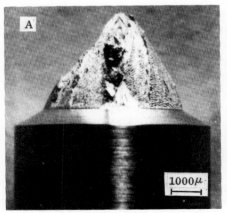 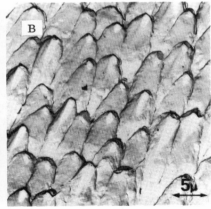

FIG. 16. Fracture appearance of Al–Al₃Ni eutectic under monotonic loading conditions: (A) macroscopic view and (B) electron fractograph revealing elongated microvoid coalescence [from Hoover and Hertzberg (1968)].

Fɪɢ. 17. Fracture appearance of Al–Al₃Ni eutectic under high cycle loading conditions: (A) macroscopic view revealing vertical fatigue crack propagation and (B) electron fractograph revealing matrix failure with evidence of interfacial failure at whisker interface [from Hoover and Hertzberg (1968)].

Initial fatigue crack propagation studies of Al–Al₃Ni notched round bars were conducted by Hoover and Hertzberg (1968). From previous studies (Hertzberg *et al.*, 1963), it was shown that the alloy contained elastic Al₃Ni whiskers embedded in a deformable Al matrix. Under high cyclic stress conditions, Al₃Ni whisker fracture and subsequent microvoid formation was observed. Complete high stress fatigue failure occurred when these microvoids coallesced on the gross fracture plane in a manner similar to that experienced under monotonic loading conditions (Figs. 16A and 16B). A dramatic change in the fracture appearance was noted under low alternating stress conditions where the notch root stress level proved too low to cause individual whisker failure; cyclic damage then concentrated exclusively in the Al matrix. With plastic damage accumulation in the matrix, stable crack propagation was initiated parallel to the loading direction and whisker axes (Fig. 17A). The vertical crack front tended to move along (111) planes in the Al matrix [analogous to stage I fatigue crack propagation (Forsyth, 1963)] and occasionally along the matrix–fiber interface (Fig. 17B). It is worth noting that no evidence could be found for whisker fracture and associated microvoid formation under low alternating stress conditions.

In a comparative study, Hoover (1967) found the aligned Al–Al$_3$Ni microstructure to be more fatigue resistant than the uncontrolled as-cast material. This was attributed to the much greater load-bearing capacity of the aligned Al$_3$Ni whiskers and their ability to effectively deflect the fatigue crack parallel to the loading axis. By contrast, it was noted that fatigue crack growth occurred predominantly along denuded colony boundaries in the as-cast samples where there were no Al$_3$Ni whiskers (e.g., see denuded regions in Fig. 14).

In another study Salkind *et al.* (1966) reported the smooth bar fatigue response of controlled Al–CuAl$_2$ specimens to be superior to that of similarly-prepared samples of Al–Al$_3$Ni. The authors reasoned that the greater strain hardening rate and consequent lower hystersis damage for a given loading cycle in the Al–CuAl$_2$ system was related to its lamellar morphology. The lamellar CuAl$_2$ reinforcing phase was considered to be more effective in restricting dislocation movement than the much smaller Al$_3$Ni whiskers. While this argument is conceptually valid, it can not be proven with the data cited above. Since the volume fraction of Al$_3$Ni whiskers is five times less than that of the CuAl$_2$ lamellae, a much greater stress (and plastic strain) amplitude would exist in the matrix of the Al–Al$_3$Ni system for any given composite stress level. Therefore, regardless of the reinforcing phase morphology, the Al–Al$_3$Ni alloy should exhibit inferior fatigue behavior compared to the Al–CuAl$_2$ system due to the lower volume fraction of the reinforcing phase.

The effect of reinforcing phase morphology and spacing was evaluated in a recent study of the notched fatigue behavior in the Mg–Mg$_2$Ni eutectic alloy. While fatigue response of this alloy was complicated by solidification rate-induced crystallographic relationship changes which affected the Mg matrix deformation mechanisms and by test environment factors, it was demonstrated that superior fatigue response could be achieved with the presence of finer and morphologically less complex whiskers of the Mg$_2$Ni reinforcing phase (Eckelmeyer, 1971; Eckelmeyer and Hertzberg, 1972). It was found that coarser and often interconnected whiskers grown at low solidification rates always fractured to form an extensive crack network; consequently, failure was controlled at all stress levels examined by failure of the reinforcing phase. By comparison, the finer, discrete Mg$_2$Ni whiskers formed at higher solidification rates did not produce a large interconnected crack front upon failure. Rather, the fatigue mechanism was similar to that found in the Al–Al$_3$Ni system: fiber fracture and void coalescence at high cyclic stress levels but crack deflection and vertical crack extension in the matrix at low stress levels. A marked increase in low stress fatigue life was observed in the Mg–Mg$_2$Ni alloy when the failure mechanism was

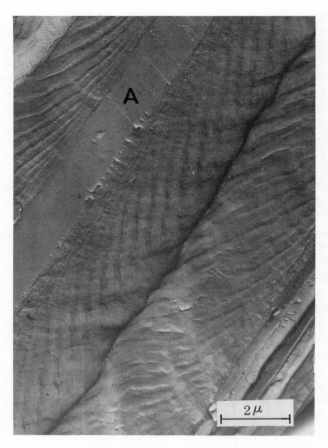

FIG. 18. Electron micrograph revealing cracked Ni_3Nb lamellae (A) and fatigue striation formation in Ni lamellae [from Hoover and Hertzberg (1971)].

changed from whisker fracture to whisker-induced crack deflection and subsequent vertical crack extension through the matrix phase.

In another investigation, Hoover and Hertzberg (1971a) examined the notched fatigue behavior of the $Ni–Ni_3Nb$ eutectic composite. This system was chosen for study in light of the ability of both phases to deform (Hoover and Hertzberg (1971b), Annarumma *et al.* (1972), and Gangloff (1972); the Ni_3Nb reinforcing phase was found to deform by twinning on {211}-type planes. In contrast to the response of the $Al–Al_3Ni$ system, fatigue cracks in the Ni–Nb alloy traversed both Ni and Ni_3Nb lamellae at all stress levels examined. At high cyclic stress levels, crack extension was promoted by Ni_3Nb twin formation and subsequent cracking ahead of the main crack front (Fig. 18). Failure in the Ni phase was by a striation formation mechanism. At intermediate stress levels, striation formation in the Ni phase and twin boundary cracking in Ni_3Nb proceeded at about the same rate. The number of Ni_3Nb deformation twins nucleated away from the main crack plane was seen to decrease sharply with decreasing net section stress. Finally, at low stress levels no additional twins were seen other than those directly involved in the advance of the crack front. It would appear that the peak stress at the advancing crack tip was sufficient to nucleate just one Ni_3Nb twin per lamellae to serve as the eventual crack path for the crack.

A marked change in the fatigue fracture mechanism was noted in the Ni phase during the low alternating stress tests. The expected fatigue striations, typical of stage II fatigue crack propagation (Forsyth, 1963), disappeared and were replaced by a faceted fracture morphology (Fig. 19) similar to that observed by Gell and Leverant (1968) in MAR-M200 nickel base alloy crystals. In both instances, the faceted growth was associated with stage I fatigue crack propagation (Forsyth, 1963).

In a comparison of notched fatigue behavior between the $Al–Al_3Ni$ and

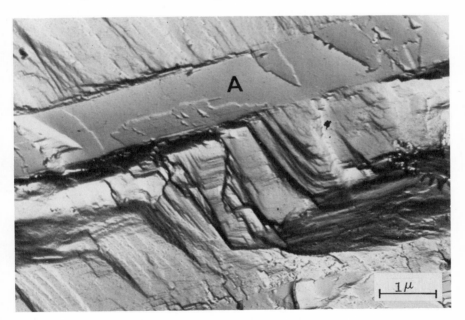

FIG. 19. Electron micrograph revealing cracked Ni_3Nb lamellae (A) and faceted (stage I) type failure in Ni lamellae [from Hoover and Hertzberg (1971)].

Ni–Ni₃Nb eutectic systems, Hoover and Hertzberg (1971a) concluded that similar fracture mechanisms were evident even though the deformation mechanisms in the Al₃Ni and Ni₃Nb reinforcing phases were dramatically different. At high stress levels, fatigue damage was controlled by fracture of the reinforcing phase while under low stress conditions, stage I fatigue crack propagation in the matrix phase was believed to be the controlling event. Assuming this to be a generally valid statement concerning the fatigue behavior of composite materials, it could prove useful in the selection of optimum fiber and matrix fatigue properties for both high and low cycle composite fatigue performance.

Encouraging elevated temperature fatigue results were reported by Thompson *et al.* (1971) for the Ni₃Nb–Ni₃Al eutectic system. This controlled eutectic alloy exhibited superior fatigue behavior at 1600°F in both smooth and notched bar configurations to that of smooth bars of B-1900, a conventional nickel-base alloy. It is worth noting that these results were obtained despite the fact that the eutectic alloy possessed inferior oxidation resistance when compared with the B-1900 alloy. While occasional interlamellar delamination was noted in the eutectic microstructure during fatigue testing, the predominant failure mode was translamellar, similar to the response of the Ni–Ni₃Nb alloy under room temperature fatigue conditions.

C. Effect of Interface on Creep Rupture Response

The main thrust of recent eutectic composite research has been related to the development of high temperature alloy systems for use in gas turbine engine applications. Such activities have given witness to a major breakthrough in eutectic composite material development; the consideration of pseudobinary and multicomponent eutectic systems as potential candidates for composite development. For example, Thompson and Lemkey (1969) investigated the mechanical response of several pseudobinary eutectic systems containing the Ni₃Al phase, known to be the key strengthening constituent in conventional nickel base alloy microstructures. In a more detailed study of the Ni₃Al–Ni₃Nb system, Thompson *et al.* (1971) reported ultimate tensile strength and total strain at 1210°C of 89 ksi and 4%, respectively. The 100 hr creep rupture strength of 90 ksi at 875°C was found to be superior to properties exhibited by any conventional nickel-based superalloy. While this system exhibited impressive strength levels, it was hampered by poor oxidation resistance. Recently, Lemkey (1971) succeeded in adding Cr and Al to the Ni–Ni₃Nb system so as to generate an aligned microstructure consisting of Ni₃Nb reinforcing platelets embedded within a corrosion-resistant Ni–Cr solid solution matrix further strengthened by the precipitation of Ni_3Al particles. Preliminary results from this investigation are very encouraging.

By proceeding in different fashions, Lemkey and Thompson (1971), Buchanan and Tarshis (1971), and Bibring and co-workers (1970, 1971) demonstrated the feasibility of preparing Ni- or Co-based pseudobinary eutectic composite systems reinforced with monocarbides of Nb, Ta, Ti, V, Zr, or Hf. Such multicomponent systems have enabled investigators to prepare eutectic microstructures with a greater range of allowable compositions and volume fraction of the constituent phases than is possible with the simple binary eutectic reaction. To wit, Bibring *et al.* (1970) modified the Ni–Cr–TaC eutectic (Ni–Cr matrix reinforced by 50 vol % TaC whiskers) with additions of Co, Ti, Al, and Ta so as to develop room temperature tensile strength and ductility of 225 ksi and 12% strain, respectively.

While intense interest has been devoted toward the development of eutectic composite systems capable of withstanding the high stress and corrosive high temperature environments of gas turbine engines, little attention has been given to an analysis of deformation and fracture mechanisms of these aligned microstructures. What little has been done appears to fit together into a consistent pattern; it is generally agreed that high temperature mechanical response of aligned eutectic composites is severely degraded by the existence of noncoherent boundaries or misaligned microstructure. In a comparative study of equiaxed and lamellar Ag–Cu alloy containing noncoherent and semicoherent boundaries, respectively, Cline and Lee (1970) reported the semicoherent boundary material to exhibit markedly superior elevated temperature behavior. In fact, the extensive grain boundary sliding associated with the incoherent boundaries rendered the equiaxed material superplastic. The high temperature colony boundary failures in the NiAl–Cr system may also reflect the inferior mechanical response of noncoherent boundaries (Walter and Cline, 1970).

Unfortunately, semicoherent boundaries may gradually degenerate to a noncoherent equiaxed structure by recrystallization and coarsening processes. This is believed to be the case in high temperature tests of the aligned Ag–Cu eutectic alloy (Cline and Lee, 1970). It is possible that slip dislocations interacted with the interface dislocations of the semicoherent boundaries to cause the formation of localized regions of disorder. Should such a mechanism prove to be common to the many aligned eutectic systems possessing semicoherent boundaries, the supposed structural thermal stability (discussed in an earlier section) would be severely limited under conditions of both high temperature and stress.

Misaligned structures have been observed to seriously degrade the ele-

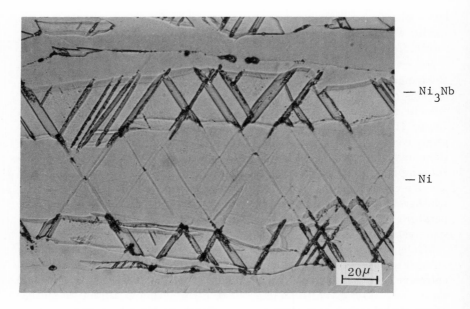

— Ni$_3$Nb

— Ni

FIG. 20. Nickel lamellae accommodation twinning with regard to Ni$_3$Nb twin formation [from Gangloff (1972)].

vated temperature mechanical response of eutectic composites (de Silva and Chadwick, 1969; Breinen *et al.*, 1972). In most instances, this is due to the existence of a resolved shear stress acting along the fiber axis producing either shear failure in the matrix or interfacial failure at the matrix–fiber (or lamellae) interface. Even at room temperature, shear failures parallel to the whisker axis have been observed when flexually-induced shear stresses were applied parallel to the Al–CuAl$_2$ interface (Hertzberg *et al.*, 1965), and when torsional shear stresses were applied to the Ni$_3$Al–Ni$_3$Nb composite (Thompson *et al.*, 1971).

Aside from participating in the fracture process, the interphase interface has served to control the deformation mechanisms in the composite. As shown by Annarumma and Turpin (1972) and Gangloff (1972), the Ni$_3$Nb reinforcing phase in the Ni–Ni$_3$Nb eutectic alloy may twin on one or more {211}-type planes. At intermediate temperature creep rupture conditions, it has been shown by Grossiord *et al.* (1971) and Gangloff (1972) that accommodation twinning is developed in adjacent Ni lamellae; both investigations have shown that only two specific {111}-type twins are ever observed from a possible total of four such twins. By considering the details of the crystallographic relationship between the aligned Ni and Ni$_3$Nb lamellae, Grossiord *et al.* (1971) clearly showed that only two specific {111}-type Ni twins would be compatible with the Ni$_3$Nb twin-generated atomic displacements. An example of this cooperative twinning is shown in Fig. 20.

D. Summary

Yielding behavior was seen to be controlled by several possible mechanisms; two that could be directly related to the unique interface in directionally-solidified eutectic alloys were the interaction of slip dislocations with interface dislocations of the semicoherent boundaries and restrictions on allowable deformation mechanisms in two contiguous phases due to orientation effects.

Evidence found thus far for elastic–plastic eutectic composites indicates that interfaces are capable of diverting fatigue cracks parallel to the loading axis.

Some concern exists for the thermal stability of preferred eutectic interfaces under stressed conditions. It is possible that semicoherent interfaces (stable) can degenerate to noncoherent ones (unstable) due to dislocation-induced strain concentrations at the boundaries.

While considerable information is available with regard to the nature of the preferred interfaces in eutectic composites, relatively little attention has been given to interface response to various stress fields and temperature combinations.

Acknowledgments

The author thanks Richard Gangloff and William Mills for the assistance provided in gathering the reference material, Professors Kraft and Chou for many helpful discussions of the technical material, and Louise Valkenburg and Barbara Hayes for their careful preparation of the manuscript. The financial support of the National Aeronautics and Space Administration under Grant NGR-39-007-007 and that of the Alcoa Foundation is sincerely appreciated.

References

Annarumma, P., and Turpin, M. (1972). *Met. Trans.* **3**, 137.
Bayles, B. J., Ford, J. A., and Salkind, M. J. (1967). *Trans. AIME* **239**, 844.
Bibring, H., Seibel, G., and Rabinovitch, M. (1970). *C. R. Acad. Sci. Paris Ser. C* **271**, 1521.

Bibring, H., Trottier, J. P., Rabinovitch, M., and Seibel, G. (1971). *Mem. Sci. Rev. Met.* **68,** 23.

Breinen, E. M., Thompson, E. R., McCarthy, G. P., and Herman, W. J. (1972). *Met. Trans.* **3,** 221.

Brooks, H. (1952). Theory of Internal Boundaries, In "Metal Interfaces," p. 20. Amer. Soc. Metals, Metals Park, Ohio.

Buchanan, E. R., and Tarshis, L. A. (1971). Carbide fiber reinforcement in a directionalyl solidified alloyed nickel eutectic, *1971 Fall Meeting of TMS, AIME, Detroit, Michigan, October.*

Cline, H. E. (1971). *Acta Met.* **19,** 481.

Cline, H. E., and Stein, D. F. (1969). *Trans. AIME* **245,** 841.

Cline, H. E., and Lee, D. (1970). *Acta Met.* **18,** 315.

Cooksey, D. J. S., Munson, D., Wilkinson, M. P., and Hellawell, A. (1964). *Phil. Mag.* **10,** 745.

Davies, I. G., and Hellawell, A. (1969). *Phil. Mag.* **19,** 1285.

Day, M. G., and Hellawell, A. (1968). *Proc. Roy. Soc. Ser. A* **305,** 473.

deSilva, A. R. T., and Chadwick, G. A. (1969). *Met. Sci. J.* **3,** 168.

Eckelmeyer, K. E. (1971). The Structure and Mechanical Behavior of the Aligned Mg–Mg$_2$Ni Eutectic Composite. Ph.D. Dissertation, Lehigh Univ. Bethlehem, Pennsylvania.

Eckelmeyer, K. E., and Hertzberg, R. W. (1972). *Met. Trans.* **3,** 609.

Fleischer, R. C. (1969). "Electron Microscopy and Strength of Crystals," p. 973. Wiley (Interscience), New York.

Fletcher, N. H. (1971). *Advan. Mater. Res.* **5,** 281.

Forsyth, P. J. E. (1963). *Acta Met.* **11,** 703.

Friedel, J. (1964). "Dislocations," p. 220. Pergamon, Oxford.

Gangloff, R. (1972). "Elevated Temperature Tensile and Creep Rupture Behavior of the Unidirectionally-Solidified Ni–Ni$_3$Nb Eutectic Composite," Master's thesis, Lehigh Univ., Bethlehem, Pennsylvania.

Gell, M., and Leverant, G. R. (1968). *Trans. AIME* **242,** 1869.

Graham, L. D., and Kraft, R. W. (1966). *Trans. AIME* **236,** 94.

Greenwood, G. W. (1956). *Acta Met.* **4,** 243.

Grossiord, C., Lesoult, G., and Turpin, M. (1971). Slip and Mechanical Twinning in Ni–Ni$_3$Nb Directionally-Solidified Eutectic Alloy, presented at *Int. Mater. Symp.* Univ. of California, Berkeley, California.

Grossman, F. W., Yue, A. S., and Vidoz, A. E. (1969). *Trans. AIME* **245,** 397.

Hertzberg, R. W., and Kraft, R. W. (1963). *Trans. AIME* **227,** 580.

Hertzberg, R. W., Lemkey, F. D., and Ford, J. A. (1965). *Trans. TMS-AIME* **233,** 342.

Hogan, L. M., Kraft, R. W., and Lemkey, F. D. (1971). *Advan. Mater. Res.* **5,** 83.

Hoover, W. R. (1967). "The Low Cycle Fatigue Behavior of Unidirectionally Solidified Al–Al$_2$Ni Eutectic Alloy." Master's thesis, Lehigh Univ., Bethlehem, Pennsylvania.

Hoover, W. R., and Hertzberg, R. W. (1968). *Trans. ASM* **61,** 769.

Hoover, W. R., and Hertzberg, R. W. (1971a). *Met. Trans.* **2,** 1289.

Hoover, W. R., and Hertzberg, R. W. (1971b). *Met. Trans.* **2,** 1283.

Hopkins, R. H., and Kossowsky, R. (1971). *Acta Met.* **19,** 203.

Jackson, K. A., and Hunt, J. D. (1966). *Trans. AIME* **236,** 1129.

Jackson, M. R. (1967). "The Structural and Magnetic Characteristics of the Unidirectionally-Solidified MnSb–Sb Eutectic System." Master's thesis, Lehigh Univ., Bethlehem, Pennsylvania.

Koss, D. A., and Copley, S. M. (1971). *Met. Trans.* **2**, 1557.
Kossowsky, R., Johnston, W. C., and Shaw, B. J. (1969). *Trans. AIME* **245**, 1219.
Kraft, R. W. (1962). *Trans. AIME* **224**, 65.
Kraft, R. W. (1963). *Trans. AIME* **227**, 393.
Kraft, R. W., and Albright, D. L. (1961). *Trans. AIME* **221**, 95.
Kraft, R. W. (1966). *J. Metals* **18**, 192.
Kraft, R. W., Albright, D. L., and Ford, J. A. (1963). *Trans. AIME* **227**, 540.
Lemkey, F. D. (1971). "Developing Directionally-Solidified Eutectics for Use up to 1235°C," United Aircraft Res. Lab., First Quart. Rep., Contract NAS 3-15562.
Lemkey, F. D., and Kraft, R. W. (1962). *Rev. Sci. Instrum.* **33**, 846.
Lemkey, F. D., and Thompson, E. R. (1971). *Met. Trans.* **2**, 1537.
Marich, S., and Jaffrey, D. (1971). *Met. Trans.* **2**, 2681.
Nakagawa, Y. G., and Weatherly, G. C. (1972). *Acta Met.* **20**, 345.
Nichols, F. A., and Mullins, W. W. (1965). *J. Appl. Phys.* **36**, 1826.
Quinn, R. T., Kraft, R. W., and Hertzberg, R. W. (1969). *Trans. ASM* **62**, 38.
Salkind, M., George, F. D., Lemkey, F. D., and Bayles, B. J. (1966). "Investigation of the Creep, Fatigue, and Transverse Properties of Al₃Ni Whiskers and CuAl₂ Platelet-Reinforced Aluminum." Contract NOw 65-0384-d, May.
Scheil, E. (1946). *Z. Metalk.* **37**, 1.
Scheil, E. (1959). *Giesserei* **24**, 1313.
Shaw, B. J. (1967). *Acta Met.* **15**, 1169.
Smartt, H. B., Tu, L. K., and Courtney, T. H. (1971). *Met. Trans.* **2**, 2717.
Stoloff, N. S., and Kim, Y. G. (1971). "Ordering and the Strength of an Aligned Ag₃Mg–AgMg Eutectic." Tech. Rep. No. 1, Contract N00014-67-A-0117-0010, NR 031-745, October 31.
Thompson, E. R., and Lemkey, F. D. (1969). *Trans. ASM* **62**, 140.
Thompson, E., and Lemkey, F. D. (1972). Review article in preparation.
Thompson, E. R., Kraft, E. H., and George, F. D. (1971). "Investigation to Develop a High Strength Eutectic for Aircraft Engine Use." Final Rep. on Contract N00019-71-C-0096, July 31.
Tiller, W. A. (1958). Polyphase solidification *in* "Liquid Metals and Solidification," p. 276. Amer. Soc. Metals, Metals Park, Ohio.
Walter, J. L., and Cline, H. E. (1970). *Met. Trans.* **1**, 1221.
Walter, J. L., Cline, H. E., and Koch, E. (1969). *Trans. AIME* **245**, 2073.
Weertman, J., and Weertman, J. R. (1964). "Elementary Dislocation Theory," Macmillan Ser. in Mater. Sci., p. 168. Macmillan, New York.
Zener, C. (1946). *Trans. AIME* **167**, 550.

10

The Effect of Impurity on Reinforcement–Matrix
Compatibility

W. BONFIELD

Department of Materials,
Queen Mary College,
London, England

I. Introduction

A. The Significance of High Temperature Compatibility

Some of the potential uses of whisker or fiber–metal matrix composites are for high temperature applications in which the superior properties of the composites will directly allow an improvement in the device performance. A prime example is the development of a nickel-based composite to replace a Nimonic alloy as the turbine blade material in the critical "hot end" of a jet engine. The substitution could be made simply on the basis of the composite having superior creep resistance at the present operating temperature, but a more exciting prospect is of obtaining equivalent properties at a temperature 200–300°C higher than the present operating temperature, which would lead to a considerably increased power output from the engine. It is for such high temperature applications that the compatibility of the reinforcement with the matrix, where compatibility implies the ability to co-exist without mutually-induced disintegration, becomes a significant controlling factor.

In order to achieve a compatible reinforcement–metal matrix system, a compromise must be effected between two opposing requirements, the desire to form a strongly-bonded interface in order to allow efficient stress transfer and maintain continuity during heating–cooling cycles and at the same time, the need to prevent a destructive reaction at the operating temperature. Hence the first requirement implies the initiation of a chemical reaction while the second requirement involves the prevention of a chemical reaction. Therefore the ideal situation is one in which the reinforcement and matrix only react to form a bond at a temperature higher than the projected operating temperature. Alternatively, a continuing reaction at the operating temperature could be acceptable if the rate of reaction is sufficiently slow to give an adequate service life (what "adequate" means will be dictated largely by economic factors).

At the moment a critical, quantitative selection of the optimum reinforcement–metal matrix system for a particular temperature, based on compatibility concepts, is hindered by the fragmentary data available on the nature and extent of reinforcement–matrix reactions. This chapter highlights the important influence of impurity as a factor controlling reinforcement–metal matrix compatibility. The three examples considered in detail to illustrate this theme are sapphire whiskers, nickel-coated carbon fibers, and nickel-coated silicon nitride whiskers, which demonstrate, respectively, the effects of impurity in the reinforcement, impurity in the matrix, and "impurity" in the surrounding environment.

B. *The Case for an Investigation of Individual Whiskers or Fibers*

In view of the complex thermomechanical history of a "bulk" composite test specimen, i.e., it has been subjected to a sequence of mixing, alignment, forming, and machining operations, it is felt to be advantageous to consider the compatibility of a composite in its simplest form, namely an individual whisker or fiber coated with a thin layer of a pure metal. This immediately eliminates the complications introduced by processing and allows the intrinsic compatibility of the reinforcement–matrix system to be determined. It also allows a direct evaluation of the nature and extent of any reaction, utilizing electron microscope techniques, and is economical in its consumption of, often expensive, whiskers or fibers. Although the applicability of such results to the case of the "infinite matrix" remains to be fully tested, some of the findings certainly suggest that compatibility is independent of the thickness of the coated layer. The points developed in this chapter are discussed with reference to the Queen Mary College program, where this individual whisker or fiber compatibility approach has been followed, and hence is not intended to represent a balanced review of compatibility in general.

II. Impurity in the Reinforcement

A. *"High Impurity" Sapphire Whiskers*

Sapphire whiskers have several intrinsic advantages as a reinforcing material for a high temperature composite system, among which are chemical inertness in an oxidizing environment, large elastic modulus and resistance to creep. However, for the effective use of sapphire in such a composite it is also necessary that the whiskers should be chemically compatible with metals such as nickel, which provide potential matrix materials over the desired operating range. In fact, it was found (Moore, 1969) for a 20% sapphire whisker–nickel composite, using both Compagnie Thomson Houston (CTH) and Thermokinetic Fibers, Inc. (TFI) whiskers, that the reinforcement was severely attacked after treatment at 1100°C in vacuum. As the potential applications of this system would be at 1100°C or higher temperatures, such a result would appear to immediately limit the usefulness of a sapphire whisker–nickel composite. However as will be discussed in this section, the apparent incompatibility at 1100°C of the sapphire whisker–nickel system is attributable to the presence of impurity at the surface of and within the "as-grown" sapphire whisker. It will be

TABLE I

SEMIQUANTITATIVE SPECTROGRAPHIC
SAPPHIRE WHISKER ANALYSES[a]

	TFI whiskers (%)	CTH whiskers (%)
Si	0.2	6
B	—	0.015
Fe	0.015	0.12
Mg	0.015	0.15
Mn	<0.001	0.001
Ga	—	0.001
Be	—	0.004
Ti	0.003	0.03
Cu	<0.001	0.001
Na	—	0.7
Ni	0.001	0.003
Ca	0.01	0.2
K	—	1
Sr	—	0.005
Cr	0.001	0.004
V	—	<0.002
Ba	—	0.01

[a] Reproduced by permission from *Journal of Materials Science*, published by Chapman & Hall, London.

demonstrated that by suitable purification the whisker breakdown at 1100°C can be prevented and a compatible sapphire whisker–nickel system obtained. This point is made to illustrate the significant influence of impurity in the reinforcement as a factor controlling reinforcement–metal matrix compatability.

Consequently a necessary prelude to a study of sapphire whisker–metal matrix compatibility is the determination of the effects of high temperature annealing on the structure of individual sapphire whiskers. Clearly it is essential to assess the nature of any morphological changes produced by such a process, particularly as the sapphire whiskers currently available contain a significant and variable concentration of impurity, which presumably results from the innoculants used in the whisker growth process (although the details are generally proprietary). As an example a typical "as grown" CTH whisker contains ~6% silicon and ~2% sodium and potassium, together with smaller concentrations of other elements, as shown in Table I. It will be demonstrated in the following sections that the presence of "high impurity" concentrations of this magnitude reduce the high temperature stability of the whiskers and as a result preclude their use, in this form, as a reinforcement at temperatures above 1000°C.

1. "As-Grown" CTH Sapphire Whiskers

The "as grown" whiskers can be prepared for electron microscope examination utilizing the replication technique developed by Andrews (1966), which allows observation of the electron transmission images of the thinner whiskers, the silhouettes of the thicker whiskers, and the corresponding replicas of all the whisker surfaces.

The width of the "as-grown" CTH whiskers measured in this manner (Bonfield and Markham, 1970) is from 0.2 to 10 μ. As the cross-sectional configuration is found in general to be either cylindrical or ribbon-like, this dimension can represent either the whisker diameter or one side of the ribbon. The width usually remains constant along the length of the whiskers, which ranges from 5μ to several mm. The thinner whiskers (i.e., $<1\mu$) are transparent to the electron beam, while for the thicker sections, examples of both transparent and opaque whiskers are present. Selected area electron diffraction through the transparent whiskers produces single-crystal alumina patterns, which indicate that most of the whiskers are oriented parallel to an "a" direction.

Particles of a second phase are often noted in the whiskers. The particles

TABLE II

SECOND-PHASE PARTICLE COARSENING[a]

Temperature (°C)	Time (hr)	Diameter (μ)			Number of particles per cm²
		Minimum	Maximum	Average	
"As-received"	0	~0.2	1.4	0.7	5.0×10^7
1100	2	~0.2	2.0	1.1	3.6×10^7
	8	~0.2	2.2	1.8	1.5×10^7
	17	~0.2	~3.7	~2.5[b]	5.8×10^6
1300	2	~0.2	2.1	1.0	3.2×10^7
	4	~0.2	2.4	1.2	2.8×10^7
	17	~0.2	~3.9	~2.5[b]	7.0×10^6

[a] Reproduced by permission from *Journal of Materials Science*, published by Chapman & Hall, London.
[b] Some particles interconnected and nonspherical.

FIG. 1. "Grown in" second-phase particles (CTH whiskers). [Reproduced by permission from *Journal of Materials Science*, published by Chapman & Hall, London.]

tend to be regularly spaced, as shown in Fig. 1, and are sometimes seen as
projections in the corresponding surface replica. The second-phase particles are always electron opaque and hence can not be directly identified
by selected area diffraction.

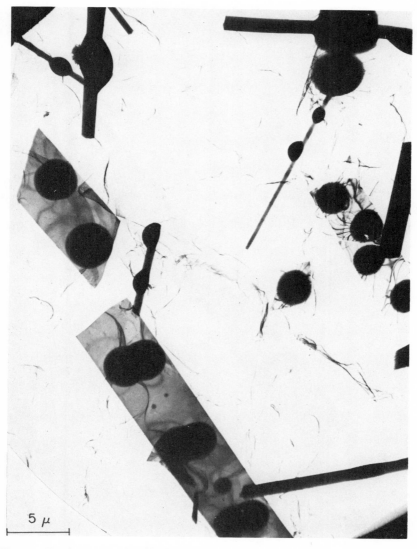

FIG. 2. Particle coarsening (CTH whiskers) after 17 hr at 1300°C. [Reproduced by permission from *Journal of Materials Science*, published by Chapman & Hall, London.]

2. Second-Phase Particle Coarsening and Whisker Breakdown

An annealing of the "as-grown" whiskers at temperatures of 800°C and 900°C under an inert atmosphere (argon), for times up to 17 hr, produces no significant changes in the morphology of the "as-grown" whiskers.

Fig. 3. Surface steps (CTH whiskers) after 8 hr at 1300°C. [Reproduced by permission from *Journal of Materials Science*, published by Chapman & Hall, London.]

However, after heat treatments at or above 1000°C, the size of the particles increases, and the number of particles per unit area decreases, with continued time, as shown in Table II for 1100°C and 1300°C treatments. As the whisker thickness is not precisely known, all the estimates in Table

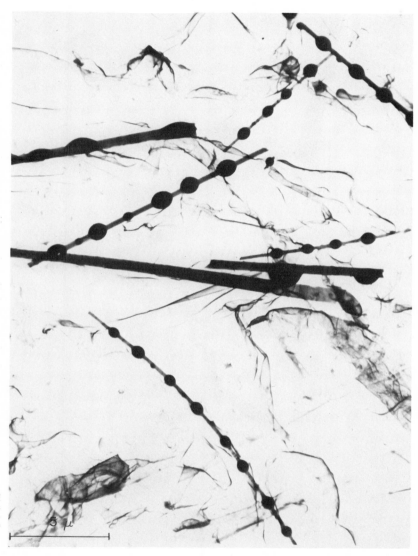

FIG. 4. Spheroidization (CTH whiskers) after 17 hr at 1100°C. [Reproduced by permission from *Journal of Materials Science*, published by Chapman & Hall, London.]

II are made on an area basis. The particles initially remain circular in outline, but tend to become elongated at longer times due to the coalescence of adjacent particles. Figure 2 shows the structure obtained after 17 hr at 1300°C and contains examples of both particle coarsening and coalescence.

Fig. 5. CTH whisker fragmentation after 4 hr at 1300°C. [Reproduced by permission from *Journal of Materials Science*, published by Chapman & Hall, London.]

In general the particle coarsening noted at 1100 and 1300°C is similar, but there are two major structural differences. First, at 1100°C, a high density of small particles (~0.2 μ diam) is produced in some whiskers, which can either be free of, or contain, some "grown-in" particles. Second,

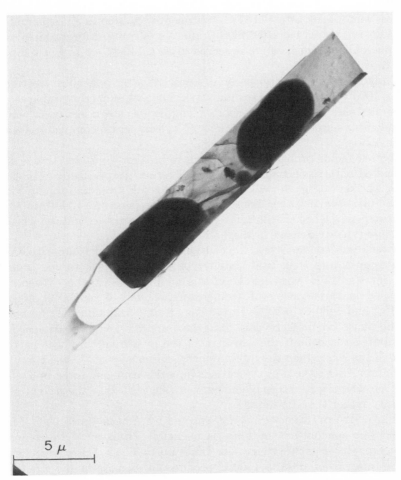

FIG. 6. CTH whisker "hole" after 17 hr at 1100°C. [Reproduced by permission from *Journal of Materials Science*, published by Chapman & Hall, London.]

at 1300°C, a pronounced diffusion zone and surface step develops around the particles, as shown in Fig. 3. (This second effect presumably could also occur at 1100°C with longer times at temperature.)

An effect related to particle coarsening is observed in the smaller whiskers after heat treatments at or above 1000°C. However, it appears to take place on the surface and results in the formation of a series of regularly spaced spheroidal projections of second phase, as shown in Fig. 4. This spheroidization is first noted after a 1000°C (for 17 hr) treatment and becomes widespread after 1100°C treatments. Examples of spheroidized whiskers are also noted after 1300°C anneals (see Fig. 2 for example), but no spheroidized whiskers are observed after a 1400°C or higher temperature heat treatment.

Whisker breakdown also becomes apparent after annealing treatments above 1000°C and occurs in several ways. Some whiskers containing second phase particles fragment into shorter lengths, as shown in Fig. 5, while a considerable number (estimated as 30%) disintegrate completely leaving a fine debris.

Another factor probably contributing to breakdown is the surface steps associated with second-phase particles. There is also evidence of holes in some whiskers (Fig. 6) which presumably result from the separation of second-phase particles. At higher annealing temperatures (1400, 1500°C), no examples of whiskers containing second phase are seen and the whiskers remaining show extensive hemispherical pits.

These results demonstrate the considerable influence of impurity on the high temperature structural stability of sapphire whiskers and indicate that for the CTH whiskers the effects of impurity can be conveniently classified in three ranges of temperature, namely 0 to 900°C, 1000 to 1300°C, and 1400 to 1500°C. In the "as grown" condition the presence of second-phase particles reveals that the impurity is not homogeneously distributed throughout the whisker. As the area fraction of the particles ($\sim$20%) is larger than the total impurity content ($\sim$8%), they are likely to contain Al_2O_3 as well as the various impurity elements listed in Table I. The "circular" appearance of the particles suggests that the second phase is glassy rather than crystalline.

The "as-grown" structure is not affected by anneals up to 900°C, from which it is concluded that there is negligible diffusion of the impurity within this temperature range (for times up to 17 hr). However, anneals in the range from 1000 to 1300°C produce an increase in the size of the "grown-in" particles and a reduction in the number of the particles per unit area. It seems reasonable then to ascribe this effect to a process in which the smaller particles redissolve and then diffuse to, and precipitate on, the larger particles. Support for this notion is provided by the reduction in the number of small particles and the more uniform particle size produced with increased time, as shown in Table II. An estimate of the activation energy Q for the initiation of particle coarsening and for the diffusion of the impurity ion(s) may be derived from the rate equation

$$\log(1/t) = \log K - Q/(RT) \tag{1}$$

in which t represents time, K a preexponential constant, R the gas con-

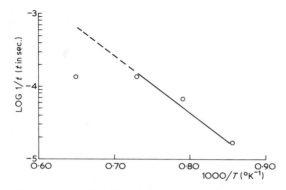

FIG. 7. Activation energy for impurity ion(s) diffusion. [Reproduced by permission from *Journal of Materials Science*, published by Chapman & Hall, London.]

stant, and T absolute temperature. From the results of Bonfield and Markham (1970) we have $T = 900, 1000, 1100,$ and $1300°C$, and $t = 17, 4, 2,$ and <2 hr, respectively, which data are shown plotted in Fig. 7 and give an approximate activation energy of 37 kcal/mole. The magnitude of the coarsening effect suggests that silicon ions, as the major impurity, must be involved in the diffusion process; this has been confirmed by some X-ray microprobe analysis on the larger particles.

The mechanism of particle coarsening appears to be by growth of a sphere to a diameter approaching half the width of the whisker, with further growth occurring by coalescence or elongation along the length of the whisker. In certain whiskers the latter process is probably promoted by the ribbon-like dimensions, with a width/thickness ratio of ~5:1. The relations between particle diameter, number of particles per unit volume, and time are shown plotted in Fig. 8. The slopes obtained are in approximate agreement with the predictions of Wagner's theory (1961) of particle coarsening by volume diffusion, which gives

$$(\bar{d}/2)^3 \propto t, \qquad N_v \propto t^{-1} \tag{2}$$

with $\bar{d}$ representing the mean particle diameter and N_v the number of particles per unit volume, but a more detailed study is required to substantiate this finding. The dependence of particle size on temperature however is not that given by Wagner for diffusion of a particular ion species ($\bar{r} \propto 1/T$). This difference is probably due to the variation in impurity type and content between whiskers. Such an effect may also account for the nucleation of "new" particles in certain whiskers apparently free from grown-in particles.

It is interesting to note that spheroidization on small whiskers takes

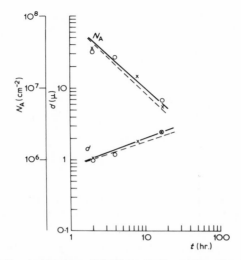

Fig. 8.   Relation between particle diameter d, number of particles per unit area N_A, and time t. Discontinuous lines represent theoretical predictions; $\times$, 1100°C measurements; and $\bigcirc$ 1300°C measurements. [Reproduced by permission from *Journal of Materials Science*, published by Chapman & Hall, London.]

place at about the same rate as particle coarsening. The likely difference between the two processes is that spheroidization is accomplished by surface diffusion, while coarsening results largely from volume diffusion. Hence the similarity of the kinetics suggests that there is little difference between surface and bulk diffusion in a material of whisker dimension.

The formation of surface steps associated with the growth of the second-phase particles may be explained in terms of the rapid diffusion of the impurity ion(s) (with $Q \sim 37$ kcal/mole) relative to the matrix Al^{+3} and $O^=$ ions [with $Q = 152$ (Oishi and Kingery, 1960), 110 (Oishi and Kingery, 1960), and 114 kcal/mole (Paladino and Kingery, 1962) for $O^=$ in single crystal Al_2O_3, $O^=$ in polycrystalline Al_2O_3, and Al^{+3} in polycrystalline Al_2O_3, respectively). Therefore, the diffusion of the impurity ions to the second phase particles at 1300°C is not compensated by counter diffusion of Al^{+3} (taking the generally accepted notion of a relatively rigid oxygen framework). Consequently a large number of vacant sites are created adjacent to the particle, which lead to the formation of a surface step.

No particles are observed after anneals at 1400°C for 4 hr or longer and, from the evidence of whisker debris, it is concluded that the particles have melted with a resultant disintegration of the whiskers. Some whisker debris is noted after all anneals above 1000°C, which suggests that melting occurs over a range of temperature. Such a result is as expected from the variable impurity content of the whiskers. The values of temperature are also in good agreement with the SiO_2 corner of the $K_2O-Al_2O_3-SiO_2$ phase diagram (Schairer and Bowen, 1955), which has a boundary line linking eutectics at 985 and 1470°C.

3. Surface Pitting

Although the disintegration of many whiskers results from melting of the second phase during anneals at 1000 to 1300°C as discussed in II,A,2., it should be stressed that a significant percentage of the CTH whiskers (~30%) are not affected by this process. However, in these whiskers localized surface pitting is noted after an 1000°C anneal, and becomes extensive after heat treatments at 1400–1500°C, when large hemispherical pits are noted. Many of these pits are associated with second-phase particles, but some pits form in impurity-free regions. The localized distribution and sharply-radiused form of the pits observed experimentally do not conform to the periodic convolutions predicted theoretically on a capillary-induced surface-diffusion model (Nichols and Mullins, 1965). Hence it is unlikely that self-diffusion of Al_2O_3 by itself would produce the surface pits. These results are consistent with the observations of Stapley and Beevers (1969) that a treatment at 1400°C for 67 hr in vacuum, i.e., longer than used in the present experiments, is required for "waist formation" in sapphire whiskers. Hence it is suggested that these surface pits result from diffusion promoted by local regions of large internal stress. The presence of such stress fields, due presumably to the limited extent of plastic flow possible in sapphire, can be experimentally demonstrated by observations under polarized light.

B. "Low Impurity" Sapphire Whiskers

On the basis of the previous section it would appear, at first sight, that an "as grown" sapphire whisker with a "low impurity" content could have improved high temperature characteristics. An example of such a whisker is that grown by Thermokinetic Fibers, Inc., which contains only ~0.2% silicon as a major impurity (Table I).

The TFI whiskers are similar in size to the CTH whiskers, but a greater percentage are electron opaque and only a few contain discrete second-phase particles. A major difference is that several "combined whiskers" and many tapered whiskers are present in the TFI samples.

There is no evidence for second-phase coalescence and only one example of spheroidization has been observed in the TFI whiskers after heat treat-

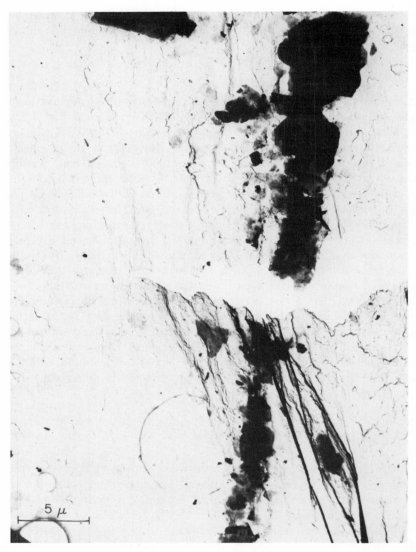

FIG. 9.   TFI whisker debris after 17 hr at 1300°C. [Reproduced by permission from *Journal of Materials Science*, published by Chapman & Hall, London.]

ment at 1100°C. However, after a 1200°C anneal some whiskers do exhibit a surface spheroidization effect, but on a much reduced scale to that noted in the CTH whiskers.

In spite of the apparently small amount of second phase present, a con-

FIG. 10.   Coherent TFI whiskers after 4 hr at 1400°C. [Reproduced by permission from *Journal of Materials Science*, published by Chapman & Hall, London.]

siderable amount of whisker debris is seen after a 1300°C anneal. Figure 9 illustrates this point and shows a whisker which had almost disintegrated.

The coherent whiskers remaining after a 1300°C or higher temperature anneal are nearly always transparent as shown in Fig. 10, which suggests

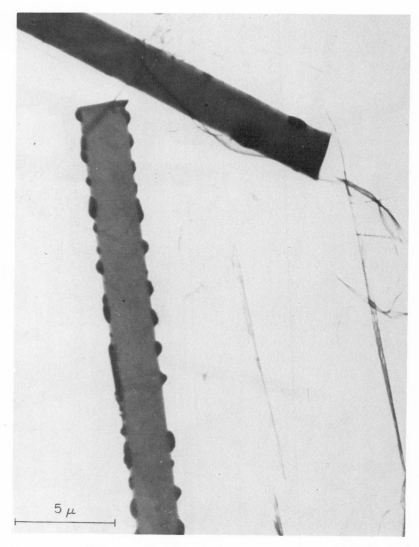

Fɪɢ. 11.   Spheroidization (TFI whiskers) after 67 hr in hydrofluoric acid and 17 hr at 1100°C. [Reproduced by permission from *Journal of Materials Science*, published by Chapman & Hall, London.]

that the breakdown is confined to the opaque specimens. It should be noted that the opaque whiskers in Fig. 10 have an irregular "globular" outline.

Fragmentation of the whiskers into shorter lengths is not observed in the TFI whiskers. Occasional surface pits can be seen on some whiskers after heat treatments above 1000°C, but at any given temperature the extent of the pitting is less than that in the CTH whiskers.

The role of impurities in the TFI whiskers (0.2% Si) is an intriguing one, as they are nominally much purer than the CTH whiskers and only a few "as grown" whiskers contain second-phase particles. However, the disintegration of many of the opaque whiskers after 1300°C heat treatment definitely appears to be associated with a melting process (see, for example, the "globular" remnants in Fig. 10). Hence, it would appear that these whiskers have a continuous surface coating of a second phase. A critical test of this hypothesis would then be to remove or reduce the thickness of the coating. Such an experiment proves singularly revealing, as after prolonged chemical polishing in phosphoric acid, there is no apparent difference in the structure of an opaque "as-grown" whisker, but definite evidence after an 1100°C anneal for a spheroidized, discontinuous coating (Fig. 11). Therefore, it may be concluded that many of the opaque TFI whiskers have a continuous second-phase coating.

From Table 1 the second phase associated with the TFI whiskers would probably consist of Al_2O_3 and the oxides of Si, Fe, and Mg. The absence of Na and K appears to account for the higher average melting point of this second phase relative to that of the CTH whiskers (1300°C versus 1100°C).

Consequently this section has demonstrated that "low impurity" sapphire whiskers also break down at relatively low temperatures due to the presence of impurity on the surface of the fibers.

C. "Purified" Sapphire Whiskers

The main conclusion to be drawn from the discussion of "high impurity" and "low impurity" sapphire whiskers is that the "as grown" whisker is not in the optimum condition for a high temperature reinforcement. In the case of sapphire whiskers the intrinsic impurity in the "as grown" condition limits the potential operating temperature of a sapphire whisker–nickel composite to ∼1000°C, a temperature which is determined exclusively by the melting of the impurity particles. Hence the potential utility of sapphire as a high temperature reinforcement, which partially is due to its high melting point, is seriously impaired.

These findings demonstrate conclusively that it is necessary to subject sapphire whiskers to a purification pretreatment before they are incorporated in a metal matrix. The objective of such a pretreatment is to reduce the concentration of the impurity contributing to whisker breakdown to a level at which it has no significant effect. Consequently in order to critically establish the form of the pretreatment it is necessary to determine, first, the particular impurity producing breakdown and, second, the level at which

TABLE III

SEMIQUANTITATIVE SPECTROGRAPHIC ANALYSES OF CTH AND TFI WHISKERS[a]

Element (%)	CTH			TFI		
	"As-grown"	Solution A	Solution B	"As-grown"	Solution A	Solution B
Si	6	0.15	10.	0.2	0.07	0.05
Na + K	1.7	—	0.75	—	—	—
Ca	0.2	0.01	0.02	0.01	0.02	0.002
Fe	0.12	0.35	1	0.01	0.07	0.04
Mg	0.15	0.06	0.15	0.01	0.04	0.02

[a] Reproduced by permission from *Journal of Materials Science*, published by Chapman & Hall, London.

it ceases to be effective. In this section, the effects of removal of impurity from the CTH and TFI whiskers, both on the "as grown" structure and on the structure after exposure to a temperature of 1100°C are discussed. Two main types of experiments which were attempted by Bonfield and Markham (1971) are described: the effect of removal of all of the impurity elements from the whiskers and the effect of selective removal of particular elements from the whiskers. In practice, it is found that the selective removal of particular elements can not be performed in the ideal manner, but a sufficient variation in impurity level is obtained to effect a reasonable evaluation of the relative contributions. In this manner, it is possible to identify the impurity producing disintegration, with the benefit that the conditions necessary for high temperature structural stability of sapphire whiskers can be clearly defined.

The effects of treatment in 20% HF–20% H_2SO_4 (solution A) for 120 hr and concentrated H_3PO_4 (solution B) on the composition of the whiskers are shown in Table III. It can be seen that, in general, solution A reduces the silicon and (sodium + potassium + calcium) to a low level, while not removing the iron. In contrast, solution B produces a partial removal of (sodium + potassium + calcium), while silicon and iron actually appear to increase (this is probably due to its selective removal of aluminum oxide).

The structure of a typical "as shown" CTH whisker, contains a number of second-phase particles, as shown in Fig. 1, with diameter d of the second phase in different whiskers ranging from approximately 0.2 to 1.1 μ. An examination of the morphology of such whiskers after treatment in solution A reveals particles of a reaction product on the surface of the whisker,

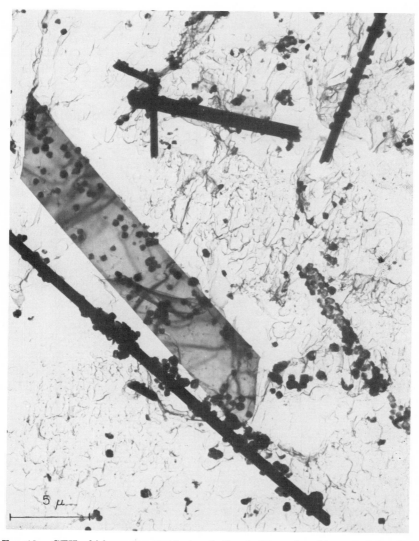

FIG. 12. CTH whiskers after 120 hr in solution A. [Reproduced by permission from *Journal of Materials Science,* published by Chapman & Hall, London.]

as shown in Fig. 12. It can be seen that the reaction product is associated with large and small whiskers and is both attached to the whiskers and also is a detached state. The reaction product has a generally square configuration with an approximate mean width of 0.4 μ and a mean density of

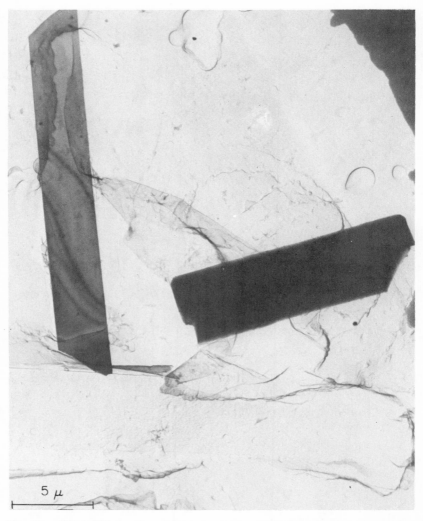

FIG. 13.   CTH whiskers after solution A treatment and 17 hr at 1100°C. [Reproduced by permission from *Journal of Materials Science*, published by Chapman & Hall, London.]

approximately 2.4 particles/μ^2. It is not possible to detect any "grown-in" second-phase particles in the whiskers after treatment in solution A.

The effect of an 1100°C/17 hr treatment on the "as-grown" whiskers (i.e., without solution A treatment) is to produce considerable coarsening of the second-phase particles (the average diameter increasing to approxi-

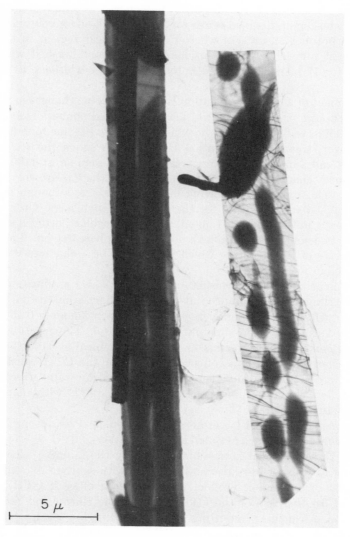

FIG. 14. Second-phase configurations formed in solution B-treated whiskers after an anneal at 1100°C for 17 hr. [Reproduced by permission from *Journal of Materials Science*, published by Chapman & Hall, London.]

mately 2.5 μ as shown in Table II). In contrast, subjecting the "solution A-treated" whiskers to a similar anneal produces an interaction between the reaction product and the whisker. The larger whiskers remain coherent but breakdown of some of the small whiskers takes place with debris formation. Consequently, it is necessary to remove the reaction product from the whisker to prevent breakdown. This can be simply done by placing the whiskers in distilled water for 24 hr, which effectively removes the reaction product and leaves whiskers which appear regular in outline. A subsequent anneal of the "solution A-treated" and washed whiskers at 1100°C for 17 hr produces no disintegration of the whiskers, as shown in Fig. 13.

Treatment of whiskers in solution B produced some changes in the shape of the sapphire whiskers. The whiskers are reduced nonuniformly in width and thickness and the ends of the whiskers tend to become rounded.

However, treatment in solution B does not remove the second-phase particles and annealing whiskers treated in this manner at 1100°C for 17 hr produces changes similar to those noted for the "as-grown" whiskers, with particle coarsening occurring in some whiskers, although on a finer scale than noted previously for the "as-grown" whiskers. Other second-phase configurations are also noted, such as irregular particles and a distinctive two-column or "duplex core" second-phase region. This duplex core feature of the structure which is found to be a characteristic one, is illustrated in Fig. 14.

Treatment of the TFI whiskers in solution A also produces a reaction product, similar in appearance but on a much finer scale to that noted on the CTH whiskers, with an average width of approximately 0.15 μ and an average density of approximately 10 particles/μ^2. The reaction product becomes embedded in the whiskers during an anneal at 1100°C for 17 hr in a similar manner to the effect described for the CTH whiskers. The reaction product can also be completely removed by water washing, after which it is possible to anneal the whiskers at 1100°C without producing any structural change.

Treatment of the TFI whiskers in solution B does not produce any second-phase effects such as noted in CTH whiskers.

From these results it is possible to identify the impurity element contributing to whisker breakdown. Treatment with solution A results in the formation of a reaction product on the surface of both CTH and TFI whiskers. In the case of the CTH whiskers the silicon and (sodium + potassium + calcium) impurities are reduced to a low level, while for the TFI whiskers the silicon impurity only is removed [there being a negligible concentration of (sodium + potassium + calcium)]. The reaction products appear similar and both are water soluble and react with the respective whiskers during an 1100°C anneal. Hence, it is likely that the reaction product contains the same impurity (i.e., silicon) in both cases. With the removal of the reaction product from the whisker surface by water washing, it is possible to anneal both CTH and TFI whiskers at 1100°C for 17 hr without any significant second-phase coalescence or whisker breakdown taking place. Hence it is concluded that for the whiskers to remain stable at 1100°C, the silicon concentration must be <0.15%. A measure of the amount of reaction product in each case can be determined on an area basis by the product of the number of particles per unit area and the average particle area. This yields the following: For CTH whiskers,

$$\text{total reaction product area} = 0.32 \ \mu^2 \text{ per } \mu^2 \text{ of whisker;}$$

for TFI whiskers,

$$\text{total reaction product area} = 0.23 \ \mu^2 \text{ per } \mu^2 \text{ of whisker.}$$

Such an estimate is only approximate as it does not take into account the particles which became detached from the whiskers, but the greater amount of reaction product on the CTH whisker is certainly consistent with the larger silicon concentration in the CTH whiskers.

Further confirmation of the detrimental effect of silicon, rather than (sodium + potassium + calcium) is provided by the results obtained in solution B. This does not reduce the silicon content in CTH whiskers and, in fact, there is an apparent increase in silicon concentration which is probably due to the removal of aluminum. However, there is partial removal of the (sodium + potassium + calcium) group. After this treatment, followed by an anneal at 1100°C for 17 hr, some of the features exhibited by the "as-grown" whiskers are observed, namely the coalescence of second-phase particles and subsequent whisker breakdown. There are also some differences in the form of the second-phase developed after annealing the "as grown" and "solution B-treated" whiskers which are attributed to the variable geometry produced during thinning and are discussed in more detail elsewhere (Bonfield and Markham, 1971).

The other major impurity in both CTH and TFI whiskers is iron but it appears that this does not contribute to whisker breakdown during 1100°C anneals, as treatment in solution A (which produces a stable whisker) does not produce any reduction in the iron concentration. Therefore by an appropriate purification treatment of both "high impurity" and "low impurity" "as-grown" sapphire whiskers, it is possible to significantly improve their high temperature stability and hence their utility as a high temperature reinforcement.

D. "Purified" Sapphire Whisker–Nickel Compatibility

It has been demonstrated in the previous sections that "as-grown" sapphire whiskers must be suitably purified to render them suitable for

FIG. 15. Nickel-coated CTH whiskers after 3 days at 1100°C.

use as a high temperature reinforcement. Two further problems then remain to be considered: firstly, the compatibility of the purified whiskers with, for example, a nickel matrix and secondly, the effectiveness of the bond between the whisker and matrix. These problems can be investigated by evaporating a thin layer of nickel (~500 Å thick) onto a sapphire whisker. If this nickel-coated whisker is then annealed at temperatures above 800°C, the continuous nickel layer breaks up to form a series of spheroidal particles, which allows the nickel–sapphire interface to be directly examined in the electron microscope. This technique provides an effective method of monitoring both compatibility and bonding, and will be returned to in more detail in later sections.

Figure 15 shows the structure observed after an anneal of "purified" nickel-coated CTH whiskers at 1100°C under argon for 3 days. It can be seen that the sapphire whiskers have remained coherent and have a number of small nickel spheres attached to the surface, although a large number of nickel spheres have also become detached from the whiskers. Observation of the profiles of some suitably oriented nickel spheres indicates that there is only point contact with the whisker surface, and hence no obvious surface wetting, while other particles have become facetted and have a larger area of contact. Therefore, it appears that at 1100°C, for times up to 3 days, the "purified" sapphire–nickel system is compatible and similar experiments are in progress for longer times and higher temperatures.

There is still only limited data available on the nature of the bonding between "purified" sapphire and nickel and hence a detailed discussion is not merited. However it is interesting to note that at 1100°C the "mechanical" bonding noted in Fig. 15 is sufficiently strong to withstand the peeling test imposed by the electron microscope replication process. Preliminary results at higher temperatures (1300°C) (Bonfield and Markahm, 1971a) have shown that some whisker surfaces become pitted and the nickel particles become irregular, indicating the possibility of a chemical reaction. If this finding is substantiated, it points to the possibility of developing a two-stage process (i.e., process at a temperature to form a chemical bond and then operate at a lower temperature where reinforcement and matrix are compatible) to ensure the optimum composite properties.

III. Impurity in the Matrix

The successful development of high strength, high modulus carbon fibers has provided an opportunity for the design of carbon fiber–metal matrix composites with high temperature mechanical properties superior to the metallic alloys currently available. Of the refractory metals available for use as the matrix material, nickel is perhaps the most suitable as it does not form a stable carbide. However, it has proved experimentally difficult to realize the theoretical values of fracture strength predicted for a carbon fiber–nickel composite, and dramatic reductions in fracture strength have resulted from comparatively low temperature annealing treatments. In recent work, Jackson and Marjoram (1968) found a significant reduction in the fracture strength of a "microcomposite" of graphitized carbon fibers (Type I) in nickel (containing approximately fifteen carbon fibers) after treatment at temperatures > 800°C, an effect which was attributed to recrystallization of the carbon fiber.

In contrast Barclay and Bonfield (1970) found that the fracture strength of individual nickel-coated (Type I) carbon fibers was not affected by anneals in a 10^{-6} torr vacuum below 1000°C; and the reduction in strength produced by treatments above this temperature was not related to fiber recrystallization, but appeared to be controlled by the formation of a suitable carbon–nickel interface.

These differences in behavior, noted for nominally the same composite system, are important, as the significant variation between the two groups of experiments is in the impurity content of the nickel matrix (and to a lesser extent in the atmosphere). Hence the effect of impurity in the matrix on reinforcement–metal matrix compatibility is considered in this section, with reference to nickel–coated carbon fibers.

A. Fracture Strength of Carbon Fibers

The average fracture strength measured at room temperature for "as-received" Type I carbon fibers is 280,000 psi (1930 MN/m²), with one standard deviation of 49,000 psi (337 MN/m²). It is found that annealing the "as-received" carbon fibers for 24 hr at temperatures up to approximately 1100°C does not significantly affect the room-temperature fracture strength (i.e., the average value is within the standard deviation), provided that the anneal is performed with a 10^{-6} torr vacuum. The atmosphere is critical, as a variable reduction in the fracture strength is obtained if the level of the vacuum is reduced to 10^{-5}–10^{-4} torr (after anneals at temperatures as low as 600°C). Under all these conditions, liberation of dissolved or adsorbed gas from the carbon fibers occurs during the "warm-up" to the annealing temperature. Thus the reduction in strength in "low" vacuum (or argon) probably results from an oxygen concentration in the furnace atmosphere which is sufficient for reaction with the carbon to form carbon monoxide. These results emphasize that, in an investigation of the effects of nickel coating on the fracture strength of carbon fibers, it is essential to maintain a 10^{-6} torr vacuum in order to avoid the complication of fiber–atmosphere reactions. Similarly, it is found difficult to obtain reproducible fracture strength results with an argon atmosphere. In all cases the fibers should be heated slowly to the annealing temperature to allow the evolution of dissolved gas. After treatment at temperatures above 1100°C the fracture strength of the carbon fibers decreases appreciably, even in a 10^{-6} torr vacuum [e.g., 147,000 psi (1020 MN/m²) after 1200°C/ 24 hr].

B. Fracture Strength of Nickel-Coated Carbon Fibers

The average room temperature fracture strength measured for the carbon fibers coated with ~400 Å of evaporated nickel is 277,000 psi (1910 MN/m²), which is similar to that of the "as-received" uncoated fibers. (The mechanical contribution of a coating of this thickness to the strength is assumed to be negligible and the fracture strength calculated on the basis of the carbon fiber area alone.)

The room temperature fracture strength of the nickel-coated carbon

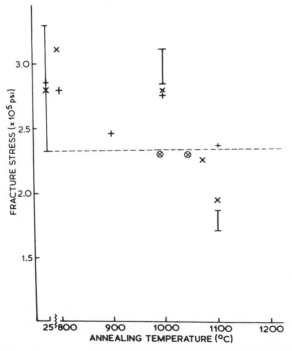

FIG. 16.   The effect of annealing on the fracture strength of uncoated and coated carbon fiber. +, 24 hr (uncoated); ×, 24 hr (~400 Å Ni); ⊗ 48 hr (~400 Å Ni); I , 24 hr (~0.7 μm Ni). [Reproduced by permission from *Journal of Materials Science*, published by Chapman & Hall, London.]

fibers is not reduced by anneals up to 1000°C for 24 hr, but as shown in Fig. 16 a possible reduction occurs after 1080°C for 24 hr, with a definite reduction after 1100°C for 24 hr. The fracture strength also appears to be time-dependent as after 1000°C for 48 hr there is a small reduction in strength, with a correspondingly larger reduction after 1100°C for 48 hr (see Fig. 16).

Some measurements have also been made on carbon fibers coated with a substantially thicker layer of nickel (approximately 0.7 μm). The fracture strength of these carbon fibers (σ_c) can be calculated in two ways; first, neglecting the contribution of the nickel, i.e., on the basis of the carbon fiber diameter alone and, second, assuming the nickel coating does contribute to the fracture strength and using the composite rule of mixtures. For example, after a treatment at 1000°C for 24 hr, we obtain by the two methods:

Method 1:

$$\sigma_c = 314{,}000 \text{ psi } (2160 \text{ MN/m}^2) \tag{3}$$

or

$$\sigma_{(c+Ni)} = \sigma_c V_c + \sigma_{Ni} V_{Ni} \tag{4}$$

$$\sigma_{Ni}/\sigma_c = E_{Ni} V_{Ni}/E_c V_c \tag{5}$$

where E is Young's modulus and V the volume fraction. By substitution in Eqs. (4) and (5) we obtain

Method 2:

$$\sigma_c = 287{,}000 \text{ psi } (1970 \text{ MN/m}^2) \tag{6}$$

Hence it is reasonable to consider these two values as the limits after the 1000°C for 24 hr treatment. This factor is important in assessing the reduction in strength produced by an 1100°C for 24 hr anneal. In this case limits of 170,000 to 185,000 psi (1170 to 1270 MN/m²) are obtained, which as shown in Fig. 16 are very near to the fracture strength measured on carbon fibers with a "thin" nickel coating after a similar anneal (193,000 psi) (1330 MN/m²).

C. Structural Changes in Nickel-Coated Carbon Fibers

1. Preferred Orientation

Type I carbon fibers are polycrystalline with graphite basal planes oriented along the fiber axis but turbostratically displaced (Johnson and Watt, 1967). This structure gives rise to a characteristic X-ray pattern as

FIG. 17.   X-ray pattern from an aligned bundle of as-received carbon fibers. [Reproduced by permission from *Journal of Materials Science*, published by Chapman & Hall, London.]

shown in Fig. 17 (taken from an aligned bundle of ∼1000 fibers). The important reflection is the arc from the (002) basal planes [a faint arc from the (004) planes can also be seen] which results from the preferred orientation of these planes along the carbon fiber axis. The angular spread of the arc (approximately 10%) is due to small deviations about the fiber axis and any misorientation of the individual fibers in the specimen holder.

The structure of the carbon fibers can be observed on a finer scale by vibrating some fibers ultrasonically and examining the detached surface fragments using transmission electron microscopy. A typical fragment from an as-received fiber reveals both the (002) plane alignment and the turbostratic arrangement.

Barclay and Bonfield (1970) found that the preferred orientation of the carbon fibers, as evidenced by both X-ray diffraction and electron microscopy, is retained after high temperature annealing of the nickel-coated carbon fibers, whereas Jackson and Marjoram (1969) observed definite recrystallization in the carbon fiber–nickel "micro" composite specimens. This point will be returned to in a later section.

2. Carbon Fiber–Nickel Adhesion

It is found that after anneals above 800°C the nickel coating breaks up into a series of spheroidal particles. As a result the replica technique can be utilized to determine the time required to form an effective bond between the carbon fiber and the nickel spheroids. A series of replicas are prepared for various times at a given temperature. The time for adhesion is then established by the transition from almost complete detachment of the nickel spheroids by the replication process to almost no nickel spheroids being present in the replica. In this manner, times for adhesion of $\sim$1, $\sim$2, $\sim$5, and $\sim$24 hr are established, respectively, for temperatures of 1100, 1080, 1050, and 1000°C.

Replicas prepared from nickel-coated fibers after adhesion has developed, reveal that the nickel spheroids tend to spread along the surface and create a continuous nickel coating.

D. Carbon Fiber–Nickel Compatibility

It has been demonstrated that a 400 Å thick nickel coating on the carbon fiber has no effect (i.e., the mean value is within one standard deviation of the as-received condition) on the fracture strength after anneals up to 1000°C for 24 hr, but that a small reduction is obtained after treatments at 1080°C for 24 hr, with a further reduction after treatment at 1100°C for 24 hr. (After higher temperature anneals (1130 to 1200°C), the fracture strength decreases markedly, but in this range some breakdown also results from the carbon–atmosphere reaction.) As the fracture strengths of the coated fibers are smaller than those measured for the uncoated fibers after equivalent treatments in the 1000 to 1100°C range, it is concluded that the reduction in strength is produced by the nickel coating. It is reasonable to conclude that the reduction in strength is a consequence of the dissolution of carbon atoms in the nickel coating, as carbon has an appreciable solubility in nickel in the 1000 to 1100°C temperature range (Dunn *et al.*, 1968) ($\sim$1.2 to 2.0 at %).

However, calculation of the rate for the diffusion of carbon into nickel suggests shorter times (t) for saturation than those measured experimentally for a reduction in strength. For example, assuming a slab configuration, we have

$$t = l^2/D \tag{7}$$

at 1000°C, as $D \sim 10^{-8}$ cm²/sec, where l is the coating thickness and D the diffusion coefficient.

$$l = 4 \times 10^{-6} \text{ cm}, \qquad t \ll 1 \text{ sec}$$

whereas for a reduction in fracture strength, we have

$$t > 24 \text{ hr}$$

This suggests that the rate-controlling process is not the diffusion of carbon into nickel, but is the establishment of a suitable carbon–nickel interface. Such a concept appears to be confirmed by the replica measurements of the times (t) required for carbon–nickel adhesion which vary from >24 hr at 1000°C to $\sim$1 hr at 1100°C. An estimate of the activation energy (Q) for the adhesion process may be derived from the rate equation (see Section II,A,2.). The data are shown plotted in Fig. 18 and give an approximate activation energy of 110 kcal mole⁻¹. This value may be compared with bond energies of 83 and 147 kcal mole⁻¹ quoted, respectively, for single and double carbon–carbon bonds. Hence, it is possible that the rate-controlling process is the breaking of a carbon–carbon bond, which is required before diffusion of carbon into the nickel can occur.

During the time period prior to adhesion, the initially homogeneous nickel coating breaks up into a series of spheroids, as has also been observed with thin coatings on sapphire whiskers. As a result, when adhesion is established and carbon diffusion into the nickel begins, there is partial nickel coverage of the fiber surface and hence only local depletion of carbon. This situation changes with time as the nickel spheroids spread over the surface, presumably due to a surface tension reduction, and eventually a continuous coating is formed which then permits diffusion of carbon into nickel from all points on the fiber.

Spheroidization and recombination of nickel coatings will only occur in relatively thin coatings. Hence it is important to establish whether a thicker nickel coating, which remains continuous at all times, has a similar effect on the fracture strength. In fact, the values of fracture strength measured for carbon fibers with a 0.7 μm nickel coating (after treatments

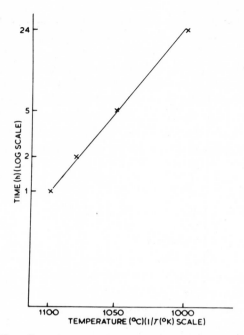

Fɪɢ. 18.   The effect of temperature T on the time t for C–Ni adhesion. [Reproduced by permission from *Journal of Materials Science*, published by Chapman & Hall, London.]

at 1000 and 1100°C for 24 hr) are reasonably similar to those of the carbon fibers with the 400 Å coating. The particular annealing treatments chosen allow two comparisons to be made. Firstly, after 1000°C for 24 hr, the 400 Å coating is spheroidized, while the 0.7 μm coating is continuous and, secondly, after 1100°C for 24 hr, both coatings are continuous. Hence, the similarity of the experimental measurements suggests that the carbon fiber fracture strength is independent of both the area of contact and the thickness of the nickel coating.

This conclusion suggests that it is reasonable to compare the results on nickel-coated carbon fibers with Jackson and Marjoram's findings on graphitized carbon fiber-nickel microcomposites. There are two major differences between the results. The first difference is the carbon fiber recrystallization noted in the microcomposites, but not observed in the nickel-coated carbon fibers. Recrystallization is attributed to the diffusion, dissolution, and precipitation of nickel within the carbon fiber, but its complete absence in the present experiments with evaporated nickel suggests that the process may be promoted by the impurities intrinsic to electroplated nickel. The second difference is the larger reduction in strength, measured in the microcomposites for any given temperature–time treatment. At first sight, it would appear that this results simply from recrystallization. However, there is an area of overlap between the results, as after a 1000°C for 24 hr treatment a reduction in strength is measured in the microcomposite, but there is no corresponding increase in crystallite size. Therefore, it is possible that in the absence of recrystallization, a similar mechanism could apply in both the nickel-coated carbon fibers and microcomposites. The larger reduction in strength in the microcomposites could then be attributed to the presence of C "impurity atom" bonds (e.g., C–S, bond energy 62 kcal mole^{-1}) which facilitate the establishment of a carbon–nickel interface.

IV. Impurity in the Environment

The gaseous environment of the reinforcement–metal matrix composite constitutes an important "impurity" factor in a consideration of compatibility. This has already been demonstrated in the example of carbon fibers, which readily disintegrate above 600°C in the presence of a small partial pressure of oxygen. Similarly, sapphire whiskers are likely to break down at high temperature under reducing conditions. Hence it is essential that the composite should be compatible with the environment, both during processing as well as in its application. Each of these aspects generally requires a different solution, i.e., a compatible carbon fiber–nickel composite could be processed under 10^{-6} torr vacuum, but its application in a jet engine requires, in addition, the incorporation of an oxygen barrier layer (a refractory metal) around the fibers. In this section the effect of gaseous environment on nickel-coated silicon nitride whiskers is considered and it is demonstrated that small variations in the oxygen and nitrogen partial pressure can significantly affect the high temperature compatibility of this system (Andrews *et al.*, 1972).

A. Compatibility in Vacuum of Nickel-Coated Silicon Nitride Whiskers

Uncoated silicon nitride whiskers (obtained from mats containing $\sim$90% of α-silicon nitride grown at the Explosives Research and Development Establishment) do not exhibit any changes of morphology after anneals in vacuum (or an inert atmosphere) in the temperature range considered.

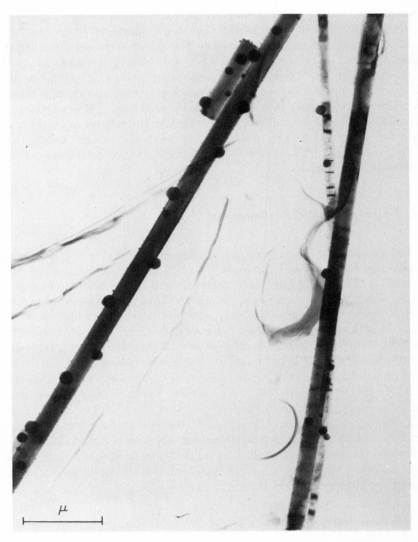

Fɪɢ. 19.   Nickel-coated silicon nitride whiskers after 50 min at 900°C in vacuum. [Reproduced by permission from *Journal of Materials Science*, published by Chapman & Hall, London.]

However, on heating nickel-coated ($\sim$200 Å thick) whiskers above 800°C, the initially continuous coating breaks up to give a series of spheroidal particles. Figure 19 shows the structure obtained after an anneal at 900°C for 50 min. It can be seen that there is a range of particle diameters from

FIG. 20. Nickel silicide platelets after 15 min at 1100°C. [Reproduced by permission from *Journal of Materials Science*, published by Chapman & Hall, London.]

approximately 0.01 μ to 0.3 μ and that some of the particles are adhering to the whiskers, although it should be emphasized that a greater number of particles have become detached from the whiskers during the replication process. An anneal at a higher temperature (950°C) for 15 min produces a similar nickel particle distribution, although some of the particles develop a faceted rather than a spheroidal form. However, with longer times at 950°C there is an increase both in the size of the particles and the average number of adherent particles per unit area of whisker. An increase is also noted in the contact area between the particle and whisker, accompanied by a general flattening or faceting of the initial spheroid, particularly for times at 950°C, greater than 4 hr. This transition is also noted after anneals at 1000°C and 1050°C, but after shorter times of approximately 90 min and approximately 30 min, respectively.

After an anneal at 1100°C for 15 min a distribution of regular "nickel" platelets is obtained, with all the platelets firmly adhering to the whisker. At this temperature, melting of some of the platelets produces holes in the whiskers, as shown, for example, in Fig. 20. Selected area diffraction from the platelets gives extra rings which are consistent with the larger interplanar spacings expected from nickel silicide as compared with pure nickel.

B. Compatibility in Nitrogen

The platelet formation and melting noted at 1100°C under vacuum is not repeated with similar temperature anneals under a high purity nitrogen atmosphere. After 15 min a few spheroidal particles are noted on the whiskers while most of the particles are detached by the replication process. This situation persists for times up to 17 hr and the whiskers remain generally coherent although there is evidence for some general irregularity and pitting of the whisker surface.

C. Compatibility in Argon

Completely different morphological changes are noted under an argon atmosphere at 1100°C from those observed under either vacuum or nitrogen. After a short time (15 min) at 1100°C under high purity argon, side growths develop on the original whiskers which displace the nickel particles away from the whisker. The side growths increase in both length and width with further time at 1100°C and after 17 hr range up to 5 and 0.5 μ, respectively, as shown in Fig. 21. It can be seen that nickel particles, of about the original size, are still attached to a number of side growths.

D. Effect of Environment on Compatibility

The results presented in Sections IV,A; IV,B; and IV,C demonstrate clearly the effect of environment on the high temperature structural stability of nickel-coated silicon nitride whiskers. At 1100°C the whiskers firstly dis-

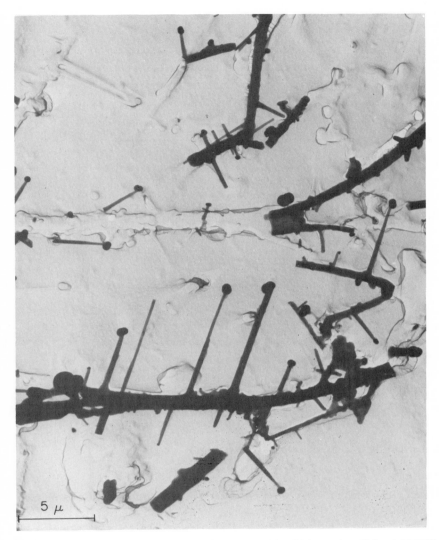

Fɪɢ. 21. Side growths on nickel-coated silicon nitride whiskers after 17 hr at 1100°C in argon. [Reproduced by permission from *Journal of Materials Science*, published by Chapman & Hall, London.]

integrate under vacuum, secondly remain essentially unchanged under nitrogen, and thirdly develop side growths under argon (while none of these changes occurred on uncoated whiskers). Further, a nickel silicide reaction product is associated with the disintegration under vacuum but cannot be

TABLE IV

NITROGEN PARTIAL PRESSURES[a]

Environment	Partial pressure of N_2 (atm)
Vacuum	$\sim 10^{-7}–10^{-8}$
Nitrogen	~ 1
Argon	$\sim 3–5 \times 10^{-5}$

[a] Reproduced by permission from *Journal of Materials Science*, published by Chapman & Hall, London.

detected under nitrogen or argon. Hence, it is concluded that the changes are all a consequence of the nickel coating, with a reaction between silicon nitride and nickel taking place under vacuum conditions to form a nickel silicide as for example

$$Si_3N_4 + 3Ni \rightarrow 3SiNi + 2N_2 \tag{8}$$

As the progress of such a reaction will depend on the partial pressure of nitrogen (P_{N_2}), it is useful to determine the approximate level required.

The free energy of the reaction in Eq. (8) at 1373°K ($\Delta G_{1373}°$) is approximately 58 kcal/mole.

Also

$$\Delta G_{1373}° = -RT \ln K \tag{9}$$

with

$$(P_{SiNi})^3 (P_{N_2})^2 / [(P_{Si_3N_4})(P_{Ni})^3] \sim (P_{N_2})^2 \tag{10}$$

Substituting in Eqs. (9) and (10) we obtain

$$P_{N_2} \sim 2.5 \times 10^{-5} \text{ atm}$$

Hence, this estimate suggests that under equilibrium conditions a nitrogen partial pressure $<2.5 \times 10^{-5}$ atm is required for reaction Eq. (8) to proceed. Under experimental conditions, the kinetics of the reaction are also important and hence the critical value can only be regarded as approximate. However, with this reservation it can be seen from Table IV, which lists the nitrogen partial pressure, that it is only under vacuum that the nitrogen partial pressure is sufficiently small for the reaction to proceed.

If the breakdown of silicon nitride, in the absence of nickel is considered, we have

$$Si_3N_4 \rightarrow 3Si + 2N_2 \tag{11}$$

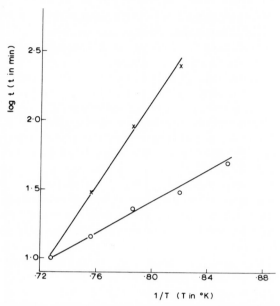

FIG. 22.   The effect of temperature on the time for nickel faceting and wetting on silicon nitride (O, faceting; ×, wetting). [Reproduced by permission from *Journal of Materials Science*, published by Chapman & Hall, London.]

with $\Delta G_{1373} \sim 70$ kcal/mole. By a similar argument we obtain

$$P_{N_2} \sim 2.8 \times 10^{-6} \text{ atm}$$

This result suggests that direct breakdown of Si_3N_4 could occur in vacuum. Hence the fact that no breakdown is observed in the uncoated condition emphasizes that the critical nitrogen partial pressure also depends on kinetic factors.

Phenomenologically, it is observed that the formation of nickel silicide, under vacuum, is preceded by three distinct stages. Firstly, above 800°C, the nickel coating breaks up to form a series of spheroidal particles, in the manner already described for nickel-coated sapphire whiskers (Section II,D) and nickel-coated carbon fibers (Section III,C,2). Secondly, the nickel particles develop facets, an event that takes place rapidly in the 900–1100°C range. Substituting estimates of the time required for faceting at the various temperatures in the rate equation (Section II,A,2), we obtain an activation energy of 26 kcal/mole (Fig. 22). It is suggested that this represents the activation energy for self diffusion in the nickel particles. Thirdly, the whiskers are wetted by the nickel particles, for which, again from the rate equation, an activation energy of 74 kcal/mole is obtained (Fig. 22). This value compares with the significantly larger activation energy of 110 kcal/mole required for the diffusion of nickel into carbon fibers under similar experimental conditions.

Although the mechanism(s) producing the side growths on nickel-coated whiskers under an argon atmosphere cannot yet be precisely identified, some of the necessary conditions can be defined. The reaction requires the presence of nickel as it does not occur on uncoated whiskers. In addition the nickel particles do not decrease significantly in size during the extension of the side growths, which suggests that the side growths do not contain nickel. Such a conclusion is supported by electron microprobe analysis, which indicates the presence of silicon but not nickel in the side growths. The nitrogen partial pressure in the argon is sufficient to inhibit the formation of silicon nitride, but the oxygen partial pressure ($\sim 10^{-5}$ atmospheres) is greater than that required for equilibrium oxidation of silicon ($\sim 3 \times 10^{-8}$ atmospheres), although this level depends critically on kinetic factors (Kaiser and Breslin, 1958). The oxidation of silicon nitride powder results in an appreciable weight gain ($\sim 1.5\%$/hr) (Colquhoun *et al.*, 1970) which could account for the magnitude of the side growths ($\sim 30\%$ in 17 hr). Consequently, it seems most likely that the effect is due to the oxidation of the whisker to form silicon. The absence of localized depletion of the whiskers further suggests that it is a vapor phase reaction which is catalyzed by nickel.

V. Conclusion

One approach to development of a reinforcement–metal matrix composite has been firstly to make it and secondly to test a sample of the bulk composite. This approach has often resulted in disappointment as the properties achieved rarely correspond to those predicted on theoretical grounds. The difficulty that then arises is which of the many variables introduced during composite production should now be manipulated to optimize its behavior. It is within this context that the single whisker or fiber compatibility outlined in this chapter is of importance, as the controlling factors can be directly evaluated for any given composite system. The "bonus" in terms of temperature and/or time which can be achieved by attention to, and control of, impurity has been demonstrated for the examples of sapphire whisker–nickel, carbon fiber–nickel, and silicon nitride whiskeṛ–nickel. Such principles then provide realistic guidelines for evaluating and setting up a production process designed to give a composite with an optimized reinforcement–matrix compatibility for each particular system. The resultant process will probably be more complex (and expensive) than hitherto, but if reinforcement breakdown and disintegration is effectively minimized, this could well be a worthwhile investment. Within this area, the deliberate development of side growths on silicon nitride whiskers to provide effective bonding has some interesting possibilities. In addition, the most useful applications of such systems in terms of temperature environment can be quantitatively assessed. For example, the data on the nickel-coated carbon fibers indicate that, even under high vacuum, a lifetime of 1000 hr requires that the temperature be less than 950°C. Consequently, it is considered that the continued study of these concepts, on a wider range of potential composite systems, will contribute significantly to the development of a comprehensive understanding of reinforcement–metal matrix compatibility.

Acknowledgments

The author is indebted to Professor E. H. Andrews, Miss A. J. Markham, and Dr. C. K. L. Davies for their contributions to the research discussed in this chapter. In addition the financial support of the Ministry of Technology and of the Science Research Council is gratefully acknowledged.

References

Andrews, E. H. (1966). *J. Mater. Sci.* **1**, 377.
Andrews, E. H., Bonfield, W., Davies, C. K. L., and Markham, A. J. (1972). *J. Mater. Sci.* (**7**, 1003.).
Barclay, R. B., and Bonfield, W. (1971). *J. Mater. Sci.* . **6**, 1076.
Bonfield, W., and Markham, A. J. (1970). *J. Mater. Sci.* **5**, 719.
Bonfield, W., and Markham, A. J. (1971). *J. Mater. Sci.* **6**, 1183.
Bonfield, W., and Markham, A. J. (1971a). Final Rep., Ministry of Technol. Agreement No. AT/2042/031.
Colquhoun, I., Grieveson, P., Jack, K. H., and Thompson, D. P. (1970). Ministry of Defense Progr. Rep.
Dunn, W. W., McLellan, R. B., and Oates, W. A. (1968). *Trans. Met. Soc. AIME* **242**, 2129.
Jackson, P. W., and Marjoram, J. R. (1970). *J. Mater. Sci.* **5**, 9.
Johnson, W., and Watt, W. (1967). *Nature (London)* **215**, 384.

Kaiser, W., and Breslin, J. (1958). *J. Appl. Phys.* **29,** 1292.
Moore, A. (1969). Ministry of Technol., Progr. Rep. (December).
Nichols, F. A., and Mullins, W. W. (1965). *J. Appl. Phys.* **36,** 1826.
Oishi, Y., and Kingery, W. D. (1960). *J. Chem. Phys.* **33,** 480.
Paladino, A. E., and Kingery, W. D. (1962). *J. Chem. Phys.* **37,** 957.
Schairer, J. F., and Bowen, N. L. (1955). *Amer. J. Sci.* **253,** 681.
Stapley, A. J., and Beevers, C. J. (1969). *J. Mater. Sci.* **4,** 65.
Wagner, C. (1961). *Z. Elektrochem.* **65,** 581.

Author Index

Numbers in italics refer to the pages on which the complete references are listed.

Subject Index

A

Activation energy, 93
  Al_2O_3/Ni, 111
  Al_2O_3/Ti, 110
  B/Ti, 96, 102
  ion diffusion, 375
  SiC/Ti, 105
Alumina filament composites,
  longitudinal strength, 322–323
Alumina-matrix reaction, 73
Aluminum-alumina, 26, 27
  reaction with liquid aluminum, 296–299
Aluminum–boron, 4, 46, 48, 54, 164–166,
    176, 223–225
  high temperature annealing, 84–85
  off-axis strength, 197–204, 208
  reaction-strength, 35, 155–160
  stability of interface, 6, 14, 15, 24, 86,
    131, 136–139
  thermal cycling, 84
  toughness, 260–267
Aluminum(6061)–boron, 19–21, 24, 25, 74,
    75, 136, 175, 208, 263, 264
  high temperature annealing, 85, 156–160
Aluminum–BN coated boron, 4, 17, 164
Aluminum–boron carbide, 117
Aluminum–carbon, 4, 6, 70, 83, 165
Aluminum–graphite, *see* Aluminum–
  carbon
Aluminum–silica, 4, 7, 161, 164–166, 309
  fatigue strength, 325–326
  fiber strength, 315–316
Aluminum–silicon carbide, 4, 6, 17
Aluminum–silicon carbide coated boron,
    17, 165, 223, 229
  off-axis strength, 197–207
Aluminum–silicon nitride, 27
Aluminum–stainless steel, 4, 6, 47, 86, 87,
    129, 160–163, 165, 176, 207–208, 229
  elastic-plastic behavior, 218–223
  fracture toughness, 268
Aluminum(2024)–stainless steel, 11, 175
Aluminum–tungsten, 68, 260, 266
Atmosphere effects,
  bonding of nickel–alumina, 313–315
  infiltration of nickel–alumina, 304–307

B

Bonding in composites, 18–25
Bonding to alumina, 301
Bond types,
  dissolution and wetting bond, 23, 67, 70
  exchange reaction bond, 23–24, 67,
    71–72
  mechanical bond, 23–24, 67–70
  mixed bond, 23–25, 68, 74–76
  oxide bond, 7, 68, 72–74, 84
  reaction bond, 23–24, 67, 70–71
Boron filaments,
  creep, 150
  ductile–brittle transition, 149
  stability, 150
  strength in Al(6061), 158
Brass–tungsten, 69, 129

C

Carbide eutectics,
  creep, 357–358
Cell structure in matrix,
  creep, 230, 239
Carbon fibers,
  adhesion of nickel, 394
  fracture strength, 390–391
Carbon fiber strength,
  effect of nickel coating, 391–397
Cell structure in matrix, 222–223
Charpy impact test, 246–247
Chemical continuum, 37–38
Chemical discontinuum, 35–37
Coatings, 16, 17, 113–117, 299, 303–304
  boron carbide on boron, 114–115

414